Meyler's Side Effects of Herbal Medicines

Meyler's Side Effects of Herbal Medicines

Editor

J K Aronson, MA, DPhil, MBChB, FRCP, FBPharmacolS, FFPM (Hon)
Oxford, United Kingdom

ELSEVIER

AMSTERDAM • BOSTON • HEIDELBERG • LONDON • NEW YORK • OXFORD
PARIS • SAN DIEGO • SAN FRANCISCO • SINGAPORE • SYDNEY • TOKYO

Elsevier
Radarweg 29, PO Box 211, 1000 AE Amsterdam, The Netherlands
The Boulevard, Langford Lane, Kidlington, Oxford OX5 1GB, UK
525 B Street, Suite 1900, San Diego, CA 92101-4495, USA

Notice
No responsibility is assumed by the publisher for any injury and/or damage to persons
or property as a matter of products liability, negligence or otherwise, or from any use or operation
of any methods, products, instructions or ideas contained in the material herein. Because of rapid
advances in the medical sciences, in particular, independent verification of diagnoses and drug
dosages should be made

Medicine is an ever-changing field. Standard safety precautions must be followed, but as new
research and clinical experience broaden our knowledge, changes in treatment and drug therapy
may become necessary or appropriate. Readers are advised to check the most current product
information provided by the manufacturer of each drug to be administered to verify the
recommended dose, the method and duration of administrations, and contraindications. It is the
responsibility of the treating physician, relying on experience and knowledge of the patient, to
determine dosages and the best treatment for each individual patient. Neither the publisher nor the
authors assume any liability for any injury and/or damage to persons or property arising from this
publication.

British Library Cataloguing in Publication Data
A catalogue record for this book is available from the British Library

Library of Congress Catalog Number: 2008933968

ISBN: 978-044-453269-5

For information on all Elsevier publications
visit our web site at http://www.elsevierdirect.com

Typeset by Integra Software Services Pvt. Ltd, Pondicherry, India www.integra-india.com
Printed and bound in the USA

08 09 10   10 9 8 7 6 5 4 3 2 1

Working together to grow
libraries in developing countries

www.elsevier.com | www.bookaid.org | www.sabre.org

ELSEVIER   BOOK AID International   Sabre Foundation

Contents

Preface

This volume covers the adverse effects of herbal medicines. The material has been collected from *Meyler's Side Effects of Drugs: The International Encyclopedia of Adverse Drug Reactions and Interactions* (15th edition, 2006, in six volumes), which was itself based on previous editions of *Meyler's Side Effects of Drugs*, and from the *Side Effects of Drugs Annuals* (SEDA) 28, 29, and 30. The main contributors of this material were JK Aronson, PAGM de Smet, E Ernst, and M Pittler.

A brief history of the Meyler series

Leopold Meyler was a physician who was treated for tuberculosis after the end of the Nazi occupation of The Netherlands. According to Professor Wim Lammers, writing a tribute in Volume VIII (1975), Meyler got a fever from para-aminosalicylic acid, but elsewhere Graham Dukes has written, based on information from Meyler's widow, that it was deafness from dihydrostreptomycin; perhaps it was both. Meyler discovered that there was no single text to which medical practitioners could look for information about unwanted effects of drug therapy; Louis Lewin's text "Die Nebenwirkungen der Arzneimittel" ("The Untoward Effects of Drugs") of 1881 had long been out of print (SEDA-27, xxv-xxix). Meyler therefore determined to make such information available and persuaded the Netherlands publishing firm of Van Gorcum to publish a book, in Dutch, entirely devoted to descriptions of the adverse effects that drugs could cause. He went on to agree with the Elsevier Publishing Company, as it was then called, to prepare and issue an English translation. The first edition of 192 pages (*Schadelijke Nevenwerkingen van Geneesmiddelen*) appeared in 1951 and the English version (*Side Effects of Drugs*) a year later.

The book was a great success, and a few years later Meyler started to publish what he called surveys of unwanted effects of drugs. Each survey covered a period of two to four years. They were labelled as volumes rather than editions, and after Volume IV had been published Meyler could no longer handle the task alone. For subsequent volumes he recruited collaborators, such as Andrew Herxheimer. In September 1973 Meyler died unexpectedly, and Elsevier invited Graham Dukes to take over the editing of Volume VIII.

Dukes persuaded Elsevier that the published literature was too large to be comfortably encompassed in a four-yearly cycle, and he suggested that the volumes should be produced annually instead. The four-yearly volume could then concentrate on providing a complementary critical encyclopaedic survey of the entire field. The first *Side Effects of Drugs Annual* was published in 1977. The first encyclopaedic edition of *Meyler's Side Effects of Drugs*, which appeared in 1980, was labelled the ninth edition, and since then a new encyclopaedic edition has appeared every four years. The 15th edition was published in 2006, in both hard and electronic versions.

Monograph structure

This volume starts with a general section on the adverse effects of herbal medicines, which is followed by monographs on individual herbal products, each of which has the following structure:

- Family: each monograph is organized under a family of plants (for example Liliaceae).
- Genera: the various genera that are included under the family name are tabulated (for example the family Liliaceae contains 94 genera); the major source of information on families and genera is the Plants National Database (http://plants.usda.gov/index.html).
- Species: in each monograph some species are dealt with separately. For example, in the monograph on Liliaceae, four species are included under their Latin names and major common names—*Sassafras albidum* (sassafras), *Allium sativum* (garlic), *Colchicum autumnale* (autumn crocus), and *Ruscus aculeatus* (butcher's broom).

Each monograph includes the following information in varying amounts:

- Alternative common names; the major sources of this information are *A Modern Herbal* by Mrs M Grieve (1931; http://www.botanical.com/botanical/mgmh/mgmh.html) and *The Desktop Guide to Complementary and Alternative Medicine: an Evidence-Based Approach* by E Ernst, MH Pittler, C Stevenson, and A White (Mosby, 2001).
- Active ingredients; the major source of this information is the *Dictionary of Plants Containing Secondary Metabolites* by John S Glasby (Taylor & Francis, 1991).
- Uses, including traditional and modern uses.
- Adverse effects.
- References.

The families of plants and their species that are the subjects of monographs are listed in Table 2 (p. 14) by alphabetical order of family. The same data are listed in Table 3 (p. 17) by alphabetical order of species. Other monographs cover one of the Basidiomycetes (*Lentinus edodes*, shiitake) and algae. Table 4 (p. 20) gives the Latin equivalents of the common names. To locate a plant by its common name, convert the common name into the Latin name using Table 4 and then find out to which family it belongs by consulting Table 3.

Drug names

Drugs have usually been designated by their recommended or proposed International Non-proprietary Names (rINN or pINN); when these are not available, chemical names have been used. In some cases brand names have been used.

Spelling

For indexing purposes, American spelling has been used, e.g. anemia, estrogen rather than anaemia, oestrogen.

Cross-references

The various editions of *Meyler's Side Effects of Drugs* are cited in the text as SED-l3, SED-14, etc; the *Side Effects of Drugs Annuals* are cited as SEDA-1, SEDA-2, etc.

J K Aronson
Oxford, May 2008

Herbal Medicines – Introduction and General Information

Herbal Medicines

Introduction

In principle, herbal medicines have the potential to elicit the same types of adverse reactions as synthetic drugs; the body has no way of distinguishing between "natural" and man-made compounds. Herbal medicines consist of whole extracts of plant parts (for example roots, leaves) and contain numerous potentially active molecules. Synergy is normally assumed to play a part in the medicinal effects of plant extracts, and medical herbalists have always claimed that whole plant extracts have superior effects over single isolated constituents. Similarly, it is also claimed that combinations of herbs have synergistic effects. There is in vitro and/or in vivo evidence to support the occurrence of synergism between constituents in certain herbal extracts (1,2); however, clinical evidence is lacking, and it is in any case uncertain how far the principle extends. Synergy is also taken to mean an attenuation of undesirable effects, another key tenet of herbalism being that the toxicity of plant extracts is less than that of a single isolated constituent. However, theoretically, plant constituents could also interact to render a herbal preparation more toxic than a single chemical constituent. Virtually no evidence is available to substantiate either hypothesis. It is also important to determine whether herbal treatments that have been shown to be as effective as conventional drugs have a better safety profile. Contrary to the belief of most herbalists, long-standing experience is by no means a reliable yardstick when it comes to judging the risk of adverse reactions (3).

Uses

Herbal medicine continues to be a growth area. In the UK, retail sales of complementary medicines (licensed herbal medicines, homoeopathic remedies, essential oils used in aromatherapy) were estimated to be £72 million in 1996, an increase of 36% in real terms since 1991 (4). This, however, is likely to be a gross underestimate as popular products sold as food supplements, including *Ginkgo biloba* and garlic, were not included. According to a detailed analysis of the herbal medicines market in Germany and France, total sales of herbal products in those countries in 1997 were US$1.8 billion and US$1.1 billion respectively (5). In 1994, annual retail sales of botanical medicines in the USA were estimated to be around US$1.6 billion; in 1998, the figure was closer to US$4 billion (6).

Some of the most useful data on trends in the use of herbal medicines come from two surveys of US adults carried out in 1991 and 1997/98, which involved over 1500 and over 2000 individuals respectively (7,8). The use of at least one form of complementary therapy in the 12 months preceding the survey increased significantly from 34% in 1990 to 42% in 1997. Herbal medicine was one of the therapies showing the most increase over this time period: there was a statistically significant increase in self-medication with herbal medicine from 2.5% of the sample in 1990 to 13% in 1997 (8). Disclosure rates to physicians of complementary medicine use were below 40% in both surveys (8). Furthermore, 18% of prescription medicine users took prescription medicines concurrently with herbal remedies and/or high-dose vitamins. These aspects of user behaviour clearly have implications for safety.

Several herbal medicines pose serious problems for surgical patients, for example through an increased bleeding tendency (9,10). Vulnerable populations also include children (11), and too few safety data are available to recommend herbal medicines during pregnancy or lactation (12). Several investigators have pointed out the potential of herbal medicines to harm certain organs, for example the liver (13) or the skin (14). Laxatives are often based on herbal extracts, and the risks of herbal laxatives have been emphasized (15). Many authors have reviewed the risks of herbal medicines in general terms (16,17).

The reasons for the popularity of herbal medicine are many and diverse. It appears that complementary medicine is not usually used because of an outright rejection of conventional medicine, but more because users desire to control their own health (5) and because they find complementary medicine to be more congruent with their own values, beliefs, and philosophical orientations toward health and life (18). Also, users may consult different practitioners for different reasons (5). An important reason for the increase in use is that consumers (often motivated by the lay press) consider complementary medicine to be "natural" and assume it is "safe". However, this notion is dangerously misleading; adverse effects have been associated with the use of complementary therapies (19). Furthermore, complementary therapies may not only be directly harmful (for example adverse effects of a herbal formulation), but like other medical treatments have the potential to be indirectly harmful (for example through being applied incompetently, by delaying appropriate effective treatment, or by causing needless expense) (20).

The inadequacy of regulation of herbal medicines has repeatedly been stressed (176). The need for adequate pharmacovigilance of herbal medicine is obvious (177) and guidance for safety assessment has been provided (178).

Adverse effects

Most of the data on adverse effects associated with herbal medicines is anecdotal, and assessment and classification of causality is often not possible. Likewise, there have been few attempts to determine systematically the incidence of adverse effects of non-orthodox therapies.

Several reviews have focused on herbal medicines and have covered:

- the toxicity of medicinal plants (21–23);
- the safety of herbal products in general (24–35; 179–183);
- adverse effects in specific countries, for example the USA (36) and Malaysia (37);
- adverse effects on specific organs (38), such as the cardiovascular system (39; 184), the nervous system (185) the liver (40,41,186–190), the hidneys (191–193) and the skin (42,43);

- the safety of herbal medicines in vulnerable populations: children and adolescents (194–196) elderly patients (44,197), during pregnancy and lactation (45,198–200), patients with cancers (206) and surgical patients (46,47);
- carcinogenicity (48);
- the adverse effects of herbal antidepressants (49);
- the adverse effects of Chinese herbal medicaments (50,51);
- the adverse effects of medicines used in other traditions (52,202–206);
- herb–drug interactions (53–64;207–218);
- pharmacovigilance of herbal medicines (65).

Direct effects associated with herbal medicines can occur in several ways:

- hypersusceptibility reactions
- collateral reactions
- toxic reactions
- drug interactions
- deliberate adulteration and accidental contamination (219)
- false authentication
- lack of quality control.

Some of these effects relate to product quality. While there are some data on certain of these aspects, information on other aspects is almost entirely lacking. For example, there are isolated case reports of interactions between conventional medicines and complementary (usually herbal) remedies (20,66,67), although further information is largely theoretical (68).

Even a perfectly safe remedy (mainstream or unorthodox) can become unsafe when used incompetently. Medical competence can be defined as doing everything in the best interest of the patient according to the best available evidence. There are numerous circumstances, both in orthodox and complementary medicine, when competence is jeopardized:

- missed diagnosis
- misdiagnosis
- disregarding contraindications
- preventing/delaying more effective treatments (for example misinformation about effective therapies; loss of herd immunity through a negative attitude toward immunization)
- clinical deterioration not diagnosed
- adverse reaction not diagnosed
- discontinuation of prescribed drugs
- self-medication.

The attitude of consumers toward herbal medicines can also constitute a risk. When 515 users of herbal remedies were interviewed about their behavior vis a vis adverse effects of herbal versus synthetic over-the-counter drugs, a clear difference emerged. While 26% would consult their doctor for a serious adverse effect of a synthetic medication, only 0.8% would do the same in relation to herbal remedies (69).

The only way to minimize incompetence is by proper education and training, combined with responsible regulatory control. While training and control are self-evident features of mainstream medicine they are often not fully incorporated in complementary medicine. Thus the issue of indirect health risk is particularly pertinent to complementary medicine. Whenever complementary practitioners take full responsibility for a patient, this should be matched with full medical competence; if on the other hand, competence is not demonstrably complete, the practitioner in question should not assume full responsibility (70).

Most reports of adverse effects associated with herbal remedies relate to Chinese herbal medicines (71). This is an issue of growing concern, particularly because in many Western countries the popularity of Chinese herbalism is increasing. This is happening in the almost complete absence of governmental control (72) or of systematic research into the potential hazards of Chinese herbal formulations (73).

There have been several reports of adverse effects associated with Asian herbal formulations (74–94). Most of the serious adverse effects of Chinese herbal remedies are associated with formulations containing aconitine, anticholinergic compounds, aristolochic acid, or podophyllin, contaminating substances (95). Problems with Chinese herbal formulations are intensified because of nomenclature, since common, botanical, and Chinese names exist side by side, making confusion likely.

Frequencies of adverse reactions to herbal medicines

Survey data have repeatedly shown that adverse effects of herbal medicines are more frequent than is generally appreciated. A large US study involving 11 poison centers showed that one-third of the events reported in association with intake of dietary supplements were "greater than mild severity" (220). They included myocardial infarction, liver failure, bleeding, and death. Most frequently implicated were ma huang (*Ephedra*), guaraná, ginseng, and St John's wort. A survey of 2203 Australian therapists suggested that only 17% of these professionals had ever noted an adverse reaction to a herbal treatment (221).

A survey of all patients seen in a German hospital specializing in traditional Chinese medicine included 1036 patients, in 6.5% of whom adverse events (mostly of minor severity) were recorded (222). Several severe adverse events were also noted; they included paralysis of the tibialis anterior muscle, paresthesia, and raised liver enzymes.

A small survey of adverse events after routine practice of Chinese herbal medicine included the data of 144 patients (223). A total of 20 patients reported 32 adverse events associated with Chinese herbal medicines during the 4 weeks after treatment. There were no serious adverse events. The most commonly reported adverse events were diarrhea, fatigue, and nausea.

Of 1701 patients admitted to two general wards of a Hong Kong hospital, 3 (0.2%) had adverse reactions to Chinese herbal drugs; two of the three were serious (96). In a retrospective study of all 2695 patients admitted to a Taiwan department of medicine during 10 months 4% were admitted because of drug-related problems, and herbal remedies ranked third amongst the categories of medicines responsible (97). In an active surveillance adverse drug reaction reporting program conducted in a family medicine ward of the National Taiwan University

Hospital, Chinese crude drugs were responsible for five hospital admissions (22% of the total) or 12% of all adverse reactions observed in the study (98). This is a part of the world where the herbal tradition is particularly strong; the figures do not apply elsewhere.

The incidence of contact sensitization associated with topical formulations containing plant extracts was significant when evaluated in 1032 consecutive or randomly selected patients visiting patch test clinics in The Netherlands (99).

In a 5-year toxicological study of traditional remedies and food supplements carried out by the Medical Toxicology Unit at Guy's and St. Thomas' Hospital, London, 1297 symptomatic enquiries by medical professionals were evaluated (100). Of these, an association was considered to have been confirmed, probable, or possible in 12, 35, and 738 cases respectively. Ten of the confirmed cases were related to Chinese or Indian herbal remedies. As a result of these findings, in October 1996 the UK Committee on Safety of Medicines extended its yellow card scheme for adverse drug reaction reporting to include unlicensed herbal remedies, which are marketed mostly as food supplements in the UK (the scheme had always applied to licensed herbal medicines) (101,102). This was an important milestone in herbal pharmacovigilance.

A report from the Uppsala Monitoring Centre of the WHO has summarized all suspected adverse reactions to herbal medicaments reported from 55 countries worldwide over 20 years (103). A total of 8985 case reports were on record. Most originated from Germany (20%), followed by France (17%), the USA (17%), and the UK (12%). Allergic reactions were the most frequent serious adverse events and there were 21 deaths. The authors pointed out that adverse reactions to herbal medicaments constitute only about 0.5% of all adverse reactions on record.

A hospital-based study from Oman has suggested that 15% of all cases of self-poisoning seen in this setting are with traditional medicines (104). A case series from Thailand has suggested that in patients with oral squamous cell carcinoma the use of herbal medicines before the first consultation with a healthcare professional increases the risk of an advanced stage almost six-fold (105) and survey data from the USA have suggested

that herb–drug interactions may be a significant problem in a sizeable proportion of patients (106).

In a German hospital specializing in Chinese herbalism of 145 patients who had been treated within 1 year 53% reported having had at least one adverse effect attributable to Chinese herbal medicines (107). Nausea, vomiting, and diarrhea were the most common complaints. It should be noted that causality in these cases can only be suspected and not proven. In the same institution about 1% of 1507 consecutive patients treated with Chinese herbal mixtures had clinically relevant rises in liver enzymes (108,109). *Glycyrrhiza* radix and *Atractylodis macrocephalae* rhizome were most consistently associated with such problems. In most of these cases there were no associated clinical signs and the abnormalities tended to normalize without specific therapy and in spite of continued treatment with the Chinese herbal mixtures.

When 1100 Australian practitioners of traditional Chinese medicine were asked to complete questionnaires about the adverse effects of Chinese herbal mixtures, they reported 860 adverse events, including 19 deaths (110). It was calculated that each practitioner had encountered an average of 1.4 adverse events during each year of full-time practice.

A physician prospectively monitored all 1265 patients taking traditional Chinese medicines at his clinic during 33 months (111). Liver enzymes were measured before the start of therapy and 3 and 10 weeks later. Alanine transaminase activity was raised in 107 patients (8.5%) who initially had normal values. Of these patients, about 25% reported symptoms such as abdominal discomfort, looseness of bowels, loss of appetite, or fatigue.

A retrospective analysis of all adverse events related to herbal medicines and dietary supplements reported to the California Poison Control System has given data on the risks of the adverse effects of herbal medicines (112). Between January 1997 and June 1998, 918 calls relating to such supplements were received. Exposures resulting in adverse reactions occurred most often at recommended doses. There were 233 adverse events, of which 29% occurred in children. The products most frequently implicated were zinc (38%), *Echinacea* (8%), witch hazel (6%), and chromium picolinate (6%). Most of the adverse events were not severe and required no treatment; hospitalization was required in only three cases.

Table 1 Potential adulterants that should be taken into account in the quality control of herbal medicines

Type of adulterant	Examples
Allopathic drugs	Analgesic and anti-inflammatory agents (for example aminophenazone, indometacin, phenylbutazone), benzodiazepines, glucocorticoids, sulfonylureas, thiazide diuretics, thyroid hormones
Botanicals	*Atropa belladonna*, *Digitalis* species, *Colchicum*, *Rauwolfia serpentina*, pyrrolizidine-containing plants (see separate monograph on pyrrolizidines)
Fumigation agents	Ethylene oxide, methyl bromide, phosphine
Heavy metals	Arsenic, cadmium, lead, mercury
Micro-organisms	*Escherichia coli* (certain strains), *Pseudomonas aeruginosa*, *Salmonella*, *Shigella*, *Staphylococcus aureus*
Microbial toxins	Aflatoxins, bacterial endotoxins
Pesticides	Carbamate insecticides and herbicides, chlorinated pesticides (for example aldrin, dieldrin, heptachlor, DDT, DDE, HCB, HCH isomers), dithiocarbamate fungicides, organic phosphates, triazine herbicides
Radionuclides	^{134}Cs, ^{137}Cs, ^{103}Ru, ^{131}I, ^{90}Sr

Adulteration and contamination

Quality control for herbal medicaments that are sold as dietary supplements in most countries is poor (115,116). Concerns about the quality and safety of herbal remedies are justified, and there have been repeated calls for greater control and regulation (75,113,114). Considerable variations in the contents of active ingredients have been reported, with lot-to-lot variations of up to 1000% (117). In most countries, the sale and supply of herbal remedies is to a large extent uncontrolled and unregulated; most herbal remedies are sold as unlicensed food supplements and their safety, efficacy, and quality have therefore not been assessed by licensing authorities. Adulteration and contamination of herbal remedies with other plant material and conventional drugs have been documented (72,118) (Table 1).

The authors of a survey of German importers of Chinese herbs concluded that "only rarely" had herbal drugs to be returned because of contamination (119). The authors also stated that "a 100% check for all possible contaminants is not possible." However, there have been many reports of adulteration and contamination relate to Chinese herbal remedies (120). Instances include adulteration/contamination with conventional drugs (121,122), heavy metals (123–129), and other substances (130,131).

When 27 samples of commercially available camomile formulations were tested in Brazil, it was found that all of them contained adulterants and only 50% had the essential oils needed to produce anti-inflammatory activity (132).

The Botanical Lab in the USA has manufactured the herbal medicines PC-SPES and SPES (133). These two products are marketed as "herbal dietary supplements" for "prostate health" and for "strengthening the immune system" respectively. They are sold through the internet, by mail order, by phone order, and through various distributors and health care professionals. An analytical report from the California Department of Health in 2002 showed that samples of PC-SPES and SPES have been contaminated with alprazolam and warfarin. The Canadian Medicines Regulatory Authority also reported similar contaminations. In view of these reports Health Canada, the Irish Medicines Board, and the State Health Director of California all warned consumers to stop using these two products immediately and to consult their healthcare practitioners. Botanic Lab also informed consumers of these laboratory findings and issued a product recall of all lots of PC-SPES, pending further reports from additional testing of PC-SPES in both commercial and academic laboratories.

In 2002 the Medicines Safety Authority of the Ministry of Health in New Zealand (Medsafe) ordered the withdrawal of several traditional Chinese medicines sold as herbal remedies, since they contained scheduled medicines and toxic substances (134). The products included the following allopathic drugs:

- Wei Ge Wang tablets, which contained sildenafil
- Sang Ju Gan Mao Pian tablets, which contained diclofenac and chlorphenamine
- Yen Qiao Jie Du Pian capsules, which contained chlorphenamine, diclofenac, and paracetamol

- Xiaoke Wan pills, which contained glibenclamide
- Shuen Feng cream, which contained ketoconazole
- Dezhong Rhinitis drops, which contained ephedrine hydrochloride.

The New Zealand Director General of Health issued a Public Statement asking people to stop taking these products and to seek medical advice. Medsafe asked all importers and distributors of traditional Chinese medicines to cease all distribution and sale of these products, to withdraw them from retail outlets, and to ensure that other products they sell do not contain scheduled medicines.

Benzodiazepines

In 2001 the California State Health Director warned consumers to stop using the herbal product Anso Comfort capsules immediately, because the product contains the undeclared prescription drug chlordiazepoxide. Chlordiazepoxide is a benzodiazepine that is used for anxiety and as a sedative and can be dangerous if not taken under medical supervision (135). Anso Comfort capsules, available by mail or telephone order from the distributor in 60-capsule bottles, were clear with dark green powder inside. The label was yellow with green English printing and a picture of a plant. An investigation by the California Department of Health Services Food and Drug Branch and Food and Drug Laboratory showed that the product contained chlordiazepoxide. The ingredients for the product were imported from China and the capsules were manufactured in California. Advertising for the product claimed that the capsules were useful for the treatment of a wide variety of illnesses, including high blood pressure and high cholesterol, in addition to claims that it was a natural herbal dietary supplement. The advertising also claimed that the product contained only Chinese herbal ingredients and that consumers could reduce or stop their need for prescribed medicines. No clear medical evidence supported any of these claims. The distributor, NuMeridian (formerly known as Top Line Project), voluntarily recalled the product nationwide.

A San Francisco woman with a history of diabetes and high blood pressure was hospitalized in January 2001 with life-threatening hypoglycemia after she consumed Anso Comfort capsules. This may have been due to an interaction of chlordiazepoxide with other unspecified medications that she was taking.

Fenfluramine

There have been reports of herbal remedies adulterated with fenfluramine. Fenfluramine is an appetite suppressant that was banned globally in 1997 because of concerns about its effect on the heart, while nitrosofenfluramine is toxic to the liver.

In 2004 the UK Medicines and Healthcare products Regulatory Agency (MHRA) alerted herbal interest groups and consumers about the presence of fenfluramine and nitrosofenfluramine in an unlicensed traditional Chinese medicine formulation, Shubao Slimming Capsules, supplied illegally as a slimming agent in the UK (224).

There was public health concern in the UK after the referral of a 44-year-old woman with new-onset hypertension, palpitation, anxiety, and a body mass index of 19 kg/m^2. It became apparent that an alarming number of the local population had been attending a particular Chinese herbalist for weight loss remedies. Most had been taking multiple formulations and described "spectacular" results. Several reported considerable cardiovascular symptoms, but they were reassured that Chinese medicines are natural and can cause no harm. Analysis by gas chromatography showed a high concentration of fenfluramine in two of the products (sold as Qian Er and Ma Zin Dol, presumably mimicking the brand name Mazindol). Fenfluramine was also found in the patients' urine. Subsequently, a student nurse was admitted with severe fenfluramine toxicity which developed 2 hours after her first dose of a herbal slimming remedy (136). Following an investigation of this case the Medicines Control Agency published a report on traditional ethnic medicines and the current law (137). Stringent regulation of traditional medicines, at least to the standards of conventional practice, is urgently needed. The hazards associated with the use of dietary supplements containing ma huang, a herbal source of ephedrine, have also been reported (138).

Glucocorticoids

Unregulated Chinese herbal products adulterated with glucocorticoids have been detected (139). Dexamethasone was present in eight of 11 Chinese herbal creams analysed by UK dermatologists. The creams contained dexamethasone in concentrations inappropriate for use on the face or in children (64–1500 micrograms/g). The cream with the highest concentration of dexamethasone was prescribed to treat facial eczema in a 4-month-old baby. In all cases, it had been assumed that the creams did not contain glucocorticoids. The authors were concerned that these patients received both unlabelled and unlicensed topical glucocorticoids. They wrote that "greater regulation and restriction needs to be imposed on herbalists, and continuous monitoring of side effects of these medications is necessary."

When British dermatologists analysed 24 Chinese herbal creams used by their patients for eczema they found that all but four of these samples were adulterated with glucocorticoids (225). Wau Wa cream is marketed in several countries as a herbal cream for eczema. After repeatedly observing surprising therapeutic successes, UK doctors analysed three samples of this cream given to them by three patients (140). The samples contained 0.013% clobetasol propionate, a powerful glucocorticoid that does not occur naturally.

Betamethasone, 0.1–0.3 mg per capsule, has been detected in Cheng Kum and Shen Loon, two herbal medicines that are popular for their benefits in joint pain, skin problems, colds, menopausal symptoms, and dysmenorrhea (141). Over-exposure to betamethasone can result in typical signs of glucocorticoid excess, such as moon face, hypertension, easy bruising, purple abdominal striae, truncal obesity, and hirsutism. The recommended daily adult dose of Cheng Kum is 1–3 capsules per day, and there have been reports of glucocorticoid-induced adverse effects in patients taking Cheng Kum and Shen Loon, even in the absence of other exogenous corticosteroid consumption.

- Two patients, a 29-year-old woman and a 10-year-old girl, developed Cushingoid features after taking Shen Loon for 4 and 5 months respectively (142). Their morning plasma cortisol concentrations were increased and adrenal suppression was confirmed by a short Synacthen test. Both recovered after withdrawal of the remedy and treatment with prednisone.

The New Zealand Medicines and Medical Devices Safety Authority (Medsafe) has notified that the further importation of these herbal products into New Zealand will be stopped at Customs. However, because of the risk of adrenal suppression from glucocorticoid, consumers have been sent a letter advising them against abruptly discontinuing these products. They should continue with the treatment and see their general practitioner as soon as possible for instructions on how they can be safely weaned off the product. Medsafe has issued a letter to doctors advising them to determine whether patients taking Cheng Kum or Shen Loon are at risk of adrenal suppression by estimating the potential total dose of glucocorticoid (from Cheng Kum or Shen Loon plus any exogenous steroids) and the duration of use, by examining the patient for signs of glucocorticoid excess, and by ascertaining if other risk factors for adrenal suppression are present (such as Addison's disease and AIDS).

The Norwegian Medicines Agency (NoMA) has banned the sale of two herbal medicines, Phu Chee and Lin Chee/Active Rheuma plus, which were found to contain high doses of undeclared dexamethasone (Phu Chee) and prednisolone (Lin Chee/Active Rheuma plus, 226). Physicians from a hospital in northern Norway reported that several patients taking Phu Chee or Lin Chee/Active Rheuma Plus developed symptoms similar to those observed with prolonged use, or high doses, of glucocorticoids, along with subsequent withdrawal symptoms. Laboratory analysis showed that Phu Chee contains dexamethasone 0.4–0.5 mg per tablet and that Lin Chee/Active Rheuma Plus contains an unknown quantity of prednisolone. As the recommended dosage of Phu Chee was 3–9 tablets/day, patients could have been exposed to a daily dose of dexamethasone of 1.2–4.5 mg. NoMA sent a letter to all of the distributors' customers, with a warning about use and rapid discontinuation of the herbal medicines, as well as advice to see a doctor.

Hypoglycemic drugs

Herbal medicines, particularly Chinese ones, are sometimes contaminated with conventional synthetic drugs (143). In 2000 the California Department of Health Services Food and Drug Branch issued a warning to consumers that they should immediately stop using five herbal products because they contained two prescription drugs that were not listed as ingredients and that are unsafe without monitoring by a physician (144). The products were Diabetes Hypoglucose Capsules, Pearl Hypoglycemic Capsules, Tongyi Tang Diabetes Angel Pearl Hypoglycemic Capsules, Tongyi Tang Diabetes Angel Hypoglycemic Capsules, and Zhen Qi Capsules. The products were available by mail order and could be purchased by telephone or via the Internet. Their manufacturers claimed that they contained only natural Chinese herbal ingredients. However, after a diabetic patient in Northern California had had several episodes of hypoglycemia after taking Diabetes Hypoglucose Capsules, an investigation by the Department showed that they contain the antidiabetic drugs glibenclamide (glyburide) and phenformin.

- A 48-year-old diabetic patient from London was initially managed with metformin and gliclazide, but because of poor metabolic control treatment was switched to insulin plus oral metformin (227). His metabolic control remained poor and he developed microvascular and macrovascular complications. He returned to his native India for a visit and on returning his metabolic control was improved. He attributed this to treatment that he had received in India, three different 'herbal balls' to be taken three times a day with meals. The remedies contained chlorpropamide in a dose equivalent to 200 mg/day. It was agreed that the patient should continue with this treatment. One year later, while still taking the 'herbal balls', his metabolic control deteriorated. The herbal therapy was withdrawn and bolus insulin treatment was restarted.

Other compounds

The Chinese patent medicine Hua Fo-Vigor Max is marketed for erectile dysfunction. It was shown by Canadian authorities to contain the prescription drug tadalafil, which is used to treat erectile impotence (228). Health Canada required the importer to remove the product from the market and issued a "Customs Alert" to stop importation of the product.

The regulatory authorities in New Zealand have ordered the withdrawal of 11 Chinese medicines (229). They were adulterated either with *Aristolochia* species, which causes severe kidney damage, or prescription drugs such as sildenafil, diclofenac, chlorphenamine, paracetamol, glibenclamide, ketoconazole, clobetasol, albendazole, and ephedrine. One remedy was contaminated with arsenic.

Jackyakamcho-tang is a Korean herbal mixture of *Paeoniae radix* and *Glycyrrhizae radix*, used for its alleged analgesic and spasmolytic properties. Among 81 patients treated with this remedy there adverse effects in 11%. Indigestion, diarrhea, and edema were the most common complaints; in 3.7% of cases they were rated as severe(230).

The UK Medicines traditional Chinese medicine slimming aid called Qing zhisan tain shou, reportedly contains the prescription-only medicine sibutramine (231). The

MHRA has warned consumers that sibutramine should only be prescribed under specific circumstances and requires the supervision of a registered doctor, as it can cause increased blood pressure. Qing zhisan tain shou is supplied in a bicolored cream and a brown capsule formulation. The capsules are contained within blister packs and are presented in a white and green carton with various lettering and imagery. Two other Chinese slimming products, Li da dai dai hua and Meizitang have been seized by the Netherlands' authorities and have been found to contain sibutramine.

Heavy metals

Reports continue to be published on incidents related to the use of traditional herbal medications from China and elsewhere that contain arsenic among other toxic substances (145,146).

There are still occasional cases of patients with late effects from obsolete formulations such as Fowler's solution or neoarsphenamine (SEDA-16, 231). As late as 1998 a case of severe chronic poisoning resulting in fatal multiorgan failure including hepatic portal fibrosis and subsequent angiosarcoma was traced back to exposure to arsenical salts used for the treatment of psoriasis many years before (147) (SEDA-22, 243). In some non-Western countries arsenic is apparently still being used in dentistry for devitalization of inflamed pulp and sensitive dentine, and cases have been described in which this has resulted in arsenical necrosis of the jaws, affecting the maxilla or mandible (148). Some exposure to arsenic may still be occurring from traditional remedies of undeclared composition, and as with other metals there may be environmental contact, notably from semiconductor materials.

In 2002 the Medicines Safety Authority of the Ministry of Health in New Zealand (Medsafe) ordered the withdrawal of several traditional Chinese medicines sold as herbal remedies, since they contained scheduled medicines and toxic substances (134). The products included Niu Huang Jie Du Pian tablets, which contain 4% arsenic. Of 260 Asian patent medicines available in California, 7% contained undeclared pharmaceuticals. When 251 samples were tested for heavy metals, 24 products contained at least 10 parts per million of lead, 36 contained arsenic, and 35 contained mercury (149).

Cases of adulteration with the heavy metals mercury (150) and lead (151) have been reported.

- A 5-year-old Chinese boy developed motor and vocal tics. His parents had given him a Chinese herbal spray to treat mouth ulcers. The spray contained mercury 878 ppm. Mercury poisoning was confirmed by the blood mercury concentration (183 nmol/l, normal value for adults under 50 nmol/l).
- A 5-year-old boy of Indian origin with encephalopathy, seizures, and developmental delay developed persistent anemia. The more obvious causes were ruled out and his blood lead concentration was high (860 ng/ml). He was treated with chelation therapy and his blood lead concentration fell. For the previous 4 years his parents had given him "Tibetan Herbal Vitamins," produced in

India, which contained large amounts of lead. The investigators calculated that over that time he had ingested around 63 g of lead.

A total of 54 samples of Asian remedies, purchased in Vietnam, Hong Kong, Florida, New York, and New Jersey, were analysed for heavy metal adulteration (152). They contained concentrations of arsenic, lead, and mercury that ranged from merely exceeding published guidelines (74%) to toxic (49%).

Ayurvedic medicines have also been reportedly adulterated and contaminated (153). For example, they have repeatedly been associated with arsenic poisoning, including hyperpigmentation and hyperkeratosis (154). Of 70 unique Ayurvedic products manufactured in India or Pakistan and sold in Boston, mostly for gastrointestinal disorders, 14 contained lead, mercury, and/or arsenic, in concentrations up to about 100 mg/g (155).

In Taipei, 319 children aged 1–7 years were screened for increased blood lead concentrations (156). The consumption of Chinese herbal medicines was significantly correlated with blood lead concentrations. In 2803 subjects from Taipei a history of herbal drug taking proved to be a major risk factor for increased blood lead concentrations (157).

- A 56-year-old woman developed the signs and symptoms of lead poisoning after taking an Indian herbal medicine for many years (158). Her blood and urine lead concentrations were 1530 ng/ml and 4785 µg/day. She also had raised liver enzymes. After withdrawal of the remedy and treatment with penicillamine, she made a full recovery.
- Czech doctors reported the case of a 26-year-old woman who had taken an Ayurvedic remedy (Astrum FE Femikalp) for sterility (159). It contained lead 113 mg/kg. Her blood lead concentration was raised and normalized 1 month after withdrawal of the remedy. Lead poisoning was also confirmed by hair analysis.

There is a practice among urbanized South African blacks to replace traditional herbal ingredients of purgative enemas with sodium or potassium dichromate. This switch can result in serious toxicity, characterized by acute renal insufficiency, gastrointestinal hemorrhage, and hepatocellular dysfunction 160,161.

Toxins

When 62 samples of medicinal plant material and 11 samples of herbal tea were examined in Croatia, fungal contamination was found to be abundant (162). *Aspergillus flavus*, a known producer of aflatoxins was present in 11 and one sample respectively. Mycotoxins were found in seven of the samples analysed.

Micro-organisms

Kombucha "mushroom" is a symbiotic yeast/bacteria aggregate surrounded by a permeable membrane. An outbreak of skin lesions affecting 20 patients from a village near Tehran has been reported (163). The lesions were painless and had a central black necrotic area, marginal erythema, and severe peripheral edema. These clinical signs led to the suspicion of anthrax infection. It turned out that all patients had applied Kombucha mushroom locally as a painkiller. The skin lesions had developed 5–7 days after the application of the material. Cultures from the skin lesions confirmed the presence of *Bacillus anthracis*. Cultures of the Kombucha mushrooms were inconclusive, owing to multiple bacterial contamination and overgrowth, but it was shown that anthrax would grow on uncontaminated material. The patients all recovered with antibiotic therapy.

Pesticides

Some herbal medicaments are contaminated with pesticides (164).

Herbal mixtures containing various ingredients: Organs and systems

Numerous herbal mixtures are promoted worldwide, for example through the Internet. In many cases their herbal ingredients are not disclosed.

Cardiovascular

Severe hypotension has been attributed to a Chinese herbal mixture (232).

- A 57-year-old man developed nausea, epigastric pain, dizziness, and diarrhea 4 hours after taking a decoction made of 14 Chinese herbs. On admission his blood pressure was 77/46 and his pulse 6 per minute. He was given intravenous fluids and the hypotension normalized within hours.

The authors pointed out that seven of the 14 herbal constituents are known to have vasodilatory effects. They therefore believed that this herbal mixture synergistically caused the hypotensive crisis.

Respiratory

Sho-saiko-to is a so-called kampo medicine, a mixture of herbs, including Chinese date, ginger root, and licorice root. It is reportedly contraindicated in patients taking interferons, patients with liver cirrhosis or hepatoma, and patients with chronic hepatitis and a platelet count of 100×10^9/l (http://www.kamponews.com). Sho-saiko-to has repeatedly been implicated in interstitial or eosinophilic pneumonias.

- A 45-year-old woman developed a high fever, a nonproductive cough, and severe dyspnea (171). Her chest X-ray showed bilateral alveolar infiltrates. Treatment with antibiotics was not successful and her condition deteriorated. She was finally put on mechanical ventilation and subsequently improved dramatically. It turned out that she had previously taken sho-saiko-to for liver dysfunction of unknown cause.

Based on a positive lymphocyte stimulation test, the authors were confident that this herbal remedy had caused pulmonary edema.

Nervous system

Cholinergic poisoning has been attributed to a Chinese herbal mixture, Ting kung teng.

- A 73-year-old man developed a cholinergic syndrome, with dizziness, sweating, chills, lacrimation, salivation, rhinorrhea, nausea, and vomiting after taking the Chinese patent medicine Ting kung teng for arthritis (174). The herbal mixture contained tropane alkaloids with cholinergic activity. After withdrawal of the remedy he made a swift and complete recovery.

Sensory systems

Corneal opacities causing photophobia have been attributed to a Kampo medicine (172).

- A 30-year-old Japanese woman developed bilateral photophobia. There were dust-like opacities in both corneae. She had a superficial keratectomy, and electron microscopy identified the opacities as lipid-like particles. She had intermittently taken a Kampo medicine composed of 18 different herbal ingredients. Her photophobia coincided with episodes of taking this medicine. The remedy was withdrawn and her symptoms subsequently subsided. She then abstained from the Kampo medicine without recurrence.

Psychiatric

Herbalife is a complex herbal formula that is promoted for weight loss. Acute mania has been attributed to it (168).

- A 39-year-old man developed classic symptoms of mania within 4–72 hours of taking Herbalife. He continued to take it and after several days became psychotic, paranoid, and out of control, culminating in a high-speed car chase with the police. Bipolar disorder was diagnosed and treated, including withdrawal of the Herbalife, and he remained free of symptoms 3 months later.

The author thought it likely that the herbal mixture had caused the psychotic illness in a man who had no previous history of mental disturbance.

Liver

Japanese authors have reported 12 cases of acute liver damage associated with Chaso or Onshido, two Chinese herbal aids to weight loss (233). Both supposedly herbal medicines actually contained N-nitroso-fenfluramine. Most patients made a full recovery, but one died and one required liver transplantation. A health warning about these medicines was issued in Japan, and subsequently 474 further cases of hepatotoxicity induced by herbal aids to weight loss came to light. The nature of the liver injury and its exact cause is not entirely clear. The pathology was consistent with toxic hepatitis.

- A 31-year-old woman developed severe hepatotoxicity while taking the Chinese drug Onshidou- Genbi-Kounou for weight loss (234). Her condition improved after withdrawal of the remedy. Other possible causes were excluded.

The authors believed that the Chinese medicine had caused the hepatitis.

- A 52-year-old woman developed liver damage resembling chronic hepatitis while taking the herbal weight loss aid "Be Petite" (235). After withdrawal of the product her signs and symptoms improved, and at 4 months follow-up she had recovered fully.

In the absence of other risk factors, the authors believed that "Be Petite" may have caused this hepatotoxic event.

- When a 37-year-old woman was admitted for elective surgery her liver enzymes and prothrombin time were found to be abnormal (236). The most plausible reason for this was that she was taking Chinese herbal mixtures containing a total of 61 different herbal ingredients. She was operated on after withdrawal of the herbal medicines for 4 days and developed a vaginal vault hematoma postoperatively. She was discharged free of symptoms on day 17.
- A 37-year-old woman developed toxic hepatitis and a vaginal vault hematoma after hysterectomy (237). She had been taking a Chinese herbal mixture containing 61 different ingredients for 6 weeks. Her prothrombin time and liver enzymes were abnormal. No other reason for her medical problems was identified. Her herbal prescription included ingredients known to have hepatotoxic and anticoagulant properties. The remedy was withdrawn and she made an uneventful recovery.

Copaltra is a herbal tea sold in France as an adjuvant therapy for diabetes. It contains *Coutarea latiflora* (50 g) and *Centaurium erythreae* (50 g).

- A 49-year-old black woman was admitted with jaundice and raised liver enzymes 4 months after starting to take Copaltra (166). She also took fenofibrate, polyunsaturated fatty acids, metformin, benfluorex, and veralipride. Liver biopsy confirmed the diagnosis of acute, severe, cytolytic hepatitis, most likely drug-induced. She made a full recovery after withdrawal of Copaltra.

The authors mentioned that five similar cases of Copaltra-induced hepatitis have been reported to the French authorities.

Isabgol is an Italian herbal mixture that is promoted for constipation.

- Syncytial giant cell hepatitis occurred in a 26-year-old woman who used Isabgol (169). Autoimmune disease and viral infections were excluded.

The authors felt that the causative role of the Isabgol was supported by the spontaneous and dramatic clinical, biochemical, and histological improvement that followed the withdrawal of Isabgol without any further therapy.

Severe liver damage has been attributed to a Kampo medicine (173).

- A 50-year-old Japanese woman with a 20-year history of asthma was taking steroids and bronchodilators when she started self-medicating with a Kampo mixture called Saiko-Keishi-Kankyo-To. Two months later, she developed acute severe liver damage. The Kampo mixture was withdrawn and she promptly recovered.

The authors attributed the liver damage to one ingredient of the mixture, *Trichosanthes* radix, a Chinese medicament that is prepared from the root of *Trichosanthes kirilowii maxim* (Tian-hua-fen).

Tsumura, a Japanese herbal mixture has been associated with hepatotoxicity (175).

- A 49-year-old Japanese woman had taken oral Tsumura for about 6 weeks to treat internal hemorrhoids when she felt unwell. Her liver enzymes were raised and a diagnosis of drug-induced hepatic damage caused by *Angelica radix* and *Bupleuri* radix contained in the mixture was made. The liver function tests normalized 4 months after withdrawal.

Urinary tract

Nephritis can complicate the use of herbal medicines.

- A 52-year-old woman developed left-sided abdominal pain, diarrhea, and a hematoma after taking the herbal remedy CKLS for 5 days (238). Renal biopsy confirmed acute tubulointerstitial nephritis.

CKLS which is marketed as a "kidney purifier" contains multiple herbal ingredients, including aloe vera, cascara, and chaparral. The authors suggested that the nephritis was most probably due to aloe vera and cascara, both of which contain anthraquinone glycosides.

- A 36-year-old woman was admitted to hospital because of general malaise (170). She had lost 5 kg within 6 months. She was found to have interstitial renal fibrosis and irreversible renal insufficiency. She had taken a Chinese herbal mixture Jai wey guo sao to treat irregular menses. The formulation contained *Angelica sinensis* root, *Rhemanniae* root and rhizome, *Ligustici* rhizome, *Paeoniae lactiflore* root, ginseng root, *Eucommiae* cortex, and honey.

The causative agent in this case could not be identified beyond doubt.

Infection risk

The patient information leaflet for a traditional Chinese medicine named Nu Bao lists human placenta, deer antler (*Corna cervi oantotrichum*), and donkey skin (*Colla cori astini*) as ingredients of capsules of the product (239). Although information about the sources of these ingredients is limited, the Medicines and Healthcare products Regulatory Agency (MHRA) in the UK has advised that all animal and human tissue derivatives carry a risk of infectious diseases, because of transmission of infective agents. The MHRA has therefore advised that consumers should not take this product. Current users should stop taking it and should consult their doctor if they feel unwell. The MHRA has written to suppliers to stop marketing Nu Bao with immediate effect.

Body temperature

Essiac is a Canadian herbal mixture promoted as a cancer cure. It has been reported to have caused fever (167).

- A 46-year-old woman with a squamous cell carcinoma of the cervix developed neutropenic fever during radiation therapy, 10 days after taking Essiac. The fever resolved after antibiotic therapy, but it delayed her radiation therapy for 9 days and required 4 days of hospitalization.

The authors felt that Essiac had caused this problem, but causality was uncertain, not least because she also took four other herbal medicaments.

Susceptibility factor

Owing to extensive modifications of drug formulations and chemical extracts from an expanding range of natural products, herbal formulations may contain ingredients that are particularly harmful to individuals with glucose-6-phosphate dehydrogenase (G6PD) deficiency. Extra vigilance is therefore required when herbal medicinal formulations, including topical applications, are used by patients with G6PD deficiency and even their carers (165).

Herb-drug interactions

Herb-drug interactions are covered in individual monographs. Here we review interactions of various herbal medicines with the anticoagulant warfarin. When mixtures of herbs are used, as is common practice in the far East, it is not possible to be sure which component was responsible for a reported interaction.

Systematic reviews

Drug interactions of warfarin with herbal preparations have been reviewed (240) (241).

In a meta-analysis of interactions of warfarin with other drugs, herbal medicines, Chinese herbal drugs, and foods 642 citations were retrieved, of which 181 eligible articles contained original reports on 120 drugs or foods (242). Of all the reports, 72% described potentiation of the effect of warfarin, and the authors considered that 84% were of poor quality, 86% of which were single case reports. The 31 incidents of clinically significant bleeding were all single case reports. Relatively few anecdotal reports of adverse event–drug associations are followed up with formal studies (243), and reports of interactions of warfarin with herbal medicines are no exception—most are based on anecdotal reports.

In a systematic review, warfarin was the most common cardiovascular drug involved in interactions with herbal medicines (244). Medicines that resulted in increased anticoagulation include *Allium sativum* (garlic), *Angelica sinensis* (dong quai), *Carica papaya* (papaya), curbicin (from *Cucurbita pepo* seed and *Serenoa repens* fruit), *Ginkgo biloba* (maidenhair), *Harpagophytum procumbens* (devil's claw), *Lycium barbarum* (Chinese wolfberry), *Mangifera indica* (mango) (245), *Peumus boldus* (boldo) (246), *Salvia miltiorrhiza* (danshen), *Trigonella foenum graecum* (fenugreek), and PC-SPES (a patented combination of eight herbs). Medicines that resulted in reduced anticoagulation include *Camellia sinensis* (green tea), milk prepared from *Hypericum perforatum* (St. John's wort), and *Panax ginseng* (ginseng).

In a retrospective analysis of the pharmaceutical care plans of 631 patients, 170 (27%) were taking some form of complementary or alternative medicine and 99 were using a medicine that could interact with warfarin, the commonest being cod-liver oil and garlic (247).

Allium sativum (garlic)

The interaction of garlic with warfarin has been reviewed (248). Certain organosulfur components inhibit human platelet aggregation in vitro and in vivo; some garlic components have an anticoagulant effect and might thus enhance the effect of warfarin. However, there is only anecdotal evidence that this occurs. Two case reports have suggested that the combination of warfarin with garlic extract prolonged the clotting time and increased the international normalized ratio (INR). There have also been reports that garlic can cause postoperative bleeding and spontaneous spinal epidural hematoma. Garlic should be withdrawn 4–8 weeks before an operation or in those taking long-term warfarin.

Angelica sinesis (dong quai)

Although Angelica sinesis is a commonly used herbal medicine, there are no clinical data on drug interactions except for one report of a 46-year-old African–American woman with atrial fibrillation stabilized on warfarin who had a greater than two-fold increase in prothrombin time and INR after taking *Angelica sinesis* for 4 weeks (249). *Angelica sinesis* extract and its active ingredient, ferulic acid, inhibit rat platelet aggregation in vivo. In rabbits oral administration *Angelica sinesis* root extract (2 g/kg bd) significantly reduced the prothrombin time when combined with warfarin (2 mg/kg), while the pharmacokinetics of warfarin were not altered (250). However, in rats an aqueous extract of *Angelica sinesis* increased the activities of CYP2D6 and CYP3A (251), and in in vitro studies components from *Angelica sinesis* root altered CYP3A4 and CYP1A activity, indicating a potential for interactions with CYP substrates. For example, a decoction or infusion of *Angelica sinesis* root inhibited CYP3A4-catalysed testosterone 6-beta-hydroxylation in human liver microsomes, whereas ferulic acid (0.5 μmol/l) from *Angelica sinesis* root significantly inhibited ethoxyresorufin O-methylase (CYP1A) activity. All of these findings suggest that precautionary advice should be given to patients who self-medicate with *Angelica sinensis* root preparations while taking long-term warfarin. Well-designed case-control studies are needed to evaluate these effects of *Angelica sinesis* root.

Camellia sinensis (green tea)

Camellia sinensis has been anecdotally reported to reduce the effect of warfarin (252).

- A 44-year-old white man taking warfarin had an INR of 3.8. He then drank green tea 0.5–1 gallon/day (4.5 l/day) for about 1 week and the INR fell to 1.37. He stopped drinking green tea and the INR rose to 2.55.

Green tea is a source of vitamin K. Dry green leaves contain 1428 micrograms of vitamin K per 100 g of leaves compared with only 262 micrograms per 100 g of dry black tea leaves (253). The amount of vitamin K ingested will obviously depend on the dilution and amount of tea leaves used to brew the tea and the quantity of tea consumed.

Cucurbita pepo

Curbicin has been anecdotally reported to cause altered coagulation in the absence of anticoagulant therapy and to enhance the anticoagulant action of warfarin; however, the authors attributed this effect to the vitamin E that was also present in the curbicin tablets (254).

Ginkgo biloba

There are anecdotal reports of possible interactions of ginkgo with warfarin (255). However, formal, albeit small, studies in patients and healthy volunteers have not confirmed this. In an open, crossover, randomized study, 12 healthy men took a single dose of warfarin 25 mg either alone or after pretreatment with *Ginkgo biloba* for 7 days; ginkgo did not significantly affect clotting or the pharmacokinetics or pharmacodynamics of warfarin (256). In a randomized, double-blind, placebo-controlled, crossover study, oral ginkgo extract 100 mg/day for 4 weeks did not alter the INR in 24 Danish outpatients (14 women and 10 men) taking stable, long-term warfarin, and the geometric mean dosage of warfarin did not change (257).

The mechanism for this interaction, if it occurs, is unknown, but both pharmacokinetic and pharmacodynamic mechanisms may be involved, given that ginkgo extracts can modulate various CYP isoenzymes and exert antiplatelet activity. Ginkgolides are also potent inhibitors of platelet-activating factor (258). There are reports of postoperative bleeding and spontaneous hemorrhage attributed to consumption of gingko (259) (260) and interactions have been described with antiplatelet drugs. For example, spontaneous hyphema occurred when ginkgo extract was combined with aspirin (acetylsalicylic acid) (261) and fatal intracerebral bleeding was associated with the combined use of ginkgo extract and ibuprofen (262). Ginkgo extract also enhanced the antiplatelet and antithrombotic effects of ticlopidine in rats, resulting in prolongation of the bleeding time by 150%

(263). However, in a double-blind, randomized, placebo-controlled study in 32 young healthy men oral ginkgo extract 120, 240, or 480 mg/day for 14 days did not alter platelet function or coagulation (264). Bleeding attributed to ginkgo often occurs in elderly or postoperative patients who may have had impaired platelet function before the use of ginkgo.

Hypericum perforatum (St John's wort)

An interaction of St John's wort with warfarin has been reported anecdotally, including 22 spontaneous reports of reduced warfarin effect after treatment with St John's wort submitted to regulatory authorities in Europe between 1998 and 2000 (265). These interactions all resulted in unstable INR values, a reduction in INR being the most common effect. Although no thromboembolic episodes occurred, the reduction in anticoagulant activity was considered clinically significant. Anticoagulant activity was restored when St John's wort was withdrawn or the warfarin dose was increased.

In a crossover study, healthy volunteers who took hypericum extract LI 160, 900 mg/day for 11 days before a single dose of phenprocoumon had a lower AUC of the unbound fraction than when they took placebo(266).

In an open, three-way, crossover, randomized study in 12 healthy men who took a single dose of warfarin 25 mg alone or after pretreatment for 14 days with St John's wort, the apparent clearance of S-warfarin was 3.3 ml/minute before St John's wort was added and 3.7 ml/minute after (267). The respective apparent clearances of R-warfarin were 1.8 and 2.4 ml/minute. The mean ratios of the apparent clearances were 1.29 (95% CI = 1.16, 1.46) for S-warfarin and 1.23 (1.11, 1.37) for R-warfarin. St John's wort did not affect the apparent volume of distribution or protein binding of either enantiomer of warfarin. The authors concluded that St John's wort induces the clearance of both enantiomers of warfarin. INR was slightly reduced as a result, but platelet aggregation was not altered.

These observations suggest that St John's wort increases the clearance of both warfarin and phenprocoumon, possibly because of induction of CYP isozymes, particularly CYP2C9 and CYP3A4.

Lycium barbarum (Chinese wolfberry)

There has been a single anecdotal report of a possible interaction of *Lycium barbarum* with warfarin (268A).

- A 61-year-old Chinese woman, previously stabilized on warfarin (INR 2–3), drank a concentrated Chinese herbal tea made from *Lycium barbarum* fruits (3–4 glasses/day) for 4 days; her INR rose to 4.1. Warfarin was withheld for 1 day and then restarted at a lower dose. She stopped drinking the tea, and 7 days later her INR was 2.4.

In vitro studies showed that *Lycium barbarum* tea inhibited S-warfarin metabolism by CYP2C9; however, the inhibition was weak, with a dissociation constant of 3.4 g/l, suggesting that the observed interaction may have been caused by other mechanisms.

Panax ginseng

Ginseng can reduce the effect of warfarin (269), and there have been anecdotal reports of such an interaction (270) (271). There have also been several formal studies of the pharmacokinetic and pharmacodynamic effects of ginseng on warfarin.

The effects of American ginseng (*Panax quinquefolium*) have been studied in a double-blind, randomized, placebo-controlled trial in 20 young healthy subjects (272). Warfarin was given for 3 days during weeks 1 and 4, and starting in week 2 the subjects were assigned to either ginseng or placebo. The peak INR fell significantly after 2 weeks of ginseng administration compared with placebo; the difference between ginseng and placebo was −0.19 (95% CI = −0.36, −0.07).

However, in an open, three-way, crossover, randomized study in 12 healthy men who took a single dose of warfarin 25 mg alone or after pretreatment for 7 days with ginseng, there was no change in the pharmacokinetics or pharmacodynamics of either S-warfarin or R-warfarin (35).

In 20 healthy volunteers who took 100 mg of an extract of *Panax ginseng* standardized to 4% ginsenosides twice daily for 14 days there were no effects on CYP3A and this result was confirmed in in vitro studies (273C).

The discrepancies between these studies could be explained by differing susceptibilities of different populations or by different effects of ginseng from different sources.

Both pharmacokinetic and pharmacodynamic components could play a role in an interaction of ginseng with warfarin. Ginseng extracts have an antiplatelet effect. Ginsenosides Rg3 and protopanaxadiol-type saponins were platelet-activating factor antagonists with IC50 values of 49–92 μmol/l (274). Modulation of various CYP isoenzymes could also be a mechanism. In rats the pharmacokinetics and pharmacodynamics of warfarin after a single dose and at steady state were not altered by co-administered ginseng (275). However, extensive in vitro and in vivo animal studies have shown that constituents of ginseng can modulate various CYP isoenzymes that metabolize warfarin. Ginsenoside Rd was weakly inhibitory against recombinant CYP3A4, CYP2D6, CYP2C19, and CYP2C9, whereas ginsenoside Re and ginsenoside Rf (200 μmol/l) increased the activity of CYP2C9 and CYP3A4 (276). In rats, the standardized saponin of red ginseng was inhibitory on p-nitrophenol hydroxylase (CYP2E1) activity in a dose-related manner (277).

Salvia miltiorrhiza (danshen)

There have been anecdotal reports of enhanced anticoagulation and bleeding when patients taking long-term warfarin therapy consumed *Salvia miltiorrhiza* root (278) (279) (280). As these patients were also taking other medications, the contribution of *Salvia miltiorrhiza* to the interaction was difficult to determine. However, the author of a systematic review concluded that danshen should be avoided in patients taking warfarin (281).

Table 2   Families of plants and their species that are the subjects of monographs in this encyclopedia (by alphabetical order of family)

Family (common name)	Species (common name)
Acanthaceae	*Andrographis paniculata* (waterwillow)
Acoraceae (calamus)	*Acorus calamus* (calamus root)
Aloeaceae (aloe)	*Aloe capensis* (aloe)
	Aloe vera (aloe)
Amaranthaceae (amaranth)	*Pfaffia paniculata* (Brazilian ginseng)
Anacardiaceae (sumac)	*Rhus species* (sumac)
Apiaceae (carrot)	*Ammi majus* (bishop's weed)
	Ammi visnaga (toothpick weed)
	Angelica sinensis (dong quai)
	Centella asiatice (centella)
	Conium maculatum (hemlock)
	Coriandrum sativum (coriander)
	Ferula assa-foetida (asafetida)
Apocynaceae (dogbane)	*Rauwolfia serpentina* (snakeroot)
Araliaceae (ginseng)	*Eleutherococcus senticosus* (Siberian ginseng)
	Panax ginseng (Asian ginseng)
Arecaceae (palm)	*Areca catechu* (areca, betel)
	Serenoa repens (saw palmetto)
Aristolochiaceae (birthwort)	*Aristolochia species* (Dutchman's pipe)
	Asarum heterotropoides (Xu xin)
Asclepiadaceae (milkweed)	*Asclepias tuberosa* (pleurisy root)
	Xysmalobium undulatum (xysmalobium)
Asteraceae (aster)	*Achillea millefolium* (yarrow)
	Anthemis species *and Matricaria recutita* (chamomile)
	Arnica montana (arnica)
	Artemisia absinthium (wormwood)
	Artemisia annua (qinghaosu)
	Artemisia cina (wormseed)
	Artemisia vulgaris (common wormwood)
	Calendula officinalis (marigold)
	Callilepis laureola (impila, ox-eye daisy)
	Chrysanthemum vulgaris (common tansy)
	Cynara scolymus (artichoke)
	Echinacea species (coneflower)
	Eupatorium species (thoroughwort)
	Inula helenium (elecampane)
	Petasites species (butterbur)
	Senecio species (ragwort)
	Silybum marianum (milk thistle)
	Tanacetum parthenium (feverfew)
	Tussilago farfara (coltsfoot)
Berberidaceae (barberry)	*Berberis vulgaris* (European barberry)
	Caulophyllum thalictroides (blue cohosh)
	Dysosma pleianthum (bajiaolian)
	Mahonia species (barberry)
Boraginaceae (borage)	*Cynoglossum officinale* (hound's tongue)
	Symphytum officinale (black wort)
	Heliotropium species (heliotrope)
Brassicaceae (mustard)	*Armoracia rusticana* (horseradish)
	Brassica nigra (black mustard)
	Raphanus sativus var. *niger* (black radish)
	Sinapis species (mustard)
Burseraceae	*Boswellia serrata* species (frankincense)
	Commiphora species (myrrh)
Campanulaceae (bellflower)	*Lobelia inflata* (Indian tobacco)
Cannabaceae	*Cannabis sativa* (cannabis)
	Humulus lupulus (hop)
Capparaceae (caper)	*Capparis spinosa* (caper plant)
Celastraceae (bittersweet)	*Catha edulis* (khat, qat)
	Euonymus europaeus (spindle tree)
	Tripterygium wilfordii (Lei gong teng)
Chenopodiaceae (goosefoot)	*Chenopodium ambrosioides* (American wormseed)

(Continued)

Table 2 (Continued)

Family (common name)	Species (common name)
Clusiaceae (mangosteen)	*Hypericum perforatum* (St John's wort)
Convolvulaceae (morning glory)	*Convolvulus scammonia* (Mexican scammony)
	Ipomoea purga (jalap)
Coriariaceae	*Coriaria arborea* (tutu)
Cucurbitaceae (cucumber)	*Bryonia alba* (white bryony)
	Citrullus colocynthis (colocynth)
	Ecballium elaterium (squirting cucumber)
	Momordica charantia (karela fruit, bitter melon)
	Sechium edule (chayote)
Cupressaceae (cypress)	*Juniperus communis* (juniper)
Cycadaceae (cycad)	*Cycas circinalis* (false sago palm)
Droseraceae (sundew)	*Dionaea muscipula* (Venus flytrap)
Dryopteraceae (wood fern)	*Dryopteris filix-mas* (male fern)
Ephedraceae	*Ephedra* species (jointfir)
Ericaceae (heath)	*Arctostaphylos uva-ursi* (bearberry)
	Gaultheria procumbens (wintergreen)
	Ledum palustre (marsh Labrador tea)
	Vaccinium macrocarpon (cranberry)
Euphorbiaceae (spurge)	*Breynia officinalis* (Chi R Yun)
	Croton tiglium (croton)
	Ricinus communis (castor oil plant)
	Arachis species (peanut)
Fabaceae (pea)	*Cassia* species (senna)
	Crotalaria species (rattlebox)
	Cyamopsis tetragonoloba (cluster bean)
	Cytisus scoparius (Scotch broom)
	Dipteryx species (tonka beans)
	Genista tinctoria (dyer's broom)
	Glycyrrhiza glabra (liquorice)
	Lupinus species (lupin)
	Medicago sativa (alfalfa)
	Melilotus officinalis (sweet clover)
	Myroxylon species (balsam of Peru)
	Pithecollobium jiringa (jering fruit)
	Sophora falvescens (Ku shen)
	Trifolium pratense (red clover)
Gentianaceae (gentian)	*Gentiana* species (gentian)
	Swertia species (felwort)
Ginkgoaceae	*Ginkgo biloba* (maidenhair)
Hippocastanaceae (horse chestnut)	*Aesculus hippocastanum* (horse chestnut)
Illiciaceae (star anise)	*Illicium* species (star anise)
	Crocus sativus (Indian saffron)
Iridaceae (iris)	*Illicium verum/anisatum* (star anise)
Juglandaceae (walnut)	*Juglans regia* (English walnut)
Krameriaceae (Krameria)	*Krameria* species (ratany)
Lamiaceae (mint)	*Hedeoma pulegoides* (pennyroyal)
	Lavandula angustifolia (lavender)
	Mentha piperita (peppermint)
	Mentha pulegium (pennyroyal)
	Salvia miltiorrhiza (danshen)
	Salvia officinalis (sage)
	Scutellaria species (skullcap)
	Teucrium species (germander)
Lauraceae (laurel)	*Cinnamonum camphora* (camphor tree)
	Laurus nobilis (laurel)
	Sassafras albidum (sassafras)
Liliaceae (lily)	*Allium sativum* (garlic)
	Colchicum autumnale (autumn crocus)
	Ruscus aculeatus (butcher's broom)
	Veratrum species (hellebore)
Loganiaceae (Logania)	*Strychnos nux-vomica* (nux vomica)
Lycopodiaceae (club moss)	*Lycopodium serratum* (clubmoss)

(Continued)

Table 2 (Continued)

Family (common name)	Species (common name)
Malvaceae (mallow)	*Gossypium* species (cotton)
	Psoralea corylifolia (bakuchi)
Meliaceae (mahogany)	*Azadirachta indica* (bead tree)
Menispermaceae (moonseed)	*Stephania* species (Jin bu huan)
Myristicaceae (nutmeg)	*Myristica fragrans* (nutmeg)
Myrtaceae (myrtle)	*Eugenia caryophyllus* (clove)
	Eucalyptus species (eucalyptus)
	Melaleuca alternifolia (tea tree)
Onagraceae (evening primrose)	*Oenothera biennis* (evening primrose)
	Chelidonium majus (celandine)
Papaveraceae (poppy)	*Papaver somniferum* (opium poppy)
Passifloraceae (passion flower)	*Passiflora incarnata* (passion flower)
Pedaliaceae (sesame)	*Harpagophytum procumbens* (devil's claw)
Phytolaccaceae (pokeweed)	*Phytolacca americana* (pokeweed)
Piperaceae (pepper)	*Piper methysticum* (kava kava)
Plantaginaceae (plantain)	*Plantago* species (plantain)
Poaceae (grass)	*Anthoxanthum odoratum* (sweet vernal grass)
Polygonaceae (buckwheat)	*Polygonum* (knotweed)
	Rheum palmatum (rhubarb)
Ranunculaceae (buttercup)	*Aconitum napellus* (monkshood)
	Cimicifuga racemosa (black cohosh)
	Delphinium species (delphinium)
	Hydrastis canadensis (golden seal)
	Pulsatilla species (pasque flower)
	Ranunculus damascenus (buttercup)
Rhamnaceae (buckthorn)	*Rhamnus purshianus* (cascara sagrada)
	Ziziphus jujuba (dazao)
Rosaceae (rose)	*Crataegus* species (hawthorn)
	Prunus species (plum)
	Asperula odorata (sweet woodruff)
Rubiaceae (madder)	*Cephaelis ipecacuanha* (ipecac)
	Hintonia latiflora (copalchi bark)
	Morinda citrifolia (noni)
	Rubia tinctorum (madder)
	Uncaria tomentosa (cat's claw)
Rutaceae (rue)	*Agathosma betulina* (buchu)
	Citrus auranticum (bergamot)
	Citrus paradisi (grapefruit)
	Dictamnus dasycarpus (densefruit pittany)
	Pilocarpus species (pilocarpus)
	Ruta graveolens (rue)
Salicaceae (willow)	*Salix* species (willow)
Sapindaceae (soapberry)	*Blighia sapida* (akee)
	Paullinia cupana (guaraná)
Selaginellaceae (spike moss)	*Selaginella doederleinii* (spike moss)
Solanaceae (potato)	*Anisodus tanguticus* (Zangqie)
	Capsicum annum (chili pepper)
	Datura candida (angel's trumpet)
	Datura stramonium (Jimson weed)
	Datura suaveolens (angel's trumpet)
	Lycium barbarum (Chinese wolfberry)
	Mandragora species (mandrake)
	Nicotiana tabacum (tobacco)
	Scopolia species (scopola)
Sterculiaceae (cacao)	*Sterculia* species (sterculia)
Taxaceae (yew)	*Taxus* species (yew)
Theaceae (tea)	*Camellia sinensis* (green tea)
Urticaceae (nettle)	*Urtica dioica* (stinging nettle)
Valerianaceae (valerian)	*Valeriana* (valerian)
Verbenaceae (verbena)	*Vitex agnus-castus* (chaste tree)
Viscaceae (Christmas mistletoe)	*Phoradendron flavescens* (American mistletoe)
	Viscum album (mistletoe)
Zingiberaceae (ginger)	*Zingiber officinale* (ginger)
Zygophyllaceae (bean caper)	*Larrea tridentata* (chaparal ceosote bush)

Table 3   Families of plants and their genera or species that are the subjects of monographs in this encyclopedia (by alphabetical order of genus or species)

Genus or species (common name)	Family (common name)
Achillea millefolium (yarrow)	Asteraceae (aster)
Aconitum napellus (monkshood)	Ranunculaceae (buttercup)
Acorus calamus (calamus root)	Acoraceae (calamus)
Aesculus hippocastanum (horse chestnut)	Hippocastanaceae (horse chestnut)
Agathosma betulina (buchu)	Rutaceae (rue)
Allium sativum (garlic)	Liliaceae (lily)
Aloe capensis (aloe)	Aloeaceae (aloe)
Aloe vera (aloe)	Aloeaceae (aloe)
Ammi majus (bishop's weed)	Apiaceae (carrot)
Ammi visnaga (toothpick weed)	Apiaceae (carrot)
Andrographis paniculata (waterwillow)	Acanthaceae (acanthus)
Angelica sinensis (dong quai)	Apiaceae (carrot)
Anisodus tanguticus (zangqie),	Solanaceae (potato)
Anthemis species (chamomile)	Asteraceae (aster)
Anthoxanthum odoratum (sweet vernal grass)	Poaceae (grass)
Arachis species (peanut)	Fabaceae (pea)
Arctostaphylos uva-ursi (bearberry)	Ericaceae (heath)
Areca catechu (areca, betel)	Arecaceae (palm)
Aristolochia species (Dutchman's pipe)	Aristolochiaceae (birthwort)
Armoracia rusticana (horseradish)	Brassicaceae (mustard)
Arnica montana (arnica)	Asteraceae (aster)
Artemisia annua (qinghaosu)	Asteraceae (aster)
Artemisia absinthium (wormwood)	Asteraceae (aster)
Artemisia cina (wormseed)	Asteraceae (aster)
Artemisia vulgaris (common wormwood)	Asteraceae (aster)
Asarum heterotropoides (xu xin)	Aristolochiaceae (birthwort)
Asclepias tuberosa (pleurisy root)	Asclepiadaceae (milkweed)
Asperula odorata (sweet woodruff)	Rubiaceae (madder)
Azadirachta indica (bead tree)	Meliaceae (mahogany)
Berberis vulgaris (European barberry)	Berberidaceae (barberry)
Blighia sapida (akee)	Sapindaceae (soapberry)
Boswellia serata (frankincense)	Burseraceae (frankincense)
Brassica nigra (black mustard)	Brassicaceae (mustard)
Breynia officinalis (chi r yun)	Euphorbiaceae (spurge)
Bryonia alba (white bryony)	Cucurbitaceae (cucumber)
Calendula officinalis (marigold)	Asteraceae (aster)
Callilepis laureola (impila, ox-eye daisy)	Asteraceae (aster)
Camellia sinensis (green tea)	Theaceae (tea)
Cannabis sativa (cannabis)	Cannabaceae
Capparis spinosa (caper plant)	Capparaceae (caper)
Capsicum annum (chili pepper)	Solanaceae (potato)
Cassia species (senna)	Fabaceae (pea)
Catha edulis (khat, qat)	Celastraceae (bittersweet)
Caulophyllum thalictroides (blue cohosh)	Berberidaceae (barberry)
Centella asiatica (centella)	Apiaceae (carrot)
Cephaelis ipecacuanha (ipecac)	Rubiaceae (madder)
Chelidonium majus (celandine)	Papaveraceae (poppy)
Chenopodium ambrosioides (American wormseed)	Chenopodiaceae (goosefoot)
Chrysanthemum vulgaris (common tansy)	Asteraceae (aster)
Cimicifuga racemosa (black cohosh)	Ranunculaceae (buttercup)
Cinnamonum camphora (camphor tree)	Lauraceae (laurel)
Citrullus colocynthis (colocynth)	Cucurbitaceae (cucumber)
Citrus aurantium (bergamot)	Rutaceae (rue)
Citrus paradisi (grapefruit)	Rutaceae (rue)
Colchicum autumnale (autumn crocus)	Liliaceae (lily)
Commiphora species (myrrh)	Burseraceae (frankincense)
Conium maculatum (hemlock)	Apiaceae (carrot)
Convolvulus scammonia (Mexican scammony)	Convolvulaceae (morning glory)
Coriandrum sativum (coriander)	Apiaceae (carrot)
Coriaria arborea (tutu)	Coriariaceae
Crataegus species (hawthorn)	Rosaceae (rose)

(Continued)

Table 3 (Continued)

Genus or species (common name)	Family (common name)
Crocus sativus (Indian saffron)	Iridaceae (iris)
Crotalaria species (rattlebox)	Fabaceae (pea)
Croton tiglium (croton)	Euphorbiaceae (spurge)
Cyamopsis tetragonoloba (cluster bean)	Fabaceae (pea)
Cycas circinalis (false sago palm)	Cycadaceae (cycad)
Cynara scolymus (artichoke)	Asteraceae (aster)
Cynoglossum officinale (hound's tongue)	Boraginaceae (borage)
Cytisus scoparius (Scotch broom)	Fabaceae (pea)
Datura candida (angel's trumpet)	Solanaceae (potato)
Datura stramonium (Jimson weed)	Solanaceae (potato)
Datura suaveolens (angel's trumpet)	Solanaceae (potato)
Delphinium species (delphinium)	Ranunculaceae (buttercup)
Dictamnus dasycarpus (densefruit pittany)	Rutaceae (rue)
Dionaea muscipula (Venus flytrap)	Droseraceae (sundew)
Dipteryx species (tonka beans)	Fabaceae (pea)
Dryopteris filix-mas (male fern)	Dryopteraceae (wood fern)
Dysosma pleianthum (bajiaolian)	Berberidaceae (barberry)
Ecballium elaterium (squirting cucumber)	Cucurbitaceae (cucumber)
Echinacea species (coneflower)	Asteraceae (aster)
Eleutherococcus senticosus (Siberian ginseng)	Araliaceae
Ephedra species (jointfir)	Ephedraceae (Mormon-tea)
Eucalyptus species (eucalyptus)	Myrtaceae (myrtle)
Eugenia caryophyllus (clove)	Myrtaceae (myrtle)
Euonymus europaeus (spindle tree)	Celastraceae (bittersweet)
Eupatorium species (thoroughwort)	Asteraceae (aster)
Ferula assa-foetida (asafetida)	Apiaceae (carrot)
Gaultheria procumbens (wintergreen)	Ericaceae (heath)
Genista tinctoria (dyer's broom)	Fabaceae (pea)
Gentiana species (gentian)	Gentianaceae (gentian)
Ginkgo biloba (maidenhair)	Ginkgoaceae
Glycyrrhiza glabra (liquorice)	Fabaceae (pea)
Gossypium (cotton)	Malvaceae (mallow)
Harpagophytum procumbens (devil's claw)	Pedaliaceae (sesame)
Hedeoma pulegoides (pennyroyal)	Lamiaceae (mint)
Heliotropium species (heliotrope)	Boraginaceae (borage)
Hintonia latiflora (copalchi bark)	Rubiaceae (madder)
Humulus lupulus (hop)	Cannabaceae
Hydrastis canadensis (golden seal)	Ranunculaceae (buttercup)
Hypericum perforatum (St John's wort)	Clusiaceae (mangosteen)
Illicium species (star anise)	Illiciaceae (star anise)
Inula helenium (elecampane)	Asteraceae (aster)
Ipomoea purga (jalap)	Convolvulaceae (morning glory)
Juglans regia (English walnut)	Juglandaceae (walnut)
Juniperus communis (juniper)	Cupressaceae (cypress)
Krameria species (ratany)	Krameriaceae (Krameria)
Larres tridentata (chaparral, creosote bush)	Zygophyllaceae (bean caper)
Laurus nobilis (laurel)	Lauraceae (laurel)
Lavandula angustifolia (lavender)	Lamiaceae
Ledum palustre (marsh Labrador tea)	Ericaceae (heath)
Lobelia inflata (Indian tobacco)	Campanulaceae (bellflower)
Lupinus species (lupin)	Fabaceae (pea)
Lycium barbarum (Chinese wolfberry)	Solanaceae (potato)
Lycopodium serratum (clubmoss)	Lycopodiaceae (club moss)
Mahonia species (barberry)	Berberidaceae (barberry)
Mandragora species (mandrake)	Solanaceae (potato)
Matricaria recutita (chamomile)	Asteraceae (aster)
Medicago sativa (alfalfa)	Fabaceae (pea)
Melaleuca alternifolia (tea tree)	Myrtaceae (myrtle)
Melilotus officinalis (sweet clover)	Fabaceae (pea)
Mentha piperita (peppermint)	Lamiaceae
Mentha pulegium (pennyroyal)	Lamiaceae (mint)
Momordica charantia (karela fruit, bitter melon)	Cucurbitaceae (cucumber)

(Continued)

Table 3 (Continued)

Genus or species (common name)	Family (common name)
Morinda citrifolia (noni)	Rubiaceae (madder)
Myristica fragrans (nutmeg)	Myristicaceae (nutmeg)
Myroxylon species (balsam of Peru)	Fabaceae (pea)
Nicotiana tabacum (tobacco)	Solanaceae (potato)
Oenothera biennis (evening primrose)	Onagraceae (evening primrose)
Panax ginseng (Asian ginseng)	Araliaceae (ginseng)
Papaver somniferum (opium poppy)	Papaveraceae (poppy)
Passiflora incarnata (passion flower)	Passifloraceae (passion flower)
Paullinia cupana (guaraná)	Sapindaceae (soapberry)
Petasites species (butterbur)	Asteraceae (aster)
Pfaffia paniculata (Brazilian ginseng)	Amaranthaceae (amaranth)
Phoradendron flavescens (American mistletoe)	Viscaceae (Christmas mistletoe)
Phytolacca americana (pokeweed)	Phytolaccaceae (pokeweed)
Pilocarpus species (pilocarpus)	Rutaceae (rue)
Piper methysticum (kava kava)	Piperaceae (pepper)
Pithecollobium jiringa (jering fruit)	Fabaceae (pea)
Plantago species (plantain)	Plantaginaceae (plantain)
Polygonum (knotweed)	Polygonaceae (buckwheat)
Prunus species (plum)	Rosaceae (rose)
Psoralea corylifolia (bakuchi)	Malvaceae (mallow)
Pulsatilla species (pasque flower)	Ranunculaceae (buttercup)
Ranunculus damascenus (buttercup)	Ranunculaceae (buttercup)
Raphanus sativus var. niger (black radish)	Brassicaceae (mustard)
Rauwolfia serpentina (snakeroot)	Apocynaceae (dogbane)
Rhamnus purshianus (cascara sagrada)	Rhamnaceae (buckthorn)
Rheum palmatum (rhubarb)	Polygonaceae (buckwheat)
Rhus species (sumac)	Anacardiaceae (sumac)
Ricinus communis (castor oil plant)	Euphorbiaceae (spurge)
Rubia tinctorum (madder)	Rubiaceae (madder)
Ruscus aculeatus (butcher's broom)	Liliaceae (lily)
Ruta graveolens (rue)	Rutaceae (rue)
Salix species (willow)	Salicaceae (willow)
Salvia miltiorrhiza (danshen)	Lamiaceae (mint)
Salvia officinalis (sage)	Lamiaceae (mint)
Sassafras albidum (sassafras)	Lauraceae (laurel)
Scopolia species (scopola)	Solanaceae (potato)
Scutellaria species (skullcap)	Lamiaceae (mint)
Sechium edule (chayote)	Cucurbitaceae (cucumber)
Selaginella doederleinii (spike moss)	Selaginellaceae (spike moss)
Senecio species (ragwort)	Asteraceae (aster)
Serenoa repens (saw palmetto)	Arecaceae (palm)
Silybum marianum (milk thistle)	Asteraceae (aster)
Sinapis species (mustard)	Brassicaceae (mustard)
Sophora falvescens (ku shen)	Fabaceae (pea)
Stephania species (jin bu huan)	Menispermaceae (moonseed)
Sterculia species (sterculia)	Sterculiaceae (cacao)
Strychnos nux-vomica (nux vomica)	Loganiaceae (Logania)
Swertia species (felwort)	Gentianaceae (gentian)
Symphytum officinale (black wort)	Boraginaceae (borage)
Tanacetum parthenium (feverfew)	Asteraceae (aster)
Taxus species (yew)	Taxaceae (yew)
Teucrium species (germander)	Lamiaceae (mint)
Trifolium pratense (red clover)	Fabaceae
Tripterygium wilfordii (lei gong teng)	Celastraceae (bittersweet)
Tussilago farfara (coltsfoot)	Asteraceae (aster)
Uncaria tomentosa (cat's claw)	Rubiaceae (madder)
Urtica dioica (stinging nettle)	Urticaceae (nettle)
Vaccinium macrocarpon (cranberry)	Ericaceae (heath)
Valeriana (valerian)	Valerianaceae (valerian)
Veratrum species (hellebore)	Liliaceae (lily)
Viscum album (mistletoe)	Viscaceae (Christmas mistletoe)
Vitex agnus-castus (chaste tree)	Verbenaceae (verbena)
Xysmalobium undulatum (xysmalobium)	Asclepiadaceae (milkweed)
Zingiber officinale (ginger)	Zingiberaceae (ginger)
Ziziphus jujuba (dazao)	Rhamnaceae (buckthorn)

Table 4 Conversion of common names of plants to Latin names

Common name	Latin name (genus or species)
Akee	Blighia sapida
Alfalfa	Medicago sativa
Aloe	Aloe capensis
Aloe	Aloe vera
American mistletoe	Phoradendron flavescens
American wormseed	Chenopodium ambrosioides
Angel's trumpet	Datura candida
Angel's trumpet	Datura suaveolens
Areca, betel	Areca catechu
Arnica	Arnica montana
Artichoke	Cynara scolymus
Asafetida	Ferula assa-foetida
Asian ginseng	Panax ginseng
Autumn crocus	Colchicum autumnale
Bajiaolian	Dysosma pleianthum
Bakuchi	Psoralea corylifolia
Balsam of Peru	Myroxylon species
Barberry	Mahonia species
Bead tree	Azadirachta indica
Bearberry	Arctostaphylos uva-ursi
Bergamot	Citrus aurantium
Bishop's weed	Ammi majus
Black cohosh	Cimicifuga racemosa
Black mustard	Brassica nigra
Black radish	Raphanus sativus var. niger
Black wort	Symphytum officinale
Blue cohosh	Caulophyllum thalictroides
Brazilian ginseng	Pfaffia paniculata
Buchu	Agathosma betulina
Butcher's broom	Ruscus aculeatus
Butterbur	Petasites species
Buttercup	Ranunculus damascenus
Calamus root	Acorus calamus
Camphor tree	Cinnamonum camphora
Caper plant	Capparis spinosa
Cascara sagrada	Rhamnus purshianus
Castor oil plant	Ricinus communis
Cat's claw	Uncaria tomentosa
Celandine	Chelidonium majus
Centella	Centella asiatica
Chamomile	Anthemis species
Chamomile	Matricaria recutita
Chaste tree	Vitex agnus-castus
Chayote	Sechium edule
Chi r yun	Breynia officinalis
Chili pepper	Capsicum annum
Chinese wolfberry	Lycium barbarum
Clove	Eugenia caryophyllus
Clubmoss	Lycopodium serratum
Cluster bean	Cyamopsis tetragonoloba
Colocynth	Citrullus colocynthis
Coltsfoot	Tussilago farfara
Common tansy	Chrysanthemum vulgaris
Common wormwood	Artemisia vulgaris
Coneflower	Echinacea species
Copalchi bark	Hintonia latiflora
Coriander	Coriandrum sativum
Cotton	Gossypium species
Cranberry	Vaccinium macrocarpon
Croton	Croton tiglium
Danshen	Salvia miltiorrhiza
Dazao	Ziziphus jujuba

(Continued)

Table 4 (Continued)

Common name	Latin name (genus or species)
Delphinium	*Delphinium* species
Densefruit pittany	*Dictamnus dasycarpus*
Devil's claw	*Harpagophytum procumbens*
Dong quai	*Angelica sinensis*
Dutchman's pipe	*Aristolochia* species
Dyer's broom	*Genista tinctoria*
Elecampane	*Inula helenium*
English walnut	*Juglans regia*
Eucalyptus	*Eucalyptus* species
European barberry	*Berberis vulgaris*
Evening primrose	*Oenothera biennis*
False sago palm	*Cycas circinalis*
Felwort	*Swertia* species
Feverfew	*Tanacetum parthenium*
Frankincense	*Boswellia serrata*
Garlic	*Allium sativum*
Gentian	*Gentiana* species
Germander	*Teucrium* species
Ginger	*Zingiber officinale*
Golden seal	*Hydrastis canadensis*
Grapefruit	*Citrus paradisi*
Green tea	*Camellia sinensis*
Guaraná	*Paullinia cupana*
Guar gum	*Cyamopsis tetragonoloba*
Hawthorn	*Crataegus* species
Heliotrope	*Heliotropium* species
Hellebore	*Veratrum* species
Hemlock	*Conium maculatum*
Hop	*Humulus lupulus*
Horse chestnut	*Aesculus hippocastanum*
Horseradish	*Armoracia rusticana*
Hound's tongue	*Cynoglossum officinale*
Impila, ox-eye daisy	*Callilepis laureola*
Indian saffron	*Crocus sativus*
Indian tobacco	*Lobelia inflata*
Ipecac	*Cephaelis ipecacuanha*
Jalap	*Ipomoea purga*
Jering fruit	*Pithecollobium jiringa*
Jimson weed	*Datura stramonium*
Jin bu huan	*Stephania* species
Jointfir	*Ephedra* species
Juniper	*Juniperus communis*
Karela fruit, bitter melon	*Momordica charantia*
Kava kava	*Piper methysticum*
Khat, qat	*Catha edulis*
Knotweed	*Polygonum*
Ku shen	*Sophora falvescens*
Laurel	*Laurus nobilis*
Lavender	*Lavandula angustifolia*
Lei gong teng	*Tripterygium wilfordii*
Liquorice	*Glycyrrhiza glabra*
Lupin	*Lupinus* species
Madder	*Rubia tinctorum*
Maidenhair	*Ginkgo biloba*
Male fern	*Dryopteris filix-mas*
Mandrake	*Mandragora* species
Marigold	*Calendula officinalis*
Marsh Labrador tea	*Ledum palustre*
Mexican scammony	*Convolvulus scammonia*
Milk thistle	*Silybum marianum*
Mistletoe	*Viscum album*
Monkshood	*Aconitum napellus*

(Continued)

Table 4 (Continued)

Common name	Latin name (genus or species)
Mustard	*Sinapis* species
Myrrh	*Commiphora* species
Noni	*Morinda citrifolia*
Nutmeg	*Myristica fragrans*
Nux vomica	*Strychnos nux-vomica*
Opium poppy	*Papaver somniferum*
Pasque flower	*Pulsatilla* species
Passion flower	*Passiflora incarnata*
Peanut	*Arachis* species
Pennyroyal	*Hedeoma pulegoides*
Pennyroyal	*Mentha pulegium*
Peppermint	*Mentha piperita*
Pilocarpus	*Pilocarpus* species
Plantain	*Plantago* species
Pleurisy root	*Asclepias tuberosa*
Plum	*Prunus* species
Pokeweed	*Phytolacca americana*
Qinghaosu	*Artemisia annua*
Ragwort	*Senecio* species
Ratany	*Krameria* species
Rattlebox	*Crotalaria* species
Red clover	*Trifolium pratense*
Rhubarb	*Rheum palmatum*
Rue	*Ruta graveolens*
Sage	*Salvia officinalis*
Sassafras	*Sassafras albidum*
Saw palmetto	*Serenoa repens*
Scopola	*Scopolia* species
Scotch broom	*Cytisus scoparius*
Senna	*Cassia* species
Siberian ginseng	*Eleutherococcus senticosus*
Skullcap	*Scutellaria* species
Snakeroot	*Rauwolfia serpentina*
Spike moss	*Selaginella doederleinii*
Spindle tree	*Euonymus europaeus*
Squirting cucumber	*Ecballium elaterium*
St John's wort	*Hypericum perforatum*
Star anise	*Illicium* species
Sterculia	*Sterculia* species
Stinging nettle	*Urtica dioica*
Sumac	*Rhus* species
Sweet clover	*Melilotus officinalis*
Sweet vernal grass	*Anthoxanthum odoratum*
Sweet woodruff	*Asperula odorata*
Tea tree	*Melaleuca alternifolia*
Thoroughwort	*Eupatorium* species
Tobacco	*Nicotiana tabacum*
Tonka beans	*Dipteryx* species
Toothpick weed	*Ammi visnaga*
Tutu	*Coriaria arborea*
Valerian	*Valeriana*
Venus flytrap	*Dionaea muscipula*
Waterwillow	*Andrographis paniculata*
White bryony	*Bryonia alba*
Willow	*Salix* species
Wintergreen	*Gaultheria procumbens*
Wormseed	*Artemisia cina*
Wormwood	*Artemisia absinthium*
Xu xin	*Asarum heterotropoides*
Xysmalobium	*Xysmalobium undulatum*
Yarrow	*Achillea millefolium*
Yew	*Taxus* species
Zang qie	*Anisodus tanguticus*

The direct anticoagulant activity of *Salvia miltiorrhiza* root itself may provide a partial explanation for the interactions. However, pharmacokinetic interactions may also play a role. Warfarin is mainly metabolized by CYP2C9 and to a smaller extent by CYP1A2 and CYP3A4. In mice oral, administration of an ethyl acetate extract of danshen caused a dose-related increase in liver microsomal 7-methoxyresorufin O-demethylation activity, with a three-fold increase in warfarin 7-hydroxylation (282). However, the aqueous extract had no effects. Immunoblot analysis of microsomal proteins showed that ethyl acetate extraction increased the proteins associated with CYP1A and CYP3A. At a dose corresponding to its content in the ethyl acetate extract, tanshinone IIA, the main diterpene quinone in *Salvia miltiorrhiza*, increased mouse liver microsomal 7-methoxyresorufin O-demethylation activity. These results suggest that there are inducing agents for mouse CYP1A, CYP2C, and CYP3A in ethyl acetate extracts but not in aqueous extracts of *Salvia miltiorrhiza*.

In rats treatment with *Salvia miltiorrhiza* root extract 5 g/kg bd for 3 days followed by a single oral dose of racemic warfarin increased the absorption rate constants, AUC, C_{max}, and half-life of warfarin but reduced the clearances and apparent volumes of distribution of both (R)-warfarin and (S)-warfarin (283). A similar effect was observed during steady-state warfarin administration. The anticoagulant effect of warfarin was also potentiated. *Salvia miltiorrhiza* root extract itself had no effect on prothrombin time at this dose, suggesting that altered warfarin metabolism was a possible mechanism.

After a single oral dose of racemic warfarin 2 mg/kg in rats, an oral extract of *Salvia miltiorrhiza* 5 g/kg bd for 3 days significantly altered the pharmacokinetics of both R-warfarin and S-warfarin and increased the plasma concentrations of both enantiomers over a period of 24 hours and the prothrombin time over 2 days (284). Steady-state concentrations of racemic warfarin during administration of 0.2 mg/kg/day for 5 days with extract of *Salvia miltiorrhiza* 5 g/kg bd for 3 days not only prolonged the prothrombin time but also increased the steady-state plasma concentrations of R-warfarin and S-warfarin. These results suggested that *Salvia miltiorrhiza* increases the absorption rate, exposure, and half-lives of both R-warfarin and S-warfarin, but reduces their clearances and apparent volumes of distribution.

In addition, *Salvia miltiorrhiza* root extract might change the plasma protein binding of warfarin. Both (R)-warfarin and (S)-warfarin bind to the so-called site I of albumin with high affinity. *Salvia miltiorrhiza* root extract was 50–70% bound by albumin and in vitro Salvia miltiorrhiza root extract displaced salicylate from protein binding, thereby increasing the unbound salicylate concentration (285). However, kangen-karyu, a mixture of six herbs (peony root, *Cnidium* rhizome, safflower, *Cyperus* rhizome, *Saussurea* root, and root of *Salvia miltiorrhiza*), significantly increased the plasma warfarin concentration and prothrombin time in rats, but did not alter the serum protein binding of warfarin (286). Further studies are required to explore the effects of *Salvia miltiorrhiza* root extract on the metabolism and plasma protein binding of drugs such as warfarin in humans.

Zingiber officinale (ginger)

Despite anecdotal reports of a possible interaction (287) (288), several studies in rats and humans have shown no effect of ginger on warfarin pharmacokinetics or pharmacodynamics (15) (24) (289) (290).

PC-Spes

PC-Spes is a mixture of eight herbs: *Chrysanthemum morifolium*, *Isatis indigotica*, *Glycyrrhiza glabra* (licorice), *Ganoderma lucidum*, *Panax pseudoginseng*, *Robdosia rubescens*, *Serenoa repens* (saw palmetto), and *Scutellaria baicalensis* (skullcap). It has been reported to increase the INR in a 79-year-old man with prostate cancer taking warfarin, an effect that was attributed to inhibition of warfarin metabolism (291). However, warfarin has also been found in formulations of PC-Spes (292).

Quilinggao

Quilinggao, a popular Chinese mixture that contains a multitude of herbal ingredients (including *Fritillaria cirrhosa* and other *Fritillaria* species, *Paeoniae rubra*, *Lonicera japonica*, and *Poncirus trifoliata*, in many different brands), has been anecdotally reported to enhance the actions of warfarin (293).

- A 61-year-old man taking stable warfarin therapy developed gum bleeding, epistaxis, and skin bruising 5 days after taking quilinggao. His international normalized ratio was above 6. His warfarin was withdrawn and the international normalized ratio normalized. Days later he tried taking quilinggao again, with a similar result.

The authors pointed out that several herbs in this mixture have anticoagulant activity.

References

1. Phillipson JD. Traditional medicine treatment for eczema: experience as a basis for scientific acceptance. Eur Phytotelegram 1994;6:33–40.
2. Barnes J. A close look at synergy and polyvalent action in medicinal plants. Inpharma 1999;1185:3–4.
3. Ernst E, De Smet PA, Shaw D, Murray V. Traditional remedies and the "test of time". Eur J Clin Pharmacol 1998;54(2):99–100.
4. Anonymous. In: Complementary Medicines. London: Mintel International Group, 1997:13.
5. Institute of Medical Statistics Self-Medication International. Herbals in Europe. London: IMS Self-Medication International;. 1998.
6. Brevoort P. The booming US botanical market. A new overview. Herbalgram 1998;44:33–46.
7. Eisenberg DM, Kessler RC, Foster C, Norlock FE, Calkins DR, Delbanco TL. Unconventional medicine in the United States. Prevalence, costs, and patterns of use. N Engl J Med 1993;328(4):246–52.
8. Eisenberg DM, Davis RB, Ettner SL, Appel S, Wilkey S, Van Rompay M, Kessler RC. Trends in alternative medicine use in the United States, 1990–1997: results of a follow-up national survey. JAMA 1998;280(18):1569–75.

9. Ang-Lee MK, Moss J, Yuan CS. Herbal medicines and perioperative care. JAMA 2001;286(2):208–16.
10. Ernst E. Use of herbal medications before surgery. JAMA 2001;286(20):2542–3.
11. Tomassoni AJ, Simone K. Herbal medicines for children: an illusion of safety? Curr Opin Pediatr 2001;13(2):162–9.
12. Ernst E, Pittler MH, Stevinson C, White AR, Eisenberg D. The Desktop Guide to Complementary and Alternative Medicine. Edinburgh: Mosby;. 2001.
13. Seeff LB, Lindsay KL, Bacon BR, Kresina TF, Hoofnagle JH. Complementary and alternative medicine in chronic liver disease. Hepatology 2001;34(3):595–603.
14. Mantle D, Gok MA, Lennard TW. Adverse and beneficial effects of plant extracts on skin and skin disorders. Adverse Drug React Toxicol Rev 2001;20(2):89–103.
15. Xing JH, Soffer EE. Adverse effects of laxatives. Dis Colon Rectum 2001;44(8):1201–9.
16. Elvin-Lewis M. Should we be concerned about herbal remedies. J Ethnopharmacol 2001;75(2–3):141–64.
17. Ko R. Adverse reactions to watch for in patients using herbal remedies. West J Med 1999;171(3):181–6.
18. Astin JA. Why patients use alternative medicine: results of a national study. JAMA 1998;279(19):1548–53.
19. Abbot NC, White AR, Ernst E. Complementary medicine. Nature 1996;381(6581):361.
20. De Smet PA. Health risks of herbal remedies. Drug Saf 1995;13(2):81–93.
21. Winslow LC, Kroll DJ. Herbs as medicines. Arch Intern Med 1998;158(20):2192–9.
22. Miller LG. Herbal medicinals: selected clinical considerations focusing on known or potential drug–herb interactions. Arch Intern Med 1998;158(20):2200–11.
23. Mashour NH, Lin GI, Frishman WH. Herbal medicine for the treatment of cardiovascular disease: clinical considerations. Arch Intern Med 1998;158(20):2225–34.
24. Saller R, Reichling J, Kristof O. Phytotherapie-Behandlung ohne Nebenwirkungen?. [Phytotherapy—treatment without side effects?.] Dtsch Med Wochenschr 1998;123(3):58–62.
25. Bateman J, Chapman RD, Simpson D. Possible toxicity of herbal remedies. Scott Med J 1998;43(1):7–15.
26. Ernst E. Harmless herbs? A review of the recent literature. Am J Med 1998;104(2):170–8.
27. Shaw D. Risks or remedies? Safety aspects of herbal remedies in the UK. J R Soc Med 1998;91(6):294–6.
28. Marrone CM. Safety issues with herbal products. Ann Pharmacother 1999;33(12):1359–62.
29. Ko RJ. Causes, epidemiology, and clinical evaluation of suspected herbal poisoning. J Toxicol Clin Toxicol 1999;37(6):697–708.
30. Ernst E. Phytotherapeutika. Wie harmlos sind sie wirklich? Dtsch Arzteblatt 1999;48:3107–8.
31. Calixto JB. Efficacy, safety, quality control, marketing and regulatory guidelines for herbal medicines (phytotherapeutic agents). Braz J Med Biol Res 2000;33(2):179–89.
32. De Smet PA. Herbal remedies. N Engl J Med 2002;347(25):2046–56.
33. Ernst E, Pittler MH. Risks associated with herbal medicinal products. Wien Med Wochenschr 2002;152(7–8):183–9.
34. Ali MS, Uzair SS. Natural organic toxins. Hamdard Medicus 2002;XLIV:86–93.
35. Gee BC, Wilson P, Morris AD, Emerson RM. Herbal is not synonymous with safe. Arch Dermatol 2002;138(12):1613.
36. Matthews HB, Lucier GW, Fisher KD. Medicinal herbs in the United States: research needs. Environ Health Perspect 1999;107(10):773–8.
37. Hussain SH. Potential risks of health supplements—self-medication practices and the need for public health education. Int J Risk Saf Med 1999;12:167–71.
38. Fontana RJ. Acute liver failure. Curr Opin Gastroenterol 1999;15:270–7.
39. Valli G, Giardina EG. Benefits, adverse effects and drug interactions of herbal therapies with cardiovascular effects. J Am Coll Cardiol 2002;39(7):1083–95.
40. Chitturi S, Farrell GC. Herbal hepatotoxicity: an expanding but poorly defined problem. J Gastroenterol Hepatol 2000;15(10):1093–9.
41. Haller CA, Dyer JE, Ko R, Olson KR. Making a diagnosis of herbal-related toxic hepatitis. West J Med 2002;176(1):39–44.
42. Ernst E. Adverse effects of herbal drugs in dermatology. Br J Dermatol 2000;143(5):923–9.
43. Holsen DS. Flora og efflorescenser—om planter som arsak til hudsykdom. [Plants and plant produce—about plants as cause of diseases.] Tidsskr Nor Laegeforen 2002;122(17):1665–9.
44. Ernst E. Adverse effects of unconventional therapies in the elderly: a systematic review of the recent literature. J Am Aging Assoc 2002;25:11–20.
45. Ernst E. Herbal medicinal products during pregnancy: are they safe? BJOG 2002;109(3):227–35.
46. Hodges PJ, Kam PC. The peri-operative implications of herbal medicines. Anaesthesia 2002;57(9):889–99.
47. Cheng B, Hung CT, Chiu W. Herbal medicine and anaesthesia. Hong Kong Med J 2002;8(2):123–30.
48. Bartsch H. Gefahrliche Naturprodukte: sind Karzinogene im Kräutertee?. [Hazardous natural products. Are there carcinogens in herbal teas?.] MMW Fortschr Med 2002;144(41):14.
49. Pies R. Adverse neuropsychiatric reactions to herbal and over-the-counter "antidepressants". J Clin Psychiatry 2000;61(11):815–20.
50. Tomlinson B, Chan TY, Chan JC, Critchley JA, But PP. Toxicity of complementary therapies: an eastern perspective. J Clin Pharmacol 2000;40(5):451–6.
51. Bensoussan A, Myers SP, Drew AK, Whyte IM, Dawson AH. Development of a Chinese herbal medicine toxicology database. J Toxicol Clin Toxicol 2002;40(2):159–67.
52. Ernst E. Ayurvedic medicines. Pharmacoepidemiol Drug Saf 2002;11(6):455–6.
53. Boullata JI, Nace AM. Safety issues with herbal medicine. Pharmacotherapy 2000;20(3):257–69.
54. Shapiro R. Safety assessment of botanicals. Nutraceuticals World 2000;52–63July/August.
55. Saller R, Iten F, Reichling J. Unerwünschte Wirkungen und Wechselwirkungen von Phytotherapeutika. Erfahrungsheilkunde 2000;6:369–76.
56. Pennachio DL. Drug–herb interactions: how vigilant should you be? Patient Care 2000;19:41–68.
57. Ernst E. Possible interactions between synthetic and herbal medicinal products. Part 1: a systematic review of the indirect evidence. Perfusion 2000;13:4–68.
58. Ernst E. Interactions between synthetic and herbal medicinal products. Part 2: a systematic review of the direct evidence. Perfusion 2000;13:60–70.
59. Blumenthal M. Interactions between herbs and conventional drugs: introductory considerations. Herbal Gram 2000;49:52–63.

60. Ernst E. Herb–drug interactions: potentially important but woefully under-researched. Eur J Clin Pharmacol 2000;56(8):523–4.

61. De Smet PAGM, Touw DJ. Sint-janskruid op de balans van werking en interacties. Pharm Weekbl 2000;135:455–62.

62. Abebe W. Herbal medication: potential for adverse interactions with analgesic drugs. J Clin Pharm Ther 2002;27(6):391–401.

63. Mason P. Food–drug interactions: nutritional supplements and drugs. Pharm J 2002;269:609–11.

64. Scott GN, Elmer GW. Update on natural product—drug interactions. Am J Health Syst Pharm 2002;59(4):339–47.

65. Rahman SZ, Singhal KC. Problems in pharmacovigilance of medicinal products of herbal origin and means to minimize them. Uppsala Rep 2002;17:1–4.

66. De Smet PAGM, D'Arcy PF. Drug interactions with herbal and other non-orthodox drugs. In: D'Arcy PF, McElnay JC, Welling PG, editors. Mechanisms of Drug Interactions. Heidelberg: Springer Verlag, in press.

67. Stockley I. Drug Interactions. 4th ed.. London: The Pharmaceutical Press;. 1996.

68. Newall CA, Anderson LA, Phillipson JD. Herbal medicines. A guide for health-care professionalsLondon: The Pharmaceutical Press;. 1996.

69. Barnes J, Mills SY, Abbot NC, Willoughby M, Ernst E. Different standards for reporting ADRs to herbal remedies and conventional OTC medicines: face-to-face interviews with 515 users of herbal remedies. Br J Clin Pharmacol 1998;45(5):496–500.

70. Ernst E. Competence in complementary medicine. Comp Ther Med 1995;3:6–8.

71. Bensoussan A, Myers SP. Towards a safer choice. The practice of traditional Chinese medicine in Australia Campbelltown: University of Western Sydney Macarthur;. 1996.

72. Aslam M. Asian medicine and its practice in Britain. In: Evans WC, editor. Trease and Evans' Pharmacognosy. 14th ed.. London: WB Saunders, 1996:488–504.

73. Zhu DY, Bai DL, Tang XC. Recent studies on traditional Chinese medicinal plants. Drug Dev Res 1996;39:147–57.

74. Sanders D, Kennedy N, McKendrick MW. Monitoring the safety of herbal remedies. Herbal remedies have a heterogeneous nature. BMJ 1995;311(7019):1569.

75. Itoh S, Marutani K, Nishijima T, Matsuo S, Itabashi M. Liver injuries induced by herbal medicine, syo-saiko-to (xiao-chai-hu-tang). Dig Dis Sci 1995;40(8):1845–8.

76. Perharic L, Shaw D, Leon C, De Smet PA, Murray VS. Possible association of liver damage with the use of Chinese herbal medicine for skin disease. Vet Hum Toxicol 1995;37(6):562–6.

77. Okuda T, Umezawa Y, Ichikawa M, Hirata M, Oh-i T, Koga M. A case of drug eruption caused by the crude drug Boi (Sinomenium stem/Sinomeni caulis et Rhizoma). J Dermatol 1995;22(10):795–800.

78. Homma M, Oka K, Ikeshima K, Takahashi N, Niitsuma T, Fukuda T, Itoh H. Different effects of traditional Chinese medicines containing similar herbal constituents on prednisolone pharmacokinetics. J Pharm Pharmacol 1995;47(8):687–92.

79. Centers for Disease Control and Prevention (CDC). Adverse events associated with ephedrine-containing products—Texas, December 1993–September 1995. MMWR Morb Mortal Wkly Rep 1996;45(32):689–93.

80. Doyle H, Kargin M. Herbal stimulant containing ephedrine has also caused psychosis. BMJ 1996;313(7059):756.

81. Nadir A, Agrawal S, King PD, Marshall JB. Acute hepatitis associated with the use of a Chinese herbal product, mahuang. Am J Gastroenterol 1996;91(7):1436–8.

82. Tojima H, Yamazaki T, Tokudome T. [Two cases of pneumonia caused by Sho-saiko-to.]Nihon Kyobu Shikkan Gakkai Zasshi 1996;34(8):904–10.

83. Ishizaki T, Sasaki F, Ameshima S, Shiozaki K, Takahashi H, Abe Y, Ito S, Kuriyama M, Nakai T, Kitagawa M. Pneumonitis during interferon and/or herbal drug therapy in patients with chronic active hepatitis. Eur Respir J 1996;9(12):2691–6.

84. Doi Y, Uchida K, Tamura N, et al. A case of Sho-Saiko-To induced pneumonitis followed up by DLST testing BALF (bronchoalveolar lavage fluid) findings. Jpn J Chest Dis 1996;55:147–51.

85. Pena JM, Borras M, Ramos J, Montoliu J. Rapidly progressive interstitial renal fibrosis due to a chronic intake of a herb (Aristolochia pistolochia) infusion. Nephrol Dial Transplant 1996;11(7):1359–60.

86. Schmeiser HH, Bieler CA, Wiessler M, van Ypersele de Strihou C, Cosyns JP. Detection of DNA adducts formed by aristolochic acid in renal tissue from patients with Chinese herbs nephropathy. Cancer Res 1996;56(9):2025–8.

87. Vanherweghem JL, Abramowicz D, Tielemans C, Depierreux M. Effects of steroids on the progression of renal failure in chronic interstitial renal fibrosis: a pilot study in Chinese herbs nephropathy. Am J Kidney Dis 1996;27(2):209–15.

88. Lai RS, Chiang AA, Wu MT, Wang JS, Lai NS, Lu JY, Ger LP, Roggli V. Outbreak of bronchiolitis obliterans associated with consumption of Sauropus androgynus in Taiwan. Lancet 1996;348(9020):83–5.

89. Horowitz RS, Feldhaus K, Dart RC, Stermitz FR, Beck JJ. The clinical spectrum of Jin Bu Huan toxicity. Arch Intern Med 1996;156(8):899–903.

90. Kobayashi Y, Hasegawa T, Sato M, Suzuki E, Arakawa M. [Pneumonia due to the Chinese medicine Pien Tze Huang.]Nihon Kyobu Shikkan Gakkai Zasshi 1996;34(7):810–5.

91. Nakada T, Kawai B, Nagayama K, Tanaka T. A case of hepatic injury induced by Sai-rei-to. Acta Hepatol Jap 1996;37:233–8.

92. Shiota Y, Wilson JG, Matsumoto H, Munemasa M, Okamura M, Hiyama J, Marukawa M, Ono T, Taniyama K, Mashiba H. Adult respiratory distress syndrome induced by a Chinese medicine, Kamisyoyo-san. Intern Med 1996;35(6):494–6.

93. Yoshida EM, McLean CA, Cheng ES, Blanc PD, Somberg KA, Ferrell LD, Lake JR. Chinese herbal medicine, fulminant hepatitis, and liver transplantation. Am J Gastroenterol 1996;91(12):2647–8.

94. Yeo KL, Tan VCC. Severe hyperbilirubinemia associated with Chinese herbs. A case report. Singapore Paediatr J 1996;38:180–2.

95. Chan TY, Critchley JA. Usage and adverse effects of Chinese herbal medicines. Hum Exp Toxicol 1996;15(1):5–12.

96. Chan TY, Chan AY, Critchley JA. Hospital admissions due to adverse reactions to Chinese herbal medicines. J Trop Med Hyg 1992;95(4):296–8.

97. Lin SH, Lin MS. A survey on drug-related hospitalization in a community teaching hospital. Int J Clin Pharmacol Ther Toxicol 1993;31(2):66–9.

98. Wu FL, Yang CC, Shen LJ, Chen CY. Adverse drug reactions in a medical ward. J Formos Med Assoc 1996;95(3):241–6.

99. Bruynzeel DP, van Ketel WG, Young E, van Joost T, Smeenk G. Contact sensitization by alternative topical medicaments containing plant extracts. The Dutch Contact Dermatoses Group. Contact Dermatitis 1992;27(4):278–9.

100. Shaw D, Leon C, Kolev S, Murray V. Traditional remedies and food supplements. A 5-year toxicological study (1991–1995). Drug Saf 1997;17(5):342–56.

101. Anonymous. Extension of the Yellow Card scheme to unlicensed herbal remedies. Curr Prob Pharmacovig 1996;22:10.

102. Yamey G. Government launches green paper on mental health. BMJ 1999;319(7221):1322.

103. Farah MH, Edwards R, Lindquist M, Leon C, Shaw D. International monitoring of adverse health effects associated with herbal medicines. Pharmacoepidemiol Drug Saf 2000;9:105–12.

104. Hanssens Y, Deleu D, Taqi A. Etiologic and demographic characteristics of poisoning: a prospective hospital-based study in Oman. J Toxicol Clin Toxicol 2001;39(4):371–80.

105. Kerdpon D, Sriplung H. Factors related to advanced stage oral squamous cell carcinoma in southern Thailand. Oral Oncol 2001;37(3):216–21.

106. Rogers EA, Gough JE, Brewer KL. Are emergency department patients at risk for herb–drug interactions? Acad Emerg Med 2001;8(9):932–4.

107. Melchart D, Hager S, Weidenhammer W, Liao JZ, Sollner C, Linde K. Tolerance of and compliance with traditional drug therapy among patients in a hospital for Chinese medicine in Germany. Int J Risk Saf Med 1998;11:61–4.

108. Melchart D, Linde K, Weidenhammer W, Hager S, Shaw D, Bauer R. Liver enzyme elevations in patients treated with traditional Chinese medicine. JAMA 1999;282(1):28–9.

109. Melchart D, Linde K, Hager S, Kaesmayr J, Shaw D, Bauer R, Weidenhammer W. Monitoring of liver enzymes in patients treated with traditional Chinese drugs. Complement Ther Med 1999;7(4):208–16.

110. Bensoussan A, Myers SP, Carlton AL. Risks associated with the practice of traditional Chinese medicine: an Australian study. Arch Fam Med 2000;9(10):1071–8.

111. Al-Khafaji M. Monitoring of liver enzymes in patients on Chinese medicine. J Chin Med 2000;62:6–10.

112. Yang S, Dennehy CE, Tsourounis C. Characterizing adverse events reported to the California Poison Control System on herbal remedies and dietary supplements: a pilot study. J Herb Pharmacother 2002;2(3):1–11.

113. Chan TY. Monitoring the safety of herbal medicines. Drug Saf 1997;17(4):209–15.

114. Sheerin NS, Monk PN, Aslam M, Thurston H. Simultaneous exposure to lead, arsenic and mercury from Indian ethnic remedies. Br J Clin Pract 1994;48(6):332–3.

115. Murch SJ, KrishnaRaj S, Saxena PK. Phytopharmaceuticals: problems, limitations, and solutions. Sci Rev Altern Med 2000;4:33–7.

116. Tyler VE. Product definition deficiencies in clinical studies of herbal medicines. Sci Rev Altern Med 2000;4:17–21.

117. Gurley BJ, Gardner SF, Hubbard MA. Content versus label claims in *Ephedra*-containing dietary supplements. Am J Health Syst Pharm 2000;57(10):963–9.

118. De Smet PAGM. Toxicological outlook on the quality assurance of herbal remedies. In: De Smet PAGM, Keller K, Hansel R, Chandler RF, editors. Adverse Effects of Herbal Drugs 1. Heidelberg: Springer-Verlag, 1992:1–72.

119. Wrobel A. Umfrage zu Kräutern und Kräuterprodukten bezüglich Pestizid- und Schadstoff-belastungen. Akupunktur 2002;30:38–40.

120. Huang WF, Wen KC, Hsiao ML. Adulteration by synthetic therapeutic substances of traditional Chinese medicines in Taiwan. J Clin Pharmacol 1997;37(4):344–50.

121. Abt AB, Oh JY, Huntington RA, Burkhart KK. Chinese herbal medicine induced acute renal failure. Arch Intern Med 1995;155(2):211–2.

122. Gertner E, Marshall PS, Filandrinos D, Potek AS, Smith TM. Complications resulting from the use of Chinese herbal medications containing undeclared prescription drugs. Arthritis Rheum 1995;38(5):614–7.

123. Bayly GR, Braithwaite RA, Sheehan TM, Dyer NH, Grimley C, Ferner RE. Lead poisoning from Asian traditional remedies in the West Midlands—report of a series of five cases. Hum Exp Toxicol 1995;14(1):24–8.

124. Espinoza EO, Mann MJ, Bleasdell B. Arsenic and mercury in traditional Chinese herbal balls. N Engl J Med 1995;333(12):803–4.

125. Worthing MA, Sutherland HH, al-Riyami K. New information on the composition of Bint al Dhahab, a mixed lead monoxide used as a traditional medicine in Oman and the United Arab Emirates. J Trop Pediatr 1995;41(4):246–7.

126. Wu MS, Hong JJ, Lin JL, Yang CW, Chien HC. Multiple tubular dysfunction induced by mixed Chinese herbal medicines containing cadmium. Nephrol Dial Transplant 1996;11(5):867–70.

127. Centers for Disease Control and Prevention (CDC). Mercury poisoning associated with beauty cream—Texas, New Mexico, and California, 1995–1996. MMWR Morb Mortal Wkly Rep 1996;45(19):400–3.

128. Wu TN, Yang KC, Wang CM, Lai JS, Ko KN, Chang PY, Liou SH. Lead poisoning caused by contaminated Cordyceps, a Chinese herbal medicine: two case reports. Sci Total Environ 1996;182(1–3):193–5.

129. Prpic-Majic D, Pizent A, Jurasovic J, Pongracic J, Restek-Samarzija N. Lead poisoning associated with the use of Ayurvedic metal-mineral tonics. J Toxicol Clin Toxicol 1996;34(4):417–23.

130. Lana-Moliner F, Sanchez-Cubas S. Fetal abnormalities and use of substances sold in 'herbal remedies' shops. Drug Saf 1996;14(1):68.

131. Oliver MR, Van Voorhis WC, Boeckh M, Mattson D, Bowden RA. Hepatic mucormycosis in a bone marrow transplant recipient who ingested naturopathic medicine. Clin Infect Dis 1996;22(3):521–4.

132. Brandao MGL, Freire N, Vianna-Soares CD. Vigilância de fitoterápicos em Minas Gerais. Verificação da qualidade de diferentes amostras comerciais de camomila. Cad Saude Publica Rio de Janeiro 1998;14:613–6.

133. Anonymous. Herbal dietary supplements (PC-SPES and SPES). Adulteration with prescription only medicines precipitates regulatory action. WHO Pharmaceuticals Newslett 2002;2:1–2.

134. Anonymous. Traditional medicines. Several Chinese medicines withdrawn due to presence of prescription and pharmacy-only components. WHO Pharmaceuticals Newslett 2003;1:2–3.

135. Anonymous. Herbal medicine. Warning: found to contain chlordiazepoxide. WHO Pharm Newslett 2001;1:2–3.

136. Metcalfe K, Corns C, Fahie-Wilson M, Mackenzie P. Chinese medicines for slimming still cause health problems. BMJ 2002;324(7338):679.

137. Medicines Control Agency. Traditional ethnic medicines: Public health and compliance with medicines law. London: MCA;. 2001.

138. Samenuk D, Link MS, Homoud MK, Contreras R, Theoharides TC, Wang PJ, Estes NA 3rd. Adverse cardiovascular events temporally associated with ma huang, an herbal source of ephedrine Mayo Clin Proc 2002;77(1): 12–6.

139. Keane FM, Munn SE, du Vivier AW, Taylor NF, Higgins EM. Analysis of Chinese herbal creams prescribed for dermatological conditions. BMJ 1999;318(7183):563–4.

140. Daniels J, Shaw D, Atherton D. Use of Wau Wa in dermatitis patients. Lancet 2002;360(9338):1025.

141. Anonymous. Traditional medicines. Adulterants/undeclared ingredients pose safety concerns. WHO Pharmaceuticals Newslett 2002;1:11–2.

142. Florkowski CM, Elder PA, Lewis JG, Hunt PJ, Munns PL, Hunter W, Baldwin D. Two cases of adrenal suppression following a Chinese herbal remedy: a cause for concern? NZ Med J 2002;115(1153):223–4.

143. Lau KK, Lai CK, Chan AW. Phenytoin poisoning after using Chinese proprietary medicines. Hum Exp Toxicol 2000;19(7):385–6.

144. Anonymous. Herbal medicines. Warning: found to contain antidiabetics. WHO Newslett 2000;2:4–5.

149. Ko RJ. Adulterants in Asian patent medicines. N Engl J Med 1998;339(12):847.

145. Wong ST, Chan HL, Teo SK. The spectrum of cutaneous and internal malignancies in chronic arsenic toxicity. Singapore Med J 1998;39(4):171–3.

146. Ernst E. Adverse effects of herbal drugs in dermatology. Br J Dermatol 2000;143(5):923–9.

150. Li AM, Chan MH, Leung TF, Cheung RC, Lam CW, Fok TF. Mercury intoxication presenting with tics. Arch Dis Child 2000;83(2):174–5.

147. Duenas C, Perez-Alvarez JC, Busteros JI, Saez-Royuela F, Martin-Lorente JL, Yuguero L, Lopez-Morante A. Idiopathic portal hypertension and angiosarcoma associated with arsenical salts therapy. J Clin Gastroenterol 1998;26(4):303–5.

151. Moore C, Adler R. Herbal vitamins: lead toxicity and developmental delay. Pediatrics 2000;106(3):600–2.

148. Bataineh AB, al-Omari MA, Owais AI. Arsenical necrosis of the jaws. Int Endod J 1997;30(4):283–7.

152. Garvey GJ, Hahn G, Lee RV, Harbison RD. Heavy metal hazards of Asian traditional remedies. Int J Environ Health Res 2001;11(1):63–71.

153. Fletcher J, Aslam M. Possible dangers of Ayurvedic herbal remedies. Pharm J 1991;247:456.

154. Treleaven J, Meller S, Farmer P, Birchall D, Goldman J, Piller G. Arsenic and Ayurveda. Leuk Lymphoma 1993;10(4–5):343–5.

155. Saper RB, Kales SN, Paquin J, Burns MJ, Eisenberg DM, Davis RB, Phillips RS. Heavy metal content of Ayurvedic herbal medicine products. JAMA 2004;292(23):2868–73.

156. Cheng TJ, Wong RH, Lin YP, Hwang YH, Horng JJ, Wang JD. Chinese herbal medicine, sibship, and blood lead in children. Occup Environ Med 1998;55(8):573–6.

157. Chu NF, Liou SH, Wu TN, Ko KN, Chang PY. Risk factors for high blood lead levels among the general population in Taiwan. Eur J Epidemiol 1998;14(8):775–81.

158. Ibrahim AS, Latif AH. Adult lead poisoning from a herbal medicine. Saudi Med J 2002;23(5):591–3.

159. Senft V, Kaderabkova A. [Herbal concentrates Astrum—health or intoxication with heavy metals?.]Prakt Lek 2002;82:551–3.

160. Wood R, Mills PB, Knobel GJ, Hurlow WE, Stokol JM. Acute dichromate poisoning after use of traditional purgatives. A report of 7 cases S Afr Med J 1990;77(12):640–2.

161. Dunn JP, Krige JE, Wood R, Bornman PC, Terblanche J. Colonic complications after toxic tribal enemas. Br J Surg 1991;78(5):545–8.

162. Halt M. Moulds and mycotoxins in herb tea and medicinal plants. Eur J Epidemiol 1998;14(3):269–74.

163. Sadjadi J. Cutaneous anthrax associated with the Kombucha "mushroom" in Iran. JAMA 1998;280(18):1567–8.

164. Zuin VG, Vilegas JH. Pesticide residues in medicinal plants and phytomedicines. Phytother Res 2000;14(2):73–88.

165. Li AM, Hui J, Chik KW, Li CK, Fok TF. Topical herbal medicine causing haemolysis in glucose-6-phosphate dehydrogenase deficiency. Acta Paediatr 2002;91(9):1012.

166. Wurtz AS, Vial T, Isoard B, Saillard E. Possible hepatotoxicity from Copaltra, an herbal medicine. Ann Pharmacother 2002;36(5):941–2.

167. von Gruenigen VE, Hopkins MP. Alternative medicine in gynecologic oncology: A case report. Gynecol Oncol 2000;77(1):190–2.

168. Katz JL. A psychotic manic state induced by an herbal preparation. Psychosomatics 2000;41(1):73–4.

169. Fraquelli M, Colli A, Cocciolo M, Conte D. Adult syncytial giant cell chronic hepatitis due to herbal remedy. J Hepatol 2000;33(3):505–8.

170. Ng YY, Yu S, Chen TW, Wu SC, Yang AH, Yang WC. Interstitial renal fibrosis in a young woman: association with a Chinese preparation given for irregular menses. Nephrol Dial Transplant 1998;13(8):2115–7.

171. Miyazaki E, Ando M, Ih K, Matsumoto T, Kaneda K, Tsuda T. [Pulmonary edema associated with the Chinese medicine shosaikoto.]Nihon Kokyuki Gakkai Zasshi 1998;36(9):776–80.

172. Akatsu T, Santo RM, Nakayasu K, Kanai A. Oriental herbal medicine induced epithelial keratopathy. Br J Ophthalmol 2000;84(8):934.

173. Hanawa T. A case of bronchial asthma with liver dysfunction caused by Kampo medicine, Saiko-keisi-kankyo-to, and recovered smoothly in general through natural course. Phytomed 2000;SII:123.

174. Lin CC, Chen JC. Medicinal herb Erycibe Henri Prain ("Ting Kung Teng") resulting in acute cholinergic syndrome. J Toxicol Clin Toxicol 2002;40(2):185–7.

175. Nagai K, Hosaka H, Ishii K, Shinohara M, Sumino Y, Nonaka H, Akima M, Yamamuro W. A case report: acute hepatic injury induced by Formula secundarius-haemorrhoica. J Med Soc Toho Univ 1999;46:311–7.

176. Fontanarosa PB, Rennie D, DeAngelis CD. The need for regulation of dietary supplements—lessons from ephedra. J Am Med Assoc 2003;289:1568–70.

177. Barnes J. Pharmacovigilance of herbal medicines. Drug Saf 2003;26:829–51.

178. Schilter B, Andersson C, Anton R, Constable A, Kleiner J, O'Brien J, Renwick AG, Korver O, Smit F, Walker R. Guidance for the safety assessment of botanicals and botanical preparations for use in food and food supplements. Food Chem Toxicol 2003;41:1625–49.

179. Kroes R, Walker R. Safety issues of botanical and botanical preparations in functional foods Toxicology 2004;198: 213–20.

180. Henson S. Re: Consumer Reports Lists "Dangerous supplements still at large": industry responds. HC 060245-270. HerbClip 2004: 1–6.

181. Ernst E. Risks of herbal medicinal products. Pharmacoepidemiol Drug Saf 2004;13:767–71.

182. Ernst E. Challenges for phytopharmacovigilance. Postgrad Med J 2004;80:249–50.

183. Woodward KN. The potential impact of the use of homeopathic and herbal remedies on monitoring the safety of prescription products. Hum Exp Toxicol 2005;24(5):219–33.

184. Ernst E. Cardiovascular adverse effects of herbal medicines: a systematic review of the recent literature. Can J Cardiol 2003;19:818–27.

185. Ernst E. Serious psychiatric and neurological adverse effects of herbal medicines—a systematic review. Acta Psychiatr Scand 2003;108:83–91.

186. Estes JD, Stolpman D, Olyaei A, Corless CL, Ham JM, Schwartz JM, Orloff SL. High prevalence of potentially hepatotoxic herbal supplement use in patients with fulminant hepatic failure. Arch Surg 2003;138:852–8.

187. Pittler MH, Ernst E. Systematic review: hepatotoxic events associated with herbal medicinal products. Ailment Pharmacol Ther 2003;18:451–71.

188. Peyrin-Biroulet L, Barraud H, Petit-Laurent F, Ancel D, Watelet J, Chone L, Hudziak H, Bigard MA, Bronowicki JP. Hépatotoxicité de la phytothérapie: données cliniques, biologiques, histologiques et mécanismes en cause pour quelques exemples caractéristiques. (Hepatotoxicity associated with herbal remedies: clinical, biological, histological data and the mechanisms involved in some typical examples.) Gastroenterol Clin Biol 2004;28:540–50.

189. Sata M, Hisamochi A, Nakanuma Y, Kage M, Karumu S, Okita K. Results of the secondary national survey of cases of drug-induced liver injury concerning to weight-loss supplements or health foods. Acta Hepatol Jap 2004;45:96–108.

190. Willett KL, Roth RA. Walker Workshop overview: hepatotoxicity assessment for botanical dietary supplements. Toxicol Sci 2004;79:4–9.

191. Bagnis CI, Deray G, Baumelou A, Le Quintrec M, Vanherweghem JL. Herbs and the kidney. Am J Kidney Dis 2004;44:1–11.

192. Schetz M, Dasta J, Goldstein S, Golper T. Drug-induced acute kidney injury. Curr Opin Crit Care 2005;11(6):555–65.

193. Colson CR, De Broe ME. Kidney injury from alternative medicines. Adv Chron Kidney Dis 2005;12(3):261–75.

194. Choonara I. Safety of herbal medicines in children. Arch Dis Child 2003;88:1032–3.

195. Ernst E. Serious adverse effects of unconventional therapies for children and adolescents: a systematic review of recent evidence. Eur J Pediatr 2003;162:72–80.

196. Kraft K. Verträglichkeit von pflanzenlichen Drogen für akute Erkrankungen der Atemwege im Kindesalter. (Safety of herbal medicinal drugs in acute respiratory diseases of children.) Wien Med Wochenschr 2004;154:535–8.

197. Sleeper RB, Kennedy SM. Adverse reaction to a dietary supplement in an elderly patient. Ann Pharmacother 2003;37:83–6.

198. Conover EA. Herbal agents and over-the-counter medications in pregnancy. Best Pract Res Clin Endocrinol Metab 2003;17:237–51.

199. Barnes J. Complementary therapies in pregnancy. Pharm J 2003;270:402–4.

200. Westfall RE. Use of anti-emetic herbs in pregnancy: women's choices, and the question of safety and efficacy. Complement Ther Nurs Midwifery 2004;10:30–6.

201. Boon H, Wong J. Botanical medicine and cancer: a review of the safety and efficacy. Expert Opin Pharmacother 2004;5:2485–501.

202. Ikegami F, Fujii Y, Satoh T. Toxicological considerations of Kampo medicine in clinical use. Toxicology 2004;198:221–8.

203. Bhatti HBA, Gogtay NJ, Kochar NR, Dalvi SS, Kshirsagar NA. Adverse drug reactions to Indian traditional remedies: past, present and future. Adv Drug React Bull 2004;225:1–4.

204. Fennell CW, Light ME, Sparg SG, Stafford GI, van Staden J. Assessing African medicinal plants for efficacy and safety: agricultural and storage practices. J Ethnopharmacol 2004;95:113–21.

205. Luyckx VA, Steenkamp V, Stewart MJ. Acute renal failure associated with the use of traditional folk remedies in South Africa. Ren Fail 2005;27(1):35–43.

206. Anochie IC, Eke FU. Acute renal failure in Nigerian children: Port Harcourt experience. Pediatr Nephrol 2005; 20(11):1610–4.

207. Mason P. Nutritional supplements and drugs. Pharm J 2002;269:609–11.

208. Cheng KF, Leung KS, Leung PC. Interactions between modern and Chinese medicinal drugs: a general review. Am J Chinese Med 2003;31:163–9.

209. Mason P. Food-drug interactions. Nutritional supplements and drugs. Pharm J 2003;269:609–11.

210. Ernst E. Herb-drug interactions—an update. Perfusion 2003;16:175–94.

211. Barnes J, Anderson LA, Phillipson JD. Herbal therapeutics: Herbal interactions. Pharm J 2003;270:118–21.

212. Abebe W. An overview of herbal supplement utlization with particular emphasis on possible interactions with dental drugs and oral manifestations. J Dent Hyg 2003;77:37–46.

213. Williamson EM. Drug interactions between herbal and prescription medicines. Drug Saf 2003;26:1073–92.

214. Meyer JR, Generali JA, Karpinski JL. Evaluation of herbal–drug interaction data in tertiary resources. Hosp Pharm 2004;39:149–60.

215. Izzo AA, Di Carlo G, Borrelli F, Ernst E. Cardiovascular pharmacotherapy and herbal medicines: the risk of drug interaction. Int J Cardiol 2005;98(1):1–14.

216. Singh YN. Potential for interaction of kava and St John's wort with drugs. J Ethnopharmacol 2005;100(1-2):108–13.

217. Holstege CP, Mitchell K, Barlotta K, Furbee RB. Toxicity and drug interactions associated with herbal products: Ephedra and St John's wort. Med Clin North Am 2005;89(6):1225–57.

218. Marder VJ. The interaction of dietary supplements with antithrombotic agents: scope of the problem. Thromb Res 2005;117(1-2):7–13.

219. Cole MR, Fetrow CW. Adulteration of dietary supplements. Am J Health-Syst Pharm 2003;60:1576–80.

220. Palmer ME, Haller C, McKinney PE, Klein-Schwartz W, Tschirgi A, Smolinske SC, Woolf A, Sprague BM, Ko R, Everson G, Nelson LS, Dodd-Butera T, Bartlett WD, Landzberg BR. Adverse events associated with dietary supplements: an observational study. Lancet 2003;361: 101–6.

221. Hale A. 2002 Combined survey of ATMS and ANTA acupuncturists, herbalists and naturopaths. J Aust Trad Med Soc 2003;9:9–15.

222. Melchart D, Weidenhammer W, Linde K, Saller R. "Quality profiling" for complementary medicine: the example of a hospital for Traditional Chinese Medicine. J Altern Complement Med 2003;9:193–206.

223. MacPherson H, Liu B. The safety of Chinese herbal medicine: a pilot study for a national survey. J Altern Complement Med 2005;11(4):617–26.

224. Anonymous. Shubao slimming capsules. Presence of fenfluramine and nitrosofenfluramine. WHO Pharmaceutical Newslett 2004;3:5–6.

225. Ramsay HM, Goddard W, Gill S, Moss C. Herbal creams used for atopic eczema in Birmingham, UK illegally contain potent corticosteroids. Arch Dis Child 2003;88:1056–7.

226. Anonymous. Phue Chee / Lin Chee / Active Rheuma Plus. Banned due to presence of undeclared glucocorticoids. WHO Pharmaceuticals Newslett 2004;5:2.

227. Wood DM, Athwal S, Panahloo A. The advantages and disadvantages of a 'herbal' medicine in a patient with diabetes mellitus: a case report. Diabet Med 2004;21:625–7.

228. Health Canada. Health Canada warns public not to use Hua Fo/VIGOR-MAX. http://www.hc-sc.gc.ca. 29 May 2003.

229. Ministry of Health. Director-General's privileged statement under section 98 of The Medicines Act 1981. Media release: 21 Jan 2003. http://www.medsafe.govt.nz/hot.htm (accessed 4 Aug 2004).

230. Jung WS, Moon SK, Park SU, Ko CN, Cho KH. Clinical assessment of usefulness, effectiveness and safety of Jackakamcho-tang (Shaoyaogancao-tang) on muscle spasm and pain: a case series. Am J Chinese Med 2004;32:611–20.

231. Anonymous. Qing zhisan tain shou. Presence of sibutramine. WHO Pharmaceuticals Newslett 2005;2:3.

232. Wong ALN, Chan JTS, Chan TYK. Adverse herbal interactions causing hypotension. Ther Drug Monit 2003;25:297–8.

233. Adachi M, Hidetsugu S, Kobayashi H, Horie Y, Kato S, Yoshioka M, Ishii H. Hepatic injury in 12 patients taking the herbal weight loss aids Chaso or Onshido. Ann Intern Med 2003;139:488–92.

234. Kanda T, Yokosuka O, Okada O, Suzuki Y, Saisho H. Severe hepatotoxicity associated with Chinese diet product Onshidou-Genbi-Kounou. J Gastroenterol Hepatol 2003;18:354–5.

235. Kanda T, Yokosuka O, Tada M, Kurihara T, Yoshida S, Suzuki Y, Nagao K, Saisho H. N-nitroso-fenfluramine hepatotoxicity resembling chronic hepatitis. J Gastroenterol Hepatol 2003;18:999–1000.

236. Critchley LAH, Chen DQ, Chu IT, Fok BS, Yeung C. Preoperative hepatitis in a woman treated with Chinese medicines. Anaesthesia 2003;58:1096–100.

237. Critchley LAH, Chen DQ, Chu IT, Fok BS, Yeung C. Preoperative hepatitis in a woman treated with Chinese medicines. Anaesthesia 2003;58:1096–100.

238. Adesunloye BA. Acute renal failure due to the herbal remedy CKLS. Am J Med 2003;115:506–7.

239. Anonymous. Nu Bao. Presence of animal derivatives and human tissue possess health risks. WHO Pharmaceuticals Newslett 2004;3:1–2.

240. Hu Z, Yang X, Ho PC, Chan SY, Heng PW, Chan E, Duan W, Koh HL, Zhou S. Herb-drug interactions: a literature review. Drugs 2005;65(9):1239–82.

241. Williamson EM. Interactions between herbal and conventional medicines. Expert Opin Drug Saf 2005;4(2):355–78.

242. Holbrook AM, Pereira JA, Labiris R, McDonald H, Douketis JD, Crowther M, Wells PS. Systematic overview of warfarin and its drug and food interactions. Arch Intern Med 2005;165(10):1095–106.

243. Loke YK, Price D, Derry S, Aronson JK. Case reports of suspected adverse drug reactions–systematic literature survey of follow-up. BMJ 2006;332(7537):335–9.

244. Izzo AA, Di Carlo G, Borrelli F, Ernst E. Cardiovascular pharmacotherapy and herbal medicines: the risk of drug interaction. Int J Cardiol 2005;98(1):1–14.

245. Monterrey-Rodríguez J. Interaction between warfarin and mango fruit. Ann Pharmacother 2002;36(5):940–1.

246. Lambert JP, Cormier J. Potential interaction between warfarin and boldo–fenugreek. Pharmacotherapy 2001;21(4):509–12.

247. Ramsay NA, Kenny MW, Davies G, Patel JP. Complimentary and alternative medicine use among patients starting warfarin. Br J Haematol 2005;130(5): 777–80.

248. Vaes LP, Chyka PA. Interactions of warfarin with garlic, ginger, ginkgo, or ginseng: nature of the evidence. Ann Pharmacother 2000;34(12):1478–82.

249. Page RL 2nd, Lawrence JD. Potentiation of warfarin by dong quai. Pharmacotherapy 1999;19(7):870–6.

250. Lo AC, Chan K, Yeung JH, Woo KS. Danggui (Angelica sinensis) affects the pharmacodynamics but not the pharmacokinetics of warfarin in rabbits. Eur J Drug Metab Pharmacokinet 1995;20(1):55–60.

251. Tang JC, Zhang JN, Wu YT, Li ZX. Effect of the water extract and ethanol extract from traditional Chinese medicines Angelica sinensis (Oliv.) Diels, Ligusticum chuanxiong Hort. and Rheum palmatum on rat liver cytochrome P450 activity Phytother Res 2006;20(12):1046–51.

252. Taylor JR, Wilt VM. Probable antagonism of warfarin by green tea. Ann Pharmacother 1999;33(4):426–8.

253. Cheng TO. Green tea may inhibit warfarin. Int J Cardiol 2007;115(2):236.

254. Yue QY, Jansson K. Herbal drug curbicin and anticoagulant effect with and without warfarin: possibly related to the vitamin E component. J Am Geriatr Soc 2001;49(6):838.

255. Matthews MK. Association of Ginkgo biloba with intracerebral haemorrhage. Neurology 1998;5:1933.

256. Jiang X, Williams KM, Liauw WS, Ammit AJ, Roufogalis BD, Duke CC, Day RO, McLachlan AJ. Effect of ginkgo and ginger on the pharmacokinetics and pharmacodynamics of warfarin in healthy subjects. Br J Clin Pharmacol 2005;59:425–32.

257. Engelsen J, Nielsen JD, Hansen KF. Effekten af coenzym Q10 og Ginkgo biloba pa warfarindosis hos patienter i laengerevarende warfarinbehandling. Et randomiseret, dobbeltblindt, placebokontrolleret overkrydsningsforsog. Ugeskr Laeger 2003;165(18):1868–71.

258. Koch E. Inhibition of platelet activating factor (PAF)-induced aggregation of human thrombocytes by ginkgolides: considerations on possible bleeding complications after oral intake of Ginkgo biloba extracts. Phytomedicine 2005;12(1-2):10–6.

259. Destro MW, Speranzini MB, Cavalheiro Filho C, Destro T, Destro C. Bilateral haematoma after rhytidoplasty and blepharoplasty following chronic use of Ginkgo biloba. Br J Plast Surg 2005;58(1):100–1.

260. Bebbington A, Kulkarni R, Roberts P. Ginkgo biloba: persistent bleeding after total hip arthroplasty caused by herbal self-medication. J Arthroplasty 2005;20(1):125–6.

261. Rosenblatt M, Mindel J. Spontaneous hyphema associated with ingestion of Ginkgo biloba extract. N Engl J Med 1997;336(15):1108.

262. Meisel C, Johne A, Roots I. Fatal intracerebral mass bleeding associated with Ginkgo biloba and ibuprofen. Atherosclerosis 2003;167(2):367.

263. Kim YS, Pyo MK, Park KM, Park PH, Hahn BS, Wu SJ, Yun-Choi HS. Antiplatelet and antithrombotic effects of a combination of ticlopidine and Ginkgo biloba ext (EGb 761). Thromb Res 1998;91(1):33–8.

264. Bal Dit Sollier C, Caplain H. Drouet No alteration in platelet function or coagulation induced by EGb761 in a controlled study. Clin Lab Haematol 2003;25(4):251–3.

265. Henderson L, Yue QY, Bergquist C, Gerden B, Arlett P. St John's wort (*Hypericum perforatum*): drug interactions and clinical outcomes. Br J Clin Pharmacol 2002;54(4):349–56.

266. Maurer A, Johne A, Bauer S. Interaction of St. John's wort extract with phenprocoumon. Eur J Clin Pharmacol 1999;55:A22.

267. Jiang X, Williams KM, Liauw WS, Ammit AJ, Roufogalis BD, Duke CC, Day RO, McLachlan AJ. Effect of St John's wort and ginseng on the pharmacokinetics and pharmacodynamics of warfarin in healthy subjects. Br J Clin Pharmacol 2004;57(5):592–9. Erratum 2004; 58(1):102.

268. Lam AY, Elmer GW, Mohutsky MA. Possible interaction between warfarin and *Lycium barbarum*. Ann Pharmacother 2001;35(10):1199–201.

269. Coon JT, Ernst E. *Panax ginseng*: a systematic review of adverse effects and drug interactions. Drug Saf 2002;25(5): 323–44.

270. Janetzky K, Morreale AP. Probable interaction between warfarin and ginseng. Am J Health Syst Pharm 1997;54(6): 692–3.

271. Rosado MF. Thrombosis of a prosthetic aortic valve disclosing a hazardous interaction between warfarin and a commercial ginseng product. Cardiology 2003;99(2):111.

272. Yuan CS, Wei G, Dey L, Karrison T, Nahlik L, Maleckar S, Kasza K, Ang-Lee M, Moss J. American ginseng reduces warfarin's effect in healthy patients: a randomized, controlled trial. Ann Intern Med 2004;141(1):23–7.

273. Anderson GD, Rosito G, Mohustsy MA, Elmer GW. Drug interaction potential of soy extract and *Panax ginseng*. J Clin Pharmacol 2003;43:643–8.

274. Jung KY, Kim DS, Oh SR, Lee IS, Lee JJ, Park JD, Kim SI, Lee HK. Platelet activating factor antagonist activity of ginsenosides. Biol Pharm Bull 1998;21(1):79–80.

275. Zhu M, Chan KW, Ng LS, Chang Q, Chang S, Li RC. Possible influences of ginseng on the pharmacokinetics and pharmacodynamics of warfarin in rats. J Pharm Pharmacol 1999;51(2):175–80.

276. Henderson GL, Harkey MR, Gershwin ME, Hackman RM, Stern JS, Stresser DM. Effects of ginseng components on c-DNA-expressed cytochrome P450 enzyme catalytic activity. Life Sci 1999;65(15):PL209-14.

277. Kim HJ, Chun YJ, Park JD, Kim SI, Roh JK, Jeong TC. Protection of rat liver microsomes against carbon tetrachloride-induced lipid peroxidation by red ginseng saponin through cytochrome P450 inhibition. Planta Med 1997;63(5):415–8.

278. Izzat MB, Yim APC, El–Zufari MH. A taste of Chinese medicine! Ann Thorac Surg 1998;66:941–2

279. Tam LS, Chan TYK, Leung WK, Critchley JAJH. Warfarin interactions with Chinese traditional medicines: danshen and methyl salicylate medicated oil. Aust NZ J Med 1995;25:258.

280. Yu CM, Chan JCN, Sanderson JE. Chinese herbs and warfarin potentiation by "danshen". J Int Med 1997;241:337–9.

281. Chan TY. Interaction between warfarin and danshen (*Salvia miltiorrhiza*). Ann Pharmacother 2001;35(4):501–4.

282. Kuo YH, Lin YL, Don MJ, Chen RM, Ueng YF. Induction of cytochrome P450-dependent monooxygenase by extracts of the medicinal herb *Salvia miltiorrhiza*. J Pharm Pharmacol 2006;58(4):521–7.

283. Lo AC, Chan K, Yeung JH, Woo KS. The effects of danshen (*Salvia miltiorrhiza*) on pharmacokinetics and pharmacodynamics of warfarin in rats. Eur J Drug Metab Pharmacokinet 1992;17(4):257–62.

284. Chan K, Lo AC, Yeung JH, Woo KS. The effects of danshen (*Salvia miltiorrhiza*) on warfarin pharmacodynamics and pharmacokinetics of warfarin enantiomers in rats. J Pharm Pharmacol 1995;47(5):402–6.

285. Gupta D, Jalali M, Wells A, Dasgupta A. Drug–herb interactions: unexpected suppression of free danshen concentrations by salicylate. J Clin Lab Anal 2002;16(6):290–4.

286. Makino T, Wakushima H, Okamoto T, Okukubo Y, Deguchi Y, Kano Y. Pharmacokinetic interactions between warfarin and kangen-karyu, a Chinese traditional herbal medicine, and their synergistic action. J Ethnopharmacol 2002;82(1):35–40.

287. Lesho EP, Saullo L, Udvari-Nagy S. A 76-year-old woman with erratic anticoagulation. Cleve Clin J Med 2004;71(8):651–6.

288. Anonymous. Medical mystery. A woman with too-thin blood. Why was the patient bleeding? What's her case mean to you? Heart Advis 2004;7(12):4–5.

289. Weidner MS, Sigwart K. The safety of a ginger extract in the rat. J Ethnopharmacol 2000;73(3):513–20.

290. Jiang X, Blair EY, McLachlan AJ. Investigation of the effects of herbal medicines on warfarin response in healthy subjects: a population pharmacokinetic–pharmacodynamic modeling approach. J Clin Pharmacol 2006;46(11):1370–8.

291. Davis NB, Nahlik L, Vogelzang NJ. Does PC-Spes interact with warfarin? J Urol 2002;167(4):1793.

292. Duncan GG. Re: Does PC-Spes interact with warfarin? J Urol 2003;169(1):294–5.

293. Wong ALN, Chan TYK. Interaction between warfarin and the herbal product quilinggao. Ann Pharmacother 2003;37:836–8.

Herbal Medicines – Acanthaceae – Zygophyllaceae

Acanthaceae

General information

The genera in the family Acanthaceae (Table 5) include the wild petunia and the water willow.

Table 5 The genera of Acanthaceae

Acanthus (acanthus)
Andrographis (false waterwillow)
Anisacanthus (desert honeysuckle)
Asystasia (asystasia)
Barleria (Philippine violet)
Barleriola (barleriola)
Blechum (blechum)
Carlowrightia (wrightwort)
Crossandra (crossandra)
Dicliptera (foldwing)
Dyschoriste (snakeherb)
Elytraria (scalystem)
Eranthemum (eranthemum)
Graptophyllum (graptophyllum)
Hemigraphis (hemigraphis)
Henrya (henrya)
Holographis
Hygrophila (swampweed)
Hypoestes (hypoestes)
Justicia (water-willow)
Megaskepasma (Brazilian red-cloak)
Nelsonia (nelsonia)
Nomaphila (nomaphila)
Odontonema (toothedthread)
Oplonia (oplonia)
Pachystachys (pachystachys)
Pseuderanthemum (pseuderanthemum)
Ruellia (wild petunia)
Sanchezia (sanchezia)
Siphonoglossa (tube tongue)
Stenandrium (shaggytuft)
Strobilanthes
Teliostachya (teliostachya)
Tetramerium (tetramerium)
Thunbergia (thunbergia)
Yeatesia (bractspike P)

Andrographis paniculata (false waterwillow)

Andrographis paniculata is a shrub that is found throughout India and other Asian countries and is sometimes called Indian *Echinacea*. In China it is known as Chuan Xin Lian and Kan Jang. Its constituents include diterpenoid lactones (andrographolides), paniculides, farnesols, and flavonoids. It has been used to treat viral infections, such as colds and influenza.

In a double-blind, placebo-controlled study of the use of andrographis 1200 mg/day in 61 patients there were no changes in liver or kidney function, blood counts, or other laboratory measures (1).

A systematic review of all clinical reports of adverse events resulting from treatment of upper respiratory tract infections with *Andrographis paniculata* found only mild, infrequent and reversible adverse events (2). The most frequent ones included *pruritus, fatigue, headache*, and *diarrhea*.

Adverse effects
Skin
In two double-blind placebo-controlled studies of *Andrographis paniculata* in a total of 225 patients there were two cases of urticaria (3).

References

1. Hancke J, Burgos R, Caceres D, Wikman G. A double-blind study with a new monodrug Kan Jang: decrease of symptoms and improvement in the recovery from common colds. Phytother Res 1995;9(8):559–62.
2. Coon JT, Ernst E. *Andrographis paniculata* in the treatment of upper respiratory tract infections: a systematic review of safety and efficacy. Planta Med 2004; 70: 293–8.
3. Melchior J, Spasov AA, Ostrovskij OV, Bulanov AE, Wikman G. Double-blind, placebo-controlled pilot and phase III study of activity of standardized *Andrographis paniculata Herba* extract fixed combination (Kan jang) in the treatment of uncomplicated upper-respiratory tract infection. Phytomedicine 2000; 7(5): 341–50.

Acoraceae

General Information

Acorus calamus was originally classified as a member of the arum family (Araceae), but is now designated as belonging to its own family, the Acoraceae, of which it is the only member.

Acorus calamus

Acorus calamus (calamus root, sweet flag, rat root, sweet sedge, flag root, sweet calomel, sweet myrtle, sweet cane, sweet rush, beewort, muskrat root, pine root) contains several active constituents called "asarones." The basic structure is 2,4,5-trimethoxy-1-propenylbenzene, which is related to the hallucinogen 3,4-methylenedioxyphenyliso-propylamine (MDA). The amounts of the asarones in calamus rhizomes vary considerably with the botanical variety. For example, there are high concentrations in triploid calamus from Eastern Europe but none detectable in the diploid North American variety.

Acorus calamus has been used as a hallucinogen since ancient times and it has several uses in folk medicine. It may have been one of the constituents of the Holy Oil that God commanded Moses to make (Exodus 30) and is mentioned by ancient writers on medicine, such as Hippocrates, Theophrastus, Dioscorides, and Celsus (http://www.a1b2c3.com/drugs/var002.htm).

Acorus calamus has in vitro antiproliferative and immunosuppressive actions (1).

Acorus calamus contains beta-asarone [(Z)-1,2,4-tri-methoxy-5-prop-1-enyl-benzene], which is carcinogenic (2). Commercial calamus preparations have mutagenic effects in bacteria (3), while calamus oil (Jammu variety) is carcinogenic in rats.

Walt Whitman's 39 "Calamus poems" are to be found in his well-known collection "Leaves of Grass."

References

1. Mehrotra S, Mishra KP, Maurya R, Srimal RC, Yadav VS, Pandey R, Singh VK. Anticellular and immunosuppressive properties of ethanolic extract of *Acorus calamus* rhizome. Int Immunopharmacol 2003;3(1):53–61.
2. Bertea CM, Azzolin CM, Bossi S, Doglia G, Maffei ME. Identification of an EcoRI restriction site for a rapid and precise determination of beta-asarone-free *Acorus calamus* cytotypes. Phytochemistry 2005;66(5):507–14.
3. Sivaswamy SN, Balachandran B, Balanehru S, Sivaramakrishnan VM. Mutagenic activity of south Indian food items. Indian J Exp Biol 1991;29(8):730–7.

Algae

General Information

Algae are members of the kingdom Protista. They include Bacillariophyta (unicellular diatoms), Charophyta (freshwater stoneworts), Chrysophyta (photosynthetic, unicellular organisms), Cyanobacteria (photosynthetic bacteria), Dinophyta (dinoflagellates), and the various types of seaweeds.

There are three major groups of seaweeds: the green algae (Chlorophyta), the brown algae (Phaeophyta), and the red algae (Rhodophyta).

Kelp is a general name for seaweed preparations obtained from different species of Phaeophyta (such as *Ascophyllum nodosum*, *Fucus vesiculosus*, *Fucus serratus*, *Laminaria* species, and *Macrocystis pyrifera*). As kelp contains iodine, it occasionally produces hyperthyroidism (1), hypothyroidism, or extrathyroidal reactions, such as skin eruptions. It can also contain contaminants such as arsenic, and bone marrow depression and autoimmune thrombocytopenia have been described in consequence (2).

Anaphylaxis to *Laminaria* has been described (3–5).

Laminaria has been used to induce abortion and can cause uterine rupture especially in primipara.

- A 32-year-old woman in her 25th week of gestation was hospitalized for endouterine fetal death (6). She had had two previous cesarean sections. Cervical dilatation was induced with a Hagar dilator and laminaria tents, and sudden spontaneous and strong contractions led to uterine rupture.

References

1. de Smet PA, Stricker BH, Wilderink F, Wiersinga WM. Hyperthyreoïdie tijdens het gebruik van kelptabletten. [Hyperthyroidism during treatment with kelp tablets.] Ned Tijdschr Geneeskd 1990;134(21):1058–9.
2. Pye KG, Kelsey SM, House IM, Newland AC. Severe dyserythropoiesis and autoimmune thrombocytopenia associated with ingestion of kelp supplements. Lancet 1992;339(8808):1540.
3. Knowles SR, Djordjevic K, Binkley K, Weber EA. Allergic anaphylaxis to *Laminaria*. Allergy 2002;57(4):370.
4. Cole DS, Bruck LR. Anaphylaxis after *Laminaria* insertion. Obstet Gynecol 2000;95(6 Pt 2):1025.
5. Nguyen MT, Hoffman DR. Anaphylaxis to *Laminaria*. J Allergy Clin Immunol 1995;95(1 Pt 1):138–9.
6. Menaldo G, Alemanno MG, Brizzolara M, Campogrande M. La *Laminaria* nell'induzione d'aborto al 2no trimestre. Caso di rottura d'utero. [*Laminaria* in induction of abortion in the 2nd trimester. Case of uterine rupture.] Minerva Ginecol 1981;33(6):599–601.

Aloeaceae

General Information

The family of Aloeaceae contains the single genus *Aloe*.

Aloe capensis and *Aloe vera*

Aloe capensis and *Aloe vera* (*Aloe barbadensis*) (Barbados aloe, burn plant, Curacao aloe, elephant's gall, first-aid plant, Hsiang dan, lily of the desert) contain several active constituents called aloins, including barbaloin, isobarbaloin, and aloe-emodin. Laxative anthranoid derivatives occur primarily in various laxative herbs (such as aloe, cascara sagrada, medicinal rhubarb, and senna) in the form of free anthraquinones, anthrones, dianthrones, and/or O- and C-glycosides derived from these substances. They produce harmless discoloration of the urine. Depending on intrinsic activity and dose, they can also produce abdominal discomfort and cramps, nausea, violent purgation, and dehydration. They can be distributed into breast milk, but not always in sufficient amounts to affect the suckling infant. Long-term use can result in electrolyte disturbances and in atony and dilatation of the colon. Several anthranoid derivatives (notably the aglycones aloe-emodin, chrysophanol, emodin, and physicon) are genotoxic in bacterial and/or mammalian test systems (SEDA-12, 409).

Aloe species contain laxative anthranoid derivatives, the main active ingredient being isobarbaloin. Large doses are claimed to cause nephritis and use during pregnancy is discouraged, since intestinal irritation might lead to pelvic congestion.

Aloes have been used on the skin to reduce the pain and swelling of burns, improve the symptoms of genital herpes and other skin conditions such as psoriasis, and to help heal wounds and frostbite. Aloe has been used orally to treat arthritis, asthma, diabetes, pruritus, peptic ulcers, and constipation. It may be effective in inflammatory bowel disease (1). It may reduce blood glucose in diabetes mellitus and blood lipid concentrations in hyperlipidemia (2).

Adverse effects

Aloe can reportedly cause muscle weakness, cardiac dysrhythmias, peripheral edema, bloody diarrhea, weight loss, stomach cramps, itching, redness, rash, pruritus, and a red coloration of the urine.

Liver

Hepatitis in a 57-year-old woman was linked to the ingestion of *Aloe vera*; the hepatitis resolved completely after withdrawal (3).

Gastrointestinal

- Melanosis coli occurred in a 39-year-old liver transplant patient who took an over-the-counter product containing aloe, rheum, and frangula (4). The typical brownish pigmentation of the colonic mucosa developed over 10 months. The medication was withdrawn and follow-up colonoscopy 1 year later showed normal looking mucosa. However, a sessile polypoid lesion was found in the transverse colon. Histology showed tubulovillous adenoma with extensive low-grade dysplasia.

Skin

Four patients had severe burning sensations after the application of *Aloe vera* to skin that had been subjected to chemical peel or dermabrasion (5).

Aloes can cause contact sensitivity (6,7).

Urinary tract

Acute renal insufficiency been attributed to *Aloe capensis*.

- A 47-year-old South African man developed acute oliguric renal insufficiency and liver dysfunction after taking an herbal medicine prescribed by a traditional healer to "clean his stomach" (8). After withdrawal of the remedy and dialysis he recovered slowly, but his creatinine concentration did not fully normalize. Analysis of the remedy showed that it consisted of *A. capensis* (*Aloe ferox* Miller), which contains the nephrotoxic compounds aloesin and aloesin A.

The authors also mentioned that 35% of all cases of acute renal insufficiency in Africa are due to traditional remedies.

Drug interactions

Sevoflurane

Massive intraoperative bleeding in a 35-year-old woman has been attributed to an interaction of preoperative *Aloe vera* tablets and sevoflurane, since both may inhibit platelet function (9).

References

1. Langmead L, Feakins RM, Goldthorpe S, Holt H, Tsironi E, De Silva A, Jewell DP, Rampton DS. Randomized, double-blind, placebo-controlled trial of oral *Aloe vera* gel for active ulcerative colitis. Aliment Pharmacol Ther 2004;19(7):739–47.
2. Vogler BK, Ernst E. Aloe vera: a systematic review of its clinical effectiveness. Br J Gen Pract 1999;49(447):823–8.
3. Rabe C, Musch A, Schirmacher P, Kruis W, Hoffmann R. Acute hepatitis induced by an *Aloe vera* preparation: a case report. World J Gastroenterol 2005;11(2):303–4.

4. Willems M, van Buuren HR, de Krijger R. Anthranoid self-medication causing rapid development of melanosis coli. Neth J Med 2003;61(1):22–4.

5. Hunter D, Frumkin A. Adverse reactions to vitamin E and *Aloe vera* preparations after dermabrasion and chemical peel. Cutis 1991;47(3):193–6.

6. Shoji A. Contact dermatitis to *Aloe arborescens*. Contact Dermatitis 1982;8(3):164–7.

7. Morrow DM, Rapaport MJ, Strick RA. Hypersensitivity to aloe. Arch Dermatol 1980;116(9):1064–5.

8. Luyckx VA, Ballantine R, Claeys M, Cuyckens F, Van den Heuvel H, Cimanga RK, Vlietinck AJ, De Broe ME, Katz IJ. Herbal remedy-associated acute renal failure secondary to Cape aloes. Am J Kidney Dis 2002;39(3):E13.

9. Lee A, Chui PT, Aun CS, Gin T, Lau AS. Possible interaction between sevoflurane and *Aloe vera*. Ann Pharmacother 2004;38(10):1651–4.

Amaranthaceae

General Information

The genera in the family of Amaranthaceae (Table 6) include cotton flower and rockwort.

Table 6 The genera of Amaranthaceae

Achyranthes (chaff flower)
Aerva (aerva)
Alternanthera (joyweed)
Amaranthus (pigweed)
Blutaparon (blutaparon)
Celosia (cock's comb)
Chamissoa (chamissoa)
Charpentiera (papala)
Cyathula (cyathula)
Froelichia (snake cotton)
Gomphrena (globe amaranth)
Gossypianthus (cotton flower)
Guilleminea (matweed)
Hermbstaedtia (hermbstaedtia)
Iresine (bloodleaf)
Lithophila (lithophila)
Nototrichium (rockwort)
Pfaffia (pfaffia)
Philoxerus (philoxerus)
Tidestromia (honeysweet)

Pfaffia paniculata and ginseng

Ginseng is an ambiguous vernacular term that can refer to *Panax* species such as *Panax ginseng* (Asian ginseng) and *Panax quinquefolius* (American ginseng), *Eleutherococcus senticosus* (Siberian ginseng), and *Pfaffia paniculata* (Brazilian ginseng). See ginseng in the monograph on Araliaceae.

Adverse effects
Respiratory

Asthma occurred in a patient who had been exposed to *Pfaffia paniculata* root powder used in the manufacturing of Brazil ginseng capsules (1). Airway hyper-reactivity was confirmed by a positive bronchial challenge to methacholine and sensitivity to the dust was confirmed by immediate skin test reactivity, a positive bronchial challenge (immediate response), and the presence of a specific IgE to an aqueous extract detected by ELISA. The bronchial response was inhibited by sodium cromoglicate. Unexposed subjects did not react to this ginseng extract with any of the tests referred to above. The same study performed with Korean ginseng (*Panax ginseng*) was negative.

Reference

1. Subiza J, Subiza JL, Escribano PM, Hinojosa M, Garcia R, Jerez M, Subiza E. Occupational asthma caused by Brazil ginseng dust. J Allergy Clin Immunol 1991;88(5):731–6.

Anacardiaceae

General Information

The genera in the family of Anacardiaceae (Table 7) include pistachio, poison ivy, and sumac.

Table 7 The genera of Anacardiaceae

Anacardium (anacardium)
Buchanania (buchanania)
Campnospera
Comocladia (maidenplum)
Cotinus (smoke tree)
Dracontomelon (dracontomelon)
Gluta (gluta)
Lithrea (lithrea)
Malosma (laurel sumac)
Mangifera (mango)
Melanorrhoea (melanorrhoea)
Metopium (Florida poison tree)
Pistacia (pistachio)
Rhus (sumac)
Schinus (pepper tree)
Schinopsis (schinopsis)
Sclerocarya (sclerocarya)
Semecarpus (semecarpus)
Spondias (mombin)
Toxicodendron (poison oak)

Rhus species

Rhus species are used as a traditional remedy for gastrointestinal complaints in Korea.

Adverse effects
Skin

Contact dermatitis has been attributed to *Rhus* species (1,2), including a case that followed exposure to a homeopathic remedy (3).

Oral or parenteral exposure to certain contact allergens can elicit an eczematous skin reaction in sensitized individuals. This phenomenon has been called systemic contact dermatitis (SCD) and is relatively rare compared with classical contact dermatitis.

In 42 patients with systemic contact dermatitis caused by ingestion of *Rhus* (24 men and 18 women, average age 44 years, range 24–72), 14 of whom had a history of allergy to lacquer, there were skin lesions such as generalized maculopapular eruptions (50%), erythroderma (29%), vesiculobullous lesions (14%), and erythema multiforme-like lesions (7%) (4). Many patients (57%) developed a leukocytosis with a neutrophilia (74%). In some patients (5%) there were abnormalities of liver function. The lymphocyte subsets of 12 patients studied were within the reference ranges with no differences between patients with or without a history of allergy to lacquer. The authors concluded that the skin eruptions were caused by toxic reactions to *Rhus* rather than by immunological mechanisms.

In 31 patients with *Rhus* allergy over a 10-year period the clinical manifestations included maculopapular eruptions (65%), erythema multiforme (32%), erythroderma (19%) pustules, purpura, wheals, and blisters (5). All the patients had generalized or localized pruritus, and other symptoms included gastrointestinal problems (32%), fever (26%), chills, and headache. Many developed a leukocytosis (70%) with neutrophilia (88%), and some had toxic effects on the liver or kidneys. All responded to glucocorticoids or antihistamines.

Erythema multiforme in a photodistribution has been attributed to *Rhus verniciflua*, the Japanese lacquer tree (6). The rash was reproduced by challenge with the drug and sunlight. On contact with *R. verniciflua* the patient had a flare of the eruption, which was limited to the areas previously exposed to sun. Immunohistochemical studies suggested that the keratinocytes in the skin that retain the photoactivated substances may facilitate epidermal invasion of lymphocytes by persistent expression of intercellular adhesion molecules.

References

1. Powell SM, Barrett DK. An outbreak of contact dermatitis from *Rhus verniciflua (Toxicodendron verniciluum)*. Contact Dermatitis 1986;14(5):288–9.
2. Sasseville D, Nguyen KH. Allergic contact dermatitis from *Rhus toxicodendron* in a phytotherapeutic preparation. Contact Dermatitis 1995;32(3):182–3.
3. Cardinali C, Francalanci S, Giomi B, Caproni M, Sertoli A, Fabbri P. Contact dermatitis from *Rhus toxicodendron* in a homeopathic remedy. J Am Acad Dermatol 2004;50(1):150–1.
4. Oh SH, Haw CR, Lee MH. Clinical and immunologic features of systemic contact dermatitis from ingestion of *Rhus (Toxicodendron)*. Contact Dermatitis 2003;48(5):251–4.
5. Park SD, Lee SW, Chun JH, Cha SH. Clinical features of 31 patients with systemic contact dermatitis due to the ingestion of *Rhus* (lacquer). Br J Dermatol 2000;142(5):937–42.
6. Shiohara T, Chiba M, Tanaka Y, Nagashima M. Drug-induced, photosensitive, erythema multiforme-like eruption: possible role for cell adhesion molecules in a flare induced by *Rhus* dermatitis. J Am Acad Dermatol 1990;22(4):647–50.

Apiaceae

General Information

The genera in the family of Apiaceae (formerly Umbelliferae) (Table 8) include a variety of spices and vegetables, such as angelica, anise, carrot, celery, chervil, coriander, cumin, dill, fennel, parsley, and parsnip.

Food allergy to spices accounts for 2% of all cases of food allergies but 6.4% of cases in adults. Prick tests to native spices in 589 patients with food allergies showed frequent sensitization to the Apiaceae coriander, caraway, fennel, and celery (32% of prick tests in children, 23% of prick tests in adults) (1). There were 10 cases of allergy related to the mugwort-celery-spices syndrome: coriander ($n = 1$), caraway ($n = 2$), fennel ($n = 3$), garlic ($n = 3$), and onion ($n = 1$).

Scratch tests with powdered commercial spices in 70 patients with positive skin tests to birch and/or mugwort pollens and celery were positive to aniseed, fennel, coriander, and cumin, all Apiaceae, in more than 24 patients (2). Spices from unrelated families (red pepper, white pepper, ginger, nutmeg, cinnamon) elicited positive immediate skin test reactions in only three of 11 patients. Specific serum IgE to spices (determined in 41 patients with a positive RAST to celery) up to class 3 was found, especially in patients with celery-mugwort or celery-birch-mugwort association. The celery-birch association pattern was linked to positive reactions (RAST classes 1,2) with spices from the Apiaceae family only.

Table 8 The genera of Apiaceae

Aciphylla (fierce Spaniard)	Levisticum (levisticum)
Aegopodium (goutweed)	Ligusticum (licorice-root)
Aethusa (aethusa)	Lilaeopsis (grasswort)
Aletes (Indian parsley)	Limnosciadium (dogshade)
Ammi (ammi)	Lomatium (desert parsley)
Ammoselinum (sand parsley)	Musineon (wild parsley)
Anethum (dill)	Myrrhis (myrrhis)
Angelica (angelica)	Neoparrya (neoparrya)
Anthriscus (chervil)	Oenanthe (water dropwort)
Apiastrum (apiastrum)	Oreonana (mountain parsley)
Apium (celery)	Oreoxis (oreoxis)
Arracacia (arracacia)	Orogenia (Indian potato)
Berula (water parsnip)	Osmorhiza (sweetroot)
Bifora (bishop)	Oxypolis (cowbane)
Bowlesia (bowlesia)	Pastinaca (parsnip)
Bupleurum (bupleurum)	Perideridia (yampah)
Carum (carum)	Petroselinum (parsley)
Caucalis (burr parsley)	Peucedanum (peucedanum)
Centella (centella)	Pimpinella (burnet saxifrage)
Chaerophyllum (chervil)	Podistera (podistera)
Cicuta (water hemlock)	Polytaenia (hairy moss)
	Pseudocymopterus (false spring parsley)
Cnidium (snow parsley)	
Conioselinum (hemlock parsley)	Pteryxia (wavewing)
	Ptilimnium (mock bishop weed)
Conium (poison hemlock)	Sanicula (sanicle)
Coriandrum (coriander)	Scandix (scandix)
Cryptotaenia (honewort)	Selinum (selinum)
Cuminum (cumin)	Seseli (seseli)
Cyclospermum (marsh parsley)	Shoshonea (shoshonea)
Cymopterus (spring parsley)	Sium (waterparsnip)
Cynosciadium (cynosciadium)	Smyrnium (smyrnium)
	Spermolepis (scaleseed)
	Sphenosciadium (sphenosciadium)
Daucosma (daucosma)	
Daucus (wild carrot)	Taenidia (taenidia)
Dorema (dorema)	Tauschia (umbrellawort)
Erigenia (erigenia)	Thaspium (meadowparsnip)
Eryngium (eryngo)	Tilingia (tilingia)
Eurytaenia (spreadwing)	Tordylium (tordylium)
Falcaria (falcaria)	Torilis (hedge parsley)
Ferula (asafetida)	Trachyspermum (Ajowan caraway)
Foeniculum (fennel)	
Glehnia (silvertop)	Trepocarpus (trepocarpus)
Harbouria (harbouria)	Turgenia (false carrot)
Heracleum (cow parsnip)	Yabea (yabea)
Hydrocotyle (hydrocotyle)	Zizia (zizia)

Ammi majus and *Ammi visnaga*

Ammi majus is also known as bishop's weed or large bullwort, and *Ammi visnaga* as toothpick weed. They have numerous active ingredients, including ammirin, angenomalin, kellactone, majurin, and marmesin, furanocoumarins (psoralen, bergapten, isopimpinellin, imperatorin, umbelliprenin, xanthotoxin), and flavonol triglycosides (kaempferol, isorhamnetin). In modern times they have been used to treat vitiligo, since they contain psoralens, and have several different pharmacological effects in experimental animals, including hypoglycemic effects (3), antischistosomal effects (4), and inhibition of nephrolithiasis (5).

Adverse effects
Injudicious use of the fruit of *A. majus* in combination with skin exposure to the sun can cause severe phototoxic dermatitis, owing to the presence of psoralens (6).

- IgE-mediated rhinitis and contact urticaria were caused by exposure to bishop's weed in a 31-year-old atopic female florist (7). A skin prick test with bishop's weed flowers gave an 8 mm wheal, and the bishop's weed-specific serum IgE concentration was 9.7 PRU/ml (RAST class 3).

Prolonged use or overdosing of the fruit of *Ammi visnaga* can cause nausea, dizziness, constipation, loss of appetite, headache, pruritus, and sleeping disorders.

Angelica sinensis

Angelica sinensis, known in China as "dong quai" or "dang gui," contains antioxidants (8), inhibits the growth of cancer cells in vitro (9) and stimulates immune function in experimental animals (10). It has been used to treat amenorrhea (11) and menopausal hot flushes (12), and to reduce pulmonary hypertension in patients with chronic obstructive pulmonary disease (13).

Adverse effects

Angelica sinensis can cause hypertension (14).

- A 32-year-old woman, 3 weeks post-partum, developed acute headache, weakness, light-headedness, and vomiting. Her blood pressure was 195/85 mmHg. She had taken dong quai for postpartum weakness and said that she had not been taking any other medicines. Her 3-week-old son's blood pressure was raised to 115/69. Dong quai medication of the mother and breast-feeding of the child were discontinued and the blood pressure normalized in both patients within 48 hours.

Centella asiatica

Centella asiatica (gotu kola, spadeleaf) has been used n Ayurvedic medicine to improve memory and treat several neurological disorders and may improve cognition and mood (15). It enhanced phosphorylation of cyclic AMP response element binding protein in rat neuroblastoma cells expressing amyloid beta peptide (16), and an aqueous extract inhibited phospholipase activities in rat cerebellum (17). *Centella* also contains a triterpene, madecassoside, which facilitates the healing of burn wounds in mice (18).

Adverse effects
Liver

Hepatitis has been attributed to *Centella asiatica* in three cases (19).

Skin

Centella asiatica can cause contact dermatitis (20).

- An 18-year-old woman presented with a pruritic eczematous eruption that developed after topically applying an ointment containing hydrocortisone acetate, neomycin sulfate and *Centella asiatica*. She was positive to all three ingredients of the ointment (21).

Conium maculatum

Conium maculatum (hemlock) contains a poisonous piperidine alkaloid, coniine, and related alkaloids, *N*-methyl-coniine, conhydrine, pseudoconhydrine, and gamma-coniceine. It has well-established teratogenic activity in certain animal species.

Adverse effects

The symptoms of hemlock poisoning are effects on the nervous system (stimulation followed by paralysis of motor nerve endings and nervous system stimulation and later depression), vomiting, trembling, difficulty in movement, an initially slow and weak and later a rapid pulse, rapid respiration, salivation, urination, nausea, convulsions, coma, and death (22).

Coriandrum sativum

Coriandrum sativum (coriander) has traditionally been used as a stimulant, aromatic, and carminative, and to disguise the taste of purgatives. In experimental animals it has hypolipidemic effects (23) and hypoglycemic effects (24).

Adverse effects
Respiratory

Occupational asthma has been attributed to various spices from botanically unrelated species, including coriander.

- A 27-year-old man developed rhinitis and asthma symptoms 1 year after starting to prepare a certain kind of sausage (25). He had positive immediate skin prick tests with paprika, coriander, and mace, and specific IgE antibodies to all three. He had immediate asthmatic reactions to bronchial inhalation of extracts from paprika, coriander, and mace, with maximum falls in FEV_1 of 26, 40, and 31% respectively, but no late asthmatic reactions.

Endocrine

Adrenal insufficiency has been attributed to *C. sativum* (26).

- A 28-year-old woman took an extract of *C. sativum* for 7 days to augment lactation while breastfeeding. She developed severe stomach pain and diarrhea and 15 days later resented with dark skin, depression, dehydration, and amenorrhea. A diagnosis of adrenal dysfunction was made, the herbal remedy was withdrawn, and she was treated with dexamethasone, prednisolone, and an oral contraceptive. Her symptoms resolved within 10 days.

Skin

Contact dermatitis from coriander has been described (27).

Immunologic

An anaphylactic reaction has been described in a patient who was sensitized to coriander (28).

Ferula assa-foetida

Ferula assa-foetida (asafetida) contains coumarin sesquiterpenoids, such as fukanefuromarins and kamolonol, which inhibit the production of nitric oxide and relax smooth muscle.

Adverse effects

Methemoglobinemia occurred in a 5-week-old infant treated with a gum asafetida formulation (29).

Asafetida may enhance the activity of warfarin (30).

References

1. Moneret-Vautrin DA, Morisset M, Lemerdy P, Croizier A, Kanny G. Food allergy and IgE sensitization caused by spices: CICBAA data (based on 589 cases of food allergy). Allerg Immunol (Paris) 2002;34(4):135–40.
2. Stager J, Wuthrich B, Johansson SG. Spice allergy in celery-sensitive patients. Allergy 1991;46(6):475–8.

3. Jouad H, Maghrani M, Eddouks M. Hypoglycemic effect of aqueous extract of Ammi visnaga in normal and streptozo-tocin-induced diabetic rats. J Herb Pharmacother 2002;2(4):19–29.

4. Abdulla WA, Kadry H, Mahran SG, el-Raziky EH, el-Nakib S. Preliminary studies on the anti-schistosomal effect of Ammi majus L. Egypt J Bilharz 1978;4(1):19–26.

5. Khan ZA, Assiri AM, Al-Afghani HM, Maghrabi TM. Inhibition of oxalate nephrolithiasis with Ammi visnaga (Al-Khillah). Int Urol Nephrol 2001;33(4):605–8.

6. Ossenkoppele PM, van der Sluis WG, van Vloten WA. Fototoxische dermatitis door het gebruik van de Ammi majus-vrucht bij vitiligo. [Phototoxic dermatitis following the use of Ammi majus fruit for vitiligo.] Ned Tijdschr Geneeskd 1991;135(11):478–80.

7. Kiistala R, Makinen-Kiljunen S, Heikkinen K, Rinne J, Haahtela T. Occupational allergic rhinitis and contact urticaria caused by bishop's weed (Ammi majus). Allergy 1999;54(6):635–9.

8. Wu SJ, Ng LT, Lin CC. Antioxidant activities of some common ingredients of traditional chinese medicine, Angelica sinensis, Lycium barbarum and Poria cocos. Phytother Res 2004;18(12):1008–12.

9. Cheng YL, Chang WL, Lee SC, Liu YG, Chen CJ, Lin SZ, Tsai NM, Yu DS, Yen CY, Harn HJ. Acetone extract of Angelica sinensis inhibits proliferation of human cancer cells via inducing cell cycle arrest and apoptosis. Life Sci 2004;75(13):1579–94.

10. Wang J, Xia XY, Peng RX, Chen X. [Activation of the immunologic function of rat Kupffer cells by the poly-saccharides of Angelica sinensis.]Yao Xue Xue Bao 2004;39(3):168–71.

11. He ZP, Wang DZ, Shi LY, Wang ZQ. Treating amenorrhea in vital energy-deficient patients with Angelica sinensis-astragalus membranaceus menstruation-regulating decoc-tion. J Tradit Chin Med 1986;6(3):187–90.

12. Kupfersztain C, Rotem C, Fagot R, Kaplan B. The immedi-ate effect of natural plant extract, Angelica sinensis and Matricaria chamomilla (Climex) for the treatment of hot flushes during menopause. A preliminary report. Clin Exp Obstet Gynecol 2003;30(4):203–6.

13. Xu JY, Li BX, Cheng SY. [Short-term effects of Angelica sinensis and nifedipine on chronic obstructive pulmonary disease in patients with pulmonary hypertension.]Zhongguo Zhong Xi Yi Jie He Za Zhi 1992;12(12):716–8707.

14. Nambiar S, Schwartz RH, Constantino A. Hypertension in mother and baby linked to ingestion of Chinese herbal medicine. West J Med 1999;171(3):152.

15. Wattanathorn J, Mator L, Muchimapura S, Tongun T, Pasuriwong O, Piyawatkul N, Yimtae K,

Sripanidkulchai B, Singkhoraard J. Positive modulation of cognition and mood in the healthy elderly volunteer follow-ing the administration of *Centella asiatica*. J Ethnopharmacol 2008; 116(2): 325–32.

16. Xu Y, Cao Z, Khan I, Luo Y. Gotu kola (*Centella asiatica*) extract enhances phosphorylation of cyclic amp response element binding protein in neuroblastoma cells expressing amyloid beta peptide. J Alzheimers Dis 2008; 13(3): 341–349.

17. Barbosa NR, Pittella F, Gattaz WF. *Centella asiatica* water extract inhibits iPLA(2) and cPLA(2) activities in rat cere-bellum. Phytomedicine 2008; Apr 30.

18. Liu M, Dai Y, Li Y, Luo Y, Huang F, Gong Z, Meng Q. Madecassoside isolated from Centella asiatica herbs facili-tates burn wound healing in mice. Planta Med 2008; May 16.

19. Jorge OA, Jorge AD. Hepatotoxicity associated with the ingestion of *Centella asiatica*. Rev Esp Enferm Dig 2005;97(2):115–24.

20. Bilbao I, Aguirre A, Zabala R, González R, Ratón J, Diaz Pérez JL. Allergic contact dermatitis from butoxyethyl nico-tinic acid and Centella asiatica extract. Contact Dermatitis 1995; 33(6): 435–6.

21. Oh C, Lee J. Contact allergy to various ingredients of topi-cal medicaments. Contact Dermatitis 2003; 49: 49–50.

22. Vetter J. Poison hemlock (Conium maculatum L.) Food Chem Toxicol 2004;42(9):1373–82.

23. Lal AA, Kumar T, Murthy PB, Pillai KS. Hypolipidemic effect of Coriandrum sativum L. in triton-induced hyperlipi-demic rats. Indian J Exp Biol 2004;42(9):909–12.

24. Gray AM, Flatt PR. Insulin-releasing and insulin-like activ-ity of the traditional anti-diabetic plant Coriandrum sativum (coriander). Br J Nutr 1999;81(3):203–9.

25. Sastre J, Olmo M, Novalvos A, Ibanez D, Lahoz C. Occupational asthma due to different spices. Allergy 1996;51(2):117–20.

26. Zabihi E, Abdollahi M. Endocrinotoxicity induced by Coriandrum sativa: a case report. WHO Drug Inf 2002;16:15.

27. Kanerva L, Soini M. Occupational protein contact dermati-tis from coriander. Contact Dermatitis 2001;45(6):354–5.

28. Manzanedo L, Blanco J, Fuentes M, Caballero ML, Moneo I. Anaphylactic reaction in a patient sensitized to coriander seed. Allergy 2004;59(3):362–3.

29. Kelly KJ, Neu J, Camitta BM, Honig GR. Methemoglobinemia in an infant treated with the folk remedy glycerited asafoetida. Pediatrics 1984;73(5):717–9.

30. Heck AM, DeWitt BA, Lukes AL. Potential interactions between alternative therapies and warfarin. Am J Health Syst Pharm 2000;57(13):1221–7.

Apocynaceae

General Information

The genera in the family of Apocynaceae (Table 9) include dogbane, oleander, and periwinkle.

Table 9 The genera of Apocynaceae

Adenium (desert rose)
Allamanda (allamanda)
Alstonia (alstonia)
Alyxia (alyxia)
Amsonia (blue star)
Anechites (anechites)
Angadenia (pineland golden trumpet)
Apocynum (dogbane)
Carissa (carissa)
Catharanthus (periwinkle)
Cycladenia (waxy dogbane)
Dyera (dyera)
Echites (echites)
Fernaldia (fernaldia)
Forsteronia (forsteronia)
Funtumia (funtumia)
Hancornia (hancornia)
Haplophyton (haplophyton)
Holarrhena (holarrhena)
Landolphia (landolphia)
Lepinia
Macrosiphonia (rock trumpet)

(Continued)

Nerium (oleander)
Ochrosia (yellow wood)
Pentalinon (pentalinon)
Plumeria (plumeria)
Prestonia (prestonia)
Pteralyxia (pteralyxia)
Rauvolfia (devil's pepper)
Rhabdadenia (rhabdadenia)
Saba (saba)
Strophanthus (strophanthus)
Tabernaemontana (milkwood)
Thevetia (thevetia)
Trachelospermum (trachelospermum)
Vallesia (vallesia)
Vinca (periwinkle)

Rauwolfia serpentina

The root of *Rauwolfia serpentina* (snakeroot) contains numerous alkaloids, of which reserpine and rescinnamine are said to be the most active as hypotensive agents (1).

Adverse effects

Inexpert use of *R. serpentina* can lead to serious toxicity, including symptoms such as hypotension, sedation, depression, and potentiation of other central depressants.

Reference

1. Cieri UR. Determination of reserpine and rescinnamine in Rauwolfia serpentina powders and tablets: collaborative study. J AOAC Int 1998;81(2):373–80.

Araliaceae

General Information

The genera in the family of Araliaceae (Table 10) include aralia, ginseng, and ivy.

Ginseng

Ginseng is an ambiguous vernacular term, which can refer to *Panax* species such as *Panax ginseng* (Asian ginseng) and *Panax quinquefolius* (American ginseng), *Eleutherococcus senticosus* (Siberian ginseng), *Pfaffia paniculata* (Brazilian ginseng), or unidentified material (for example Rumanian ginseng).

Asian ginseng (*P. ginseng*) contains a wide variety of flavonoids, saponins, steroids, sesquiterpenoids, and triterpenoids. These include ginsenolides and ginsenosides, protopanaxadiol, panaxadiol, panaxatriol, and panasinsene.

The botanical quality of ginseng preparations is problematic (1). For instance, when a case of neonatal androgenization was associated with maternal use of Siberian ginseng tablets during pregnancy, botanical analysis showed that the incriminated material almost certainly came from *Periploca sepium* (Chinese silk vine) (2).

Brazilian ginseng (*P. glomerata*) has been found to be contaminated with cadmium and mercury (3).

There is no good evidence that ginseng confers benefit for any indication (4).

Adverse effects

The various adverse effects that have been attributed to ginseng formulations include hypertension, pressure headaches, dizziness, estrogen-like effects, vaginal bleeding, and mastalgia. Prolonged use has been associated with a "ginseng abuse syndrome" including symptoms like hypertension, edema, morning diarrhea, skin eruptions, insomnia, depression, and amenorrhea. However, most reports are difficult to interpret, because of the absence of a control group, the simultaneous use of other agents, insufficient information about dosage, and lack of botanical authentication (1).

The authors of a review of the adverse effects associated with *P. ginseng* concluded that it is generally safe but at high doses can cause insomnia, headache, diarrhea, and cardiovascular and endocrine disorders (5). Inappropriate use and suboptimal formulations were deemed to be the most likely reason for adverse effects of ginseng. The authors of a systematic review reached similar conclusions and showed that serious adverse effects of ginseng seem to be true rarities (6).

Table 10 The genera of Araliaceae

Aralia (spikenard, angelica)
Cheirodendron (cheirodendron)
Dendropanax (dendropanax)
Eleutherococcus (ginseng)
Fatsia (fatsia)
Hedera (ivy)
Kalopanax (castor aralia)
Meryta (meryta)
Munroidendron (munroidendron)
Oplopanax (oplopanax)
Panax (ginseng)
Polyscias (aralia)
Pseudopanax (pseudopanax)
Reynoldsia (reynoldsia)
Schefflera (schefflera)
Tetraplasandra (tetraplasandra)
Tetrapanax (tetrapanax)

Cardiovascular

The WHO database contains seven cases from five countries of ginseng intake followed by arterial hypertension (7). In five of them no other medication was noted. In four cases the outcome was mentioned, which was invariably full recovery without sequelae after withdrawal of ginseng.

- A 64-year-old previously normotensive man presented with amaurosis fugax and hypertension (blood pressure 220/130 mmHg) after taking Ginseng Forte-Dietisa 500 mg/day for 13 days (8). All other tests were normal. He was advised to stop taking ginseng, and 1 week later his blood pressure was normal (140/90 mmHg).

Nervous system

- A 56-year-old woman with previous affective disorder had an episode of mania while taking ginseng (9). She was treated with neuroleptic drugs and benzodiazepines and ginseng was withdrawn. She made a rapid full recovery.

Ginseng has also been associated with a case of cerebral arteritis (10).

Psychiatric

A woman with prior episodes of depression had a manic episode several days after starting to take *P. ginseng* (11).

Skin

Stevens–Johnson syndrome occurred in a 27-year-old man who was a regular user of *P. ginseng* (12).

Musculoskeletal

Aggravation of muscle injury has been attributed to ginseng (SED-13, 1461) (13).

Drug-drug interactions

In 12 healthy volunteers the probe substrates dextromethorphan and alprazolam were given to test the effects of Siberian ginseng on the activity of CYP2D6 and CYP3A4 (14). The results showed that Siberian ginseng is unlikely to alter the disposition of co-administered medications whose metabolism primarily depends on CYP2D6 or CYP3A4.

In 20 healthy volunteers who took 100 mg of an extract of *Panax ginseng* standardized to 4% ginsenosides twice daily for 14 days no effects on CYP3A were detected and this result was confirmed in in vitro studies (15).

Digoxin

- A 74-year-old man had increased serum digoxin concentrations without signs of toxicity while taking Siberian ginseng (16). Common causes of increased serum digoxin were ruled out, and the association with ginseng use was confirmed by dechallenge and rechallenge. The rise in serum digoxin concentration could have been due to cardiac glycosides contained in *Periploca sepium*, a common substitute for *E. senticosus* (17).

Germanium

Diuretic resistance has been reported in a patient consuming a formulation containing germanium and ginseng (species unspecified) (18).

Monoamine oxidase inhibitors

Interactions of antidepressants with herbal medicines have been reported (19). Ginseng has been reported to cause mania, tremor, and headache when used in combination with conventional monoamine oxidase inhibitors.

Warfarin

In one case the concomitant use of *P. ginseng* and warfarin resulted in loss of anticoagulant activity (20). However, in an open, crossover, randomized study in 12 healthy men ginseng did not affect the pharmacokinetics or pharmacodynamics of *S*-warfarin or *R*-warfarin (21).

- A 58-year-old anticoagulated man with a mechanical bi-leaflet aortic valve prosthesis was hospitalized with an acute anteroseptal myocardial infarction and diabetic ketoacidosis (22). His international normalized ratio on admission was 1.4 and there were no other factors that could have interfered with warfarin. Thrombosis of the artificial valve was diagnosed and removed surgically.

The authors suspected that the thrombosis had been caused by an interaction of ginseng with warfarin.

References

1. Cui J, Garle M, Eneroth P, Bjorkhem I. What do commercial ginseng preparations contain? Lancet 1994;344(8915):134.
2. Awang DV. Maternal use of ginseng and neonatal androgenization. JAMA 1991;266(3):363.
3. Caldas ED, Machado LL. Cadmium, mercury and lead in medicinal herbs in Brazil. Food Chem Toxicol 2004;42(4):599–603.
4. Vogler BK, Pittler MH, Ernst E. The efficacy of ginseng. A systematic review of randomised clinical trials. Eur J Clin Pharmacol 1999;55(8):567–75.
5. Xie JT, Mehendale SR, Maleckar SA, Yuan CS. Is ginseng free from adverse effects? Oriental Pharm Exp Med 2002;2:80–6.
6. Coon JT, Ernst E. Panax ginseng: a systematic review of adverse effects and drug interactions. Drug Saf 2002;25(5):323–44.
7. Anonymous. Ginseng—hypertension. WHO SIGNAL 2002;.
8. Martínez-Mir I, Rubio E, Morales-Olivas FJ, Palop-Larrea V. Transient ischemic attack secondary to hypertensive crisis related to Panax ginseng. Ann Pharmacother 2004;38:1970.
9. Vazquez I, Aguera-Ortiz LF. Herbal products and serious side effects: a case of ginseng-induced manic episode. Acta Psychiatr Scand 2002;105(1):76–8.
10. Ryu SJ, Chien YY. Ginseng-associated cerebral arteritis. Neurology 1995;45(4):829–30.
11. Gonzalez-Seijo JC, Ramos YM, Lastra I. Manic episode and ginseng: report of a possible case. J Clin Psychopharmacol 1995;15(6):447–8.
12. Dega H, Laporte JL, Frances C, Herson S, Chosidow O. Ginseng as a cause for Stevens–Johnson syndrome? Lancet 1996;347(9011):1344.
13. Anonymous. Ginseng—muskelskada? Inf Socialstyr Läkemedelsavd 1988;4:114.
14. Donovan JL, DeVane CL, Chavin KD, Taylor RM, Markowitz JS. Siberian ginseng (*Eleutherococcus senticosus*) effects on CYP2D6 and CYP3A4 activity in normal volunteers. Drug Metab Dispos 2003;31:519–22.
15. Anderson GD, Rosito G, Mohustsy MA, Elmer GW. Drug interaction potential of soy extract and Panax ginseng. J Clin Pharmacol 2003;43:643–8.
16. McRae S. Elevated serum digoxin levels in a patient taking digoxin and Siberian ginseng. CMAJ 1996;155(3):293–5.
17. Awang DVC. Siberian ginseng toxicity may be case of mistaken identity. Can Med Assoc J 1996;155:1237.
18. Becker BN, Greene J, Evanson J, Chidsey G, Stone WJ. Ginseng-induced diuretic resistance. JAMA 1996;276(8):606–7.
19. Fugh-Berman A. Herb–drug interactions. Lancet 2000;355(9198):134–8.
20. Janetzky K, Morreale AP. Probable interaction between warfarin and ginseng. Am J Health Syst Pharm 1997;54(6):692–3.
21. Jiang X, Williams KM, Liauw WS, Ammit AJ, Roufogalis BD, Duke CC, Day RO, McLachlan AJ. Effect of St John's wort and ginseng on the pharmacokinetics and pharmacodynamics of warfarin in healthy subjects. Br J Clin Pharmacol 2004;57(5):592–9.
22. Rosado MF. Thrombosis of a prosthetic aortic valve disclosing a hazardous interaction between warfarin and a commercial ginseng product. Cardiology 2003;99:111.

Arecaceae

General Information

The genera in the family of Arecaceae (Table 11) include various types of palm.

Areca catechu

Areca catechu (areca, betel) contains piperidine alkaloids, such as guvacine, guvacoline, and isoguvacine, and pyridine alkaloids, such as arecaidine, arecolidine, and arecoline.

Many of the world's population (more than 200 million people worldwide) chew betel nut quid, a combination of areca nut, betel pepper leaf (from *Piper betle*), lime paste, and tobacco leaf. The major alkaloid of the areca nut, arecoline, can produce cholinergic adverse effects (such as bronchoconstriction) (1) as well as antagonism of anticholinergic agents (2). The lime in the betel quid causes hydrolysis of arecoline to arecaidine, a central nervous system stimulant, which accounts, together with the essential oil of the betel pepper, for the euphoric effects of chewing betel quid.

Adverse effects

Glycemia and anthropometric risk markers for type 2 diabetes were examined in relation to betel usage. Of 993 supposedly healthy Bangladeshis 12% had diabetes. A further 145 of 187 subjects at risk of diabetes (spot glucose over 6.5 mmol/l less than 2 hours after food or over 4.5 mmol/l more than 2 hours after food) had a second blood glucose sample taken; 61 were confirmed

Table 11 The genera of Arecaceae

Acoelorraphe (palm)
Acrocomia (acrocomia)
Aiphanes (aiphanes)
Archontophoenix (archontophoenix)
Calyptronoma (manac)
Caryota (fishtail palm)
Chamaedorea (chamaedorea)
Coccothrinax (silver palm)
Cocos (coconut palm)
Dypsis (butterfly palm)
Elaeis (oil palm)
Gaussia (gaussia)
Livistona (livistona)
Phoenix (date palm)
Prestoea (prestoea)
Pritchardia (pritchardia)
Pseudophoenix (pseudophoenix)
Ptychosperma (ptychosperma)
Rhapidophyllum (rhapidophyllum)
Roystonea (royal palm)
Sabal (palmetto)
Serenoa (serenoa)
Thrinax (thatch palm)
Washingtonia (fan palm)

as being at risk, and had an oral glucose tolerance test; nine new diabetics were identified. Spot blood glucose values fell with time after eating and increased independently with waist size and age. Waist size was strongly related to use of betel, and was independent of other factors such as age. The authors suggested that betel chewing may contribute to the risk of type 2 diabetes mellitus (3).

The saliva of betel nut chewers contains nitrosamines derived from areca nut alkaloids (4), and the use of areca nuts has been widely implicated in the development of oral cancers.

Serenoa repens

S. repens (American dwarf palm tree, cabbage palm, sabal, saw palmetto) has mainly been used to treat benign prostatic hyperplasia. In placebo-controlled and comparative studies its efficacy in benign prostatic hyperplasia and lower urinary tract symptoms has been demonstrated (5). Numerous mechanisms of action have been proposed, including an antiandrogenic action, an anti-inflammatory effect, and an antiproliferative influence through inhibition of growth factors.

A systematic review and meta-analysis of randomized trials of *S. repens* in men with benign prostatic hyperplasia showed that saw palmetto extracts improve urinary symptoms and flow measures to a greater extent than placebo, and similar improvements in urinary symptoms and flow measures to the 5-alpha-reductase inhibitor finasteride with fewer adverse effects (6).

Adverse effects
Hematologic

- A 53-year-old man, who had self-medicated with a saw palmetto supplement for benign prostatic hyperplasia, had profuse bleeding (estimated blood loss 2 liters) after resection of a meningioma and required 4 units of packed erythrocytes, 3 units of platelets, and 3 units of fresh frozen plasma (7). Postoperatively his bleeding time was 21 minutes (reference range 2–10 minutes), but all other coagulation tests were normal. He made an uneventful recovery.

The authors concluded that the cyclo-oxygenase inhibitory activity of saw palmetto had caused platelet dysfunction, which had resulted in abnormal bleeding.

Immunologic

Allergic contact dermatitis has been attributed to *Serenoa repens*.

- A 24-year-old woman developed allergic contact dermatitis while taking minoxidil and again while taking *Serenoa repens* solution for androgenic alopecia (8).

Drug-drug interactions

In 12 healthy volunteers *Serenoa repens* at generally recommended doses did not alter the disposition of co-administered medications whose metabolism primarily depends on CYP2D6 or CYP3A4 (9).

References

1. Taylor RF, al-Jarad N, John LM, Conroy DM, Barnes NC. Betel-nut chewing and asthma Lancet 1992;339(8802): 1134–6.

2. Deahl M. Betel nut-induced extrapyramidal syndrome: an unusual drug interaction. Mov Disord 1989;4(4):330–2.

3. Mannan N, Boucher BJ, Evans SJ. Increased waist size and weight in relation to consumption of Areca catechu (betel-nut); a risk factor for increased glycaemia in Asians in east London. Br J Nutr 2000;83(3):267–75.

4. Pickwell SM, Schimelpfening S, Palinkas LA. "Betelmania." Betel quid chewing by Cambodian women in the United States and its potential health effects West J Med 1994;160(4):326–30.

5. Buck AC. Is there a scientific basis for the therapeutic effects of Serenoa repens in benign prostatic hyperplasia? Mechanisms of action. J Urol 2004;172(5 Pt 1):1792–9.

6. Wilt T, Ishani A. Serenoa repens for treatment of benign prostatic hyperplasia (Cochrane Review). In: The Cochrane Library. Oxford: Update Software, 1999:1.

7. Cheema P, El-Mefty O, Jazieh AR. Intraoperative haemorrhage associated with the use of extract of saw palmetto herb: a case report and review of literature. J Intern Med 2001;250(2):167–9.

8. Sinclair R, Mallair R, Tate B. Sensitization to saw palmetto and minoxidil in separate topical extemporaneous treatments for androgenetic alopecia. Aust J Dermatol 2002;43:311–12.

9. Markowitz JS, Donovan JL, DeVane CL, Taylor RM, Ruan Y, Wang J-S, Chavin KD. Multiple doses of saw palmetto (Serenoa repens) did not alter cytochrome P450 2D6 and 3A4 activity in normal volunteers. Clin Pharmacol Ther 2003;74:536–42.

Aristolochiaceae

General Information

The family of Aristolochiaceae contains three genera:

1. *Aristolochia* (dutchman's pipe)
2. *Asarum* (wild ginger)
3. *Hexastylis* (heartleaf).

Aristolochia species

Plants belonging to the genus of *Aristolochia* are rich in aristolochic acids and aristolactams.

In the UK, the long-term (2 and 6 years) use of *Aristolochia* species in Chinese herbal mixtures, taken as an oral medication or herbal tea, resulted in Chinese-herb nephropathy with end-stage renal insufficiency (1). In reaction to these reports, the erstwhile Medicines Control Agency banned all *Aristolochia* species for medicinal use in the UK.

In June 2001, the FDA issued a nationwide alert, recalling 13 "Treasure of the East" herbal products containing aristolochic acid. Before this alert, the FDA had issued several warnings:

- On 4 April 2001 a "Dear Health Professional" letter was sent, drawing attention to serious renal disease associated with the use of aristolochic acid-containing dietary supplements or "traditional medicines." Health professionals were urged to review patients who had had unexplained renal disease, especially those with urothelial tract tumors and interstitial nephritis with end-stage renal insufficiency, to determine if such products had been used.
- On 9 April 2001 a letter was sent to industry associations, detailing the reported cases of renal disease associated with aristolochic acid.
- On 11 April 2001 the FDA cautioned consumers to immediately discontinue any dietary supplements or "traditional medicines" that contain aristolochic acid, including products with "Aristolochia," "Bragantia," or "Asarum" listed as their ingredients.

In a related action, Health Canada first issued a warning on aristolochic acid in November 1999 that this ingredient posed a Class I Health Hazard with a potential to cause serious health effects or death (2) and warned consumers not to use the pediatric product Tao Chih Pien. This Chinese product, sold in the form of tablets, is said to be a diuretic and a laxative. It is not labeled to contain aristolochic acid. However, the Chinese labeling says that it contains Mu Tong, a traditional term used to describe numerous herbs, including aristolochia; subsequent product analysis showed that Tao Chih Pien does indeed contain aristolochic acid. Health Canada advised individuals in possession of this product not to consume it and to return it to the place of purchase. It also issued a Customs Alert for the product to prevent the importation and sale of Tao Chih Pien and advised Canadians not to consume Longdan or Lung Tan Xi Gan products, since they may also contain aristolochic acid.

The product Longdan Qiegan Wan ("Wetness Heat" Pill) was removed from the Australian Register of Therapeutic Goods following the detection of aristolochic acid by laboratory testing by the Therapeutic Goods Administration (3).

In 2002 the Medicines Safety Authority of the Ministry of Health in New Zealand (Medsafe) ordered the withdrawal of several traditional Chinese medicines sold as herbal remedies (4). The products included Guan Xin Su He capsules, Long Dan Xie Gan Wan Pills, and Zhiyuan Xinqinkeli sachets.

In 2004 the UK's Medicines and Healthcare products Regulatory Agency, with the co-operation of Customs and Excise, seized a potentially illegal consignment of 90 000 traditional Chinese medicine tablets, Jingzhi Kesou Tanchuan, which reportedly contained *Aristolochia*.

In December 2004 the Hong Kong authorities warned the public not to take the product Shen yi Qian Lie Hui Chin, as laboratory tests showed that it contained aristolochic acid.

In 2004 China's State FDA banned two commonly used herbs containing aristolochic acid, a toxin that is linked to renal insufficiency and cancer (5). Manufacturers were directed to replace *Aristolochia fangchi* and *Aristolochia debilis* with *Staphania tetrandra* and *Inula helenium* respectively in their traditional medicine formulations by 30 September 2004. The Provincial Drug Bureau was instructed to carry out inspections to ensure compliance with the ban by 31 October. Medicines found to contain either *Aristolochia fangchi* or *Aristolochia debilis* after 30 September were to be treated as fake under Chinese law. By a previous order, special restrictions were imposed on four other potentially harmful aristolochic acid-containing herbs in China (*Fructus aristolochiae*, *Aristolochia mollissima Hance*, *Herba aristolochiae*, and *Aristolochia tuberose*); however, there was no outright ban on these products. Several countries withdrew formulations containing aristolochic acid in 1981 after the demonstration of carcinogenicity in a 3-month toxicity study in rats. A consolidated list of products whose consumption and/or sale have been banned, withdrawn, severally restricted or not approved by governments has been published (6).

Adverse effects

Several review articles have covered the toxicology of *Aristolochia* (7–10).

Liver

Hepatitis has been attributed to *Aristolochia* (11).

- A 49-year-old woman developed signs of hepatitis. All the usual causes were ruled out. The history revealed that she had recently started to use a Chinese herbal tea to treat her eczema. Examination of the herbal mixture showed that it contained *Aristolochia debilis* root and seven other medicinal plants.

Like *Aristolochia fangchi*, *A. debilis* contains the highly toxic aristolochic acid and was therefore the likely cause of the toxic hepatitis.

Urinary tract

Aristolochic acid is a potent carcinogen and can cause serious kidney damage, "Chinese herb nephropathy," which can be fatal (12). Renal fibrosis has also been reported (13).

Numerous reports from many countries have confirmed that plants from the *Aristolochia* species are the cause of the nephropathy (14,15), and the toxic agent has been confirmed to be aristolochic acid (16).

- A 46-year-old Chinese woman, living in Belgium and China, developed subacute renal insufficiency (17). Her creatinine concentration had increased from 80 μmol/l (November 1998) to 327 μmol/l (January 2000). During the preceding 6 months she had taken a patent medicine bought in China "for waste discharging and youth keeping." The package insert did not list any herbs of the *Aristolochia* species. Kidney biopsy showed extensive hypocellular interstitial fibrosis, tubular atrophy, and glomerulosclerosis. Analysis of the Chinese medicine demonstrated the presence of aristolochic acid. She required hemodialysis in June 2000 and received a renal transplant 4 months later.
- A 58-year-old Japanese woman with CREST syndrome (calcinosis, Raynaud's syndrome, esophageal sclerosis, sclerodactyly, and telangiectasia) developed progressive renal dysfunction (18). Renal biopsy showed changes typical of Chinese herb nephropathy. Analyses of Chinese herbs she had taken for several years demonstrated the presence of aristolochic acid. Oral prednisolone improved her renal function and anemia.
- A 59-year-old man developed renal insufficiency after self-medication for 5 years with a Chinese herbal remedy to treat his hepatitis (19). Renal biopsy showed signs characteristic of Chinese herb nephropathy. Analysis of the remedy proved the presence of aristolochic acids I and II.
- A 43-year-old Korean woman developed Fanconi's syndrome after taking a Chinese herbal mixture containing *Aristolochia* for 10 days, hoping to lose weight (20). Despite withdrawal of the remedy, she rapidly progressed to renal insufficiency. A renal biopsy showed typical findings of aristolochic acid-induced neuropathy.

In Belgium, an outbreak of nephropathy in about 70 individuals was attributed to a slimming formulation that supposedly included the Chinese herbs *Stephania tetrandra* and *Magnolia officinalis*. However, analysis showed that the root of *S. tetrandra* (Chinese name Fangji) had in all probability been substituted or contaminated with the root of *Aristolochia fangchi* (Chinese name Guang fangji) (21,22). The nephropathy was characterized by extensive interstitial fibrosis with atrophy and loss of the tubules (23). At least one patient had evidence suggestive of urothelial malignancies (24). The same remedy was apparently also distributed in France, and

two cases have been reported from Toulouse and one possible case from Nice (25).

Subsequently, Belgian nephrologists re-investigated 71 patients who were originally affected by this syndrome (26). Using multiple regression analysis, they showed that the original dose of *Aristolochia* was the only significant predictor of progression of renal insufficiency. The risk of end-stage renal insufficiency in these patients increased linearly with the dose of *Aristolochia*.

In Japan two cases of Chinese herb nephropathy were associated with chronic use of *Aristolochia manchuriensis* (Kan-mokutsu) (27). The diagnosis was confirmed by renal biopsy and the toxic constituents were identified as aristolochic acids I, II, and D.

Taiwanese authors have reported 12 cases of suspected Chinese herb nephropathy, confirmed by renal biopsy (28). Renal function deteriorated rapidly in most patients, despite withdrawal of the Aristolochia. Seven patients underwent dialysis and the rest had slowly progressive renal insufficiency. One patient was subsequently found to have a bladder carcinoma. Other cases have been reported from mainland China (29) and Taiwan (30).

Because of fear of malignancies the Belgian researchers who first described the condition have advocated prophylactic removal of the kidneys and ureters in patients with Chinese herb nephropathy. Of 39 patients who agreed to this, 18 (46%) had urothelial carcinoma, 19 of the others had mild to moderate urothelial dysplasia, and only two had normal urothelium (31). All tissue samples contained aristolochic acid-related DNA in adducts. The original dose of *Aristolochia* correlated positively with the risk of urothelial carcinoma.

Animal experiments have shed more light on Chinese herb nephropathy (32). Salt-depleted male Wistar rats were regularly injected with two different doses of aristolochic acid or with vehicle only for 35 days. The histological signs of Chinese herb nephropathy were demonstrated only in animals that received the high dose of 10 mg/kg. The authors presented this as an animal model for studying the pathophysiology of Chinese herb nephropathy.

An animal study has suggested that the nephrotoxicity of *Aristolochia* can be reduced by combining it with an extract of *Rhizoma coptidis* (Huanglian, 33).

Tumorigenicity

Aristolochic acid I and aristolochic acid II are mutagenic in several test systems. A mixture of these two compounds was so highly carcinogenic in rats that even homeopathic *Aristolochia* dilutions have been banned from the German market. The closely related aristolactam I and aristolactam II have not been submitted to carcinogenicity testing, but these compounds similarly show mutagenic activity in bacteria.

When 19 kidneys and urethras removed from 10 patients with Chinese herb nephropathy who required kidney transplantation were examined histologically, there were conclusive signs of neoplasms in 40% (34).

One patient who had a urothelial malignancy 6 years after the onset of Chinese herb nephropathy later developed a breast carcinoma that metastasised to the liver

(35). The urothelial malignancy contained aristolochic acid-DNA adducts and mutations in the p53 gene, and the same mis-sense mutation in codon 245 of exon 7 of p53 was found in DNA from the breast and liver tumors. However, DNA extracted from the urothelial tumor also showed a mutation in codon 139 of exon 5, which was not present in the breast and liver.

- A 69-year-old woman developed invasive urothelial cancer and later died after taking a Chinese herbal medicine containing aristolochic acid for weight loss (36). Her cancer was diagnosed 9 years after taking about 189 g of *Aristolochia fangchi* in total. Examination of tissue samples showed significant concentrations of specific aristolochic acid DNA adducts.

Unusually, the patient had not suffered from renal impairment before developing cancer.

Drug contamination

When 42 samples of Chinese herbal slimming aids were analysed in Switzerland four were found to contain the nephrotoxic aristolochic acid I and a further two were suspected to contain aristolochic acid derivatives (37). The authors called for the immediate removal of these products from the Swiss market.

Asarum heterotropoides

The Chinese herbal medicine Xu xin (38) is made from the leaves and aerial parts of *Asarum heterotropoides*. It is used for the symptomatic relief of colds, headaches, and other pains.

Adverse effects

The volatile oil of Xu xin causes the following adverse effects: vomiting, sweating, dyspnea, restlessness, fever, palpitation, and nervous system depression. Death can result from respiratory paralysis at high doses.

References

1. Lord GM, Tagore R, Cook T, Gower P, Pusey CD. Nephropathy caused by Chinese herbs in the UK. Lancet 1999;354(9177):481–2.
2. Anonymous. Aristolochic acid. Warnings on more products containing aristolochic acid. WHO Pharmaceuticals Newslett 2002;3:1.
3. Anonymous. Aristolochia. More products cancelled. WHO Pharmaceuticals Newslett 2002;1:1.
4. Anonymous. Traditional medicines. Several Chinese medicines withdrawn due to presence of prescription and pharmacy-only components. WHO Pharmaceuticals Newslett 2003;1:2–3.
5. Anonymous. Aristolochic acid. To be replaced by *Stephania tetrandra* and *Inula helenium*. WHO Pharmaceuticals Newslett 2004;5:1.
6. World Health Organization. Pharmaceuticals: restrictions in use and availability, April 2003.http://www.who.int/medicines/library/docseng_from_a_to_z.shtml#p.
7. Chen JK. Nephropathy associated with the use of Aristolochia. Herbal Gram 2000;48:44–5.
8. Pokhrel PK, Ergil KV. Aristolochic acid: a toxicological review. Clin Acupunct Orient Med 2000;1:161–6.
9. Hammes MG. Anmerkungen zu Aristolochia—eine Recherche in chinesischen Originaltexten. Dtsch Z Akupunkt 2000;3:198–200.
10. Wiebrecht A. Über die Aristolochia-Nephropathie. Dtsch Z Akupunkt 2000;3:187–97.
11. Levi M, Guchelaar HJ, Woerdenbag HJ, Zhu YP. Acute hepatitis in a patient using a Chinese herbal tea—a case report. Pharm World Sci 1998;20(1):43–4.
12. Anonymous. Aristolochia. Alert against products containing aristolochic acid. WHO Pharm Newslett 2001;2/3:1.
13. Anonymous. Aristolochic acid. Warning concerning interstitial renal fibrosis. WHO Newslett 2000;2:1.
14. Xi-wen D, Xiang-rong R, Shen LI. Current situation of Chinese-herbs-induced renal damage and its countermeasures. Chin J Integrative Med 2001;7:162–6.
15. Tamaki K, Okuda S. Chinese herbs nephropathy: a variant form in Japan. Intern Med 2001;40(4):267–8.
16. Lebeau C, Arlt VM, Schmeiser HH, Boom A, Verroust PJ, Devuyst O, Beauwens R. Aristolochic acid impedes endocytosis and induces DNA adducts in proximal tubule cells. Kidney Int 2001;60(4):1332–42.
17. Gillerot G, Jadoul M, Arlt VM, van Ypersele De Strihou C, Schmeiser HH, But PP, Bieler CA, Cosyns JP. Aristolochic acid nephropathy in a Chinese patient: time to abandon the term "Chinese herbs nephropathy?" Am J Kidney Dis 2001;38(5):E26.
18. Nishimagi E, Kawaguchi Y, Terai C, Kajiyama H, Hara M, Kamatani N. Progressive interstitial renal fibrosis due to Chinese herbs in a patient with calcinosis Raynaud esophageal sclerodactyly telangiectasia (CREST) syndrome. Intern Med 2001;40(10):1059–63.
19. Cronin AJ, Maidment G, Cook T, Kite GC, Simmonds MS, Pusey CD, Lord GM. Aristolochic acid as a causative factor in a case of Chinese herbal nephropathy. Nephrol Dial Transplant 2002;17(3):524–5.
20. Lee S, Lee T, Lee B, Choi H, Yang M, Ihm CG, Kim M. Fanconi's syndrome and subsequent progressive renal failure caused by a Chinese herb containing aristolochic acid. Nephrology 2004;9:126–9.
21. Vanherweghem JL, Depierreux M, Tielemans C, Abramowicz D, Dratwa M, Jadoul M, Richard C, Vandervelde D, Verbeelen D, Vanhaelen-Fastre R, et al. Rapidly progressive interstitial renal fibrosis in young women: association with slimming regimen including Chinese herbs. Lancet 1993;341(8842):387–91.
22. Vanhaelen M, Vanhaelen-Fastre R, But P, Vanherweghem JL. Identification of aristolochic acid in Chinese herbs. Lancet 1994;343(8890):174.
23. Depierreux M, Van Damme B, Vanden Houte K, Vanherweghem JL. Pathologic aspects of a newly described nephropathy related to the prolonged use of Chinese herbs. Am J Kidney Dis 1994;24(2):172–80.
24. Cosyns JP, Jadoul M, Squifflet JP, Van Cangh PJ, van Ypersele de Strihou C. Urothelial malignancy in nephropathy due to Chinese herbs. Lancet 1994;344(8916):188.
25. Stengel B, Jones E. Insuffisance rénale terminale associée à la consommation d'herbes chinoises en France. [End-stage renal insufficiency associated with Chinese herbal consumption in France.] Nephrologie 1998;19(1):15–20.
26. Martinez MC, Nortier J, Vereerstraeten P, Vanherweghem JL. Progression rate of Chinese herb nephropathy: impact of Aristolochia fangchi ingested dose. Nephrol Dial Transplant 2002;17(3):408–12.

27. Tanaka A, Nishida R, Maeda K, Sugawara A, Kuwahara T. Chinese herb nephropathy in Japan presents adult-onset Fanconi syndrome: could different components of aristolochic acids cause a different type of Chinese herb nephropathy? Clin Nephrol 2000;53(4):301–6.

28. Yang CS, Lin CH, Chang SH, Hsu HC. Rapidly progressive fibrosing interstitial nephritis associated with Chinese herbal drugs. Am J Kidney Dis 2000;35(2):313–8.

29. But PP, Ma SC. Chinese-herb nephropathy. Lancet 1999;354(9191):1731–2.

30. Lee CT, Wu MS, Lu K, Hsu KT. Renal tubular acidosis, hypokalemic paralysis, rhabdomyolysis, and acute renal failure—a rare presentation of Chinese herbal nephropathy. Ren Fail 1999;21(2):227–30.

31. Nortier JL, Martinez MC, Schmeiser HH, Arlt VM, Bieler CA, Petein M, Depierreux MF, De Pauw L, Abramowicz D, Vereerstraeten P, Vanherweghem JL. Urothelial carcinoma associated with the use of a Chinese herb (Aristolochia fangchi). N Engl J Med 2000;342(23):1686–92.

32. Debelle FD, Nortier JL, De Prez EG, Garbar CH, Vienne AR, Salmon IJ, Deschodt-Lanckman MM, Vanherweghem JL. Aristolochic acids induce chronic renal failure with interstitial fibrosis in salt-depleted rats. J Am Soc Nephrol 2002;13(2):431–6.

33. Hu SL, Zhang HQ, Chan K, Mei QX. Studies on the toxicity of Aristolochia manshuriensis (guanmuton). Toxicology 2004;198:195–201.

34. Cosyns JP, Jadoul M, Squifflet JP, Wese FX, van Ypersele de Strihou C. Urothelial lesions in Chinese-herb nephropathy. Am J Kidney Dis 1999;33(6):1011–7.

35. Lord GM, Hollstein M, Arlt VM, Roufosse C, Pusey CD, Cook T, Schmeiser HH. DNA adducts and p53 mutations in a patient with aristolochic acid-associated nephropathy. Am J Kidney Dis 2004;43(4):e11–7.

36. Nortier JL, Schmeiser HH, Martinez MCM, Arlt VM, Vervaet C. Invasive urothelial carcinoma after exposure to Chinese herbal medicine containing aristolochic acid may occur without severe renal failure. Nephrol Dial Transplant 2003;18:426–8.

37. Ioset JR, Raoelison GE, Hostettmann K. Detection of aristolochic acid in Chinese phytomedicines and dietary supplements used as slimming regimens. Food Chem Toxicol 2003;41:29–36.

38. Drew AK, Whyte IM, Bensoussan A, Dawson AH, Zhu X, Myers SP. Chinese herbal medicine toxicology database: monograph on Herba Asari, "xi xin". J Toxicol Clin Toxicol 2002;40(2):169–72.

Asclepiadaceae

General Information

The genera in the family of Asclepiadaceae (Table 12) include milkweed and periploca.

Table 12 The genera of Asclepiadaceae

Araujia (araujia)
Asclepias (milkweed)
Calotropis (calotropis)
Cryptostegia (rubbervine)
Cynanchum (swallow-wort)
Funastrum (twinevine)
Gonolobus (gonolobus)
Hoya (hoya)
Marsdenia (marsdenia)
Matelea (milkvine)
Metaplexis (metaplexis)
Morrenia (morrenia)
Oxypetalum (oxypetalum)
Periploca (periploca)
Stapelia (stapelia)
Xysmalobium (xysmalobium)

Asclepias tuberosa

Asclepias tuberosa (pleurisy root) contains cardenolides such as uzarigenin, coroglaucigenin, and corotoxigenin, the coumarins isorhamnetin, kaempferol, quercetin, and rutin, the steroid sitosterol, and the triterpenoids amyrin, friedelin, and lupeol.

Adverse effects

Because it contains cardenolides, *Asclepias* can have digitalis-like effects and potentiate digitalis toxicity. Interference with assays of plasma digoxin concentrations is also possible (1).

Xysmalobium undulatum

Xysmalobium undulatum (xysmalobium) contains the cardiac glycoside ascleposide. It has been used topically to treat wounds (2). Its adverse effects are likely to be those of other cardiac glycosides.

References

1. Longerich L, Johnson E, Gault MH. Digoxin-like factors in herbal teas. Clin Invest Med 1993;16(3):210–8.
2. Steenkamp V, Mathivha E, Gouws MC, van Rensburg CE. Studies on antibacterial, antioxidant and fibroblast growth stimulation of wound healing remedies from South Africa. J Ethnopharmacol 2004;95(2–3):353–7.

Asteraceae

General Information

The genera in the family of Asteraceae (Table 13) (formerly Compositae) include various types of asters (daisies), arnica, chamomile, goldeneye, marigold, snakeroot, tansy, thistle, and wormwood.

Delayed hypersensitivity reactions to the Asteraceae (Compositae) can arise from sesquiterpene lactones. To detect contact allergy to sesquiterpene lactones, a mixture of lactones (alantolactone, costunolide, and dehydrocostus lactone) is used. However, Compositae contain other sensitizers, such as polyacetylenes and thiophenes. In a prospective study, the lactone mixture was complemented with a mixture of Compositae (containing ether extracts of arnica, German chamomile, yarrow, tansy, and feverfew) to detect contact allergy to Compositae (1). Of 346 patients tested, 15 (4.3%) reacted to the mixture of Compositae, compared with eight of 1076 patients (0.7%) who gave positive results with the lactone mixture, indicating the importance of the addition of Compositae allergens to the lactone mixture. However, the authors warned that patch-testing with these mixtures can cause active sensitization.

- Compositae dermatitis occurred in a 9-year-old boy with a strong personal and family history of atopy. Positive patch test reactions were 2+ for dandelion (*Taraxacum officinale*), false ragweed (*Ambrosia acanthicarpa*), giant ragweed (*Ambrosia trifida*), short ragweed (*Ambrosia artemisifolia*), sagebrush (*Artemisia tridentata*), wild feverfew (*Parthenium hysterophorus*), yarrow (*Achillea millifolium*), and tansy (*Tanacetum vulgare*), and 1+ for *Dahlia* species and English ivy (*Hedera helix*) (2). Patch tests were negative for another 30 plants, including cocklebur (*Xanthium strumarium*), dog fennel (*Anthemis cotula*), fleabane (*Erigeron strigosus*), sneezeweed (*Helenium autumnale*), and feverfew (*Tanacetum parthenium*).

An Austrian study has re-confirmed the importance of testing with not only a mixture of Compositae and a mixture of sesquiterpene lactones, but also with additional plant extracts when there is continuing clinical suspicion of allergy to one of the Compositae (3). By using additional short ether extracts, the authors found two of five patients who had otherwise been overlooked.

Achillea millefolium

Achillea millefolium (yarrow) can cause contact dermatitis (4); a generalized eruption following the drinking of yarrow tea has also been reported (5).

- A female florist from North Germany, who ran a flower shop from 1954 to 1966 had to quit her job because of contact allergy to chrysanthemums and primrose. After a further 12 years she started to suffer occasionally from

Table 13 The genera of Asteraceae

Acamptopappus (goldenhead)
Acanthospermum (starburr)
Achillea (yarrow)
Achyrachaena (blow wives)
Acmella (spotflower)
Acourtia (desert peony)
Acroptilon (hard heads)
Adenocaulon (trail plant)
Adenostemma (medicine plant)
Adenophyllum (dogweed)
Ageratum (whiteweed)
Ageratina (snakeroot)
Agoseris (agoseris)
Almut aster (alkali marsh aster)
Amberboa (amberboa)
Amblyolepis (amblyolepis)
Amblyopappus (amblyopappus)
Ambrosia (ragweed)
Ampel aster (climbing aster)
Amphipappus (chaffbush)
Amphiachyris (broomweed)
Anacylus (anacylus)
Anaphalis (pearly everlasting)
Ancistrocarphus (nest straw)
Anisocoma (anisocoma)
Antennaria (pussytoes)
Anthemis (chamomile)
Antheropeas (e aster bonnets)
Aphanostephus (doze daisy)
Arctium (burrdock)
Arctotheca (capeweed)
Arctotis (arctotis)
Argyroxiphium (silver sword)
Argyranthemum (dill daisy)
Argyrautia (arhyrautia)
Arnica (arnica)
Arnoglossum (Indian plaintain)
Arnoseris (arnoseris)
Artemisia (sagebrush, wormwood)
Asanthus (brickell bush)
Aster (aster)
Astranthium (western daisy)
Atrichoseris (atrichoseris)
Baccharis (baccharis)
Bahia (bahia)
Baileya (desert marigold)
Balduina (honeycombhead)
Balsamorhiza (balsam root)
Balsamita (balsamita)
Baltimora (baltimora)
Barkleyanthus (willow ragwort)
Bartlettia (bartlettia)
Bartlettina (bartlettina)
Bebbia (sweetbush)
Bellis (bellis)
Benitoa (benitoa)
Berkheya (berkheya)
Berlandiera (green eyes)
Bidens (beggar ticks)
Bigelowia (rayless goldenrod)
Blennosperma (sticky seed)
Blepharipappus (blepharipappus)
Blepharizonia (blepharizonia)
Blumea (false ox tongue)

(Continued)

Table 13 (Continued)

Boltonia (doll's daisy)
Borrichia (seaside tansy)
Brickellia (brickell bush)
Brickelliastrum (brickell bush)
Buphthalmum (ox eye)
Cacaliopsis (cacaliopsis)
Calendula (marigold)
Callilepis (ox-eye daisy)
Callistephus (callistephus)
Calotis (calotis)
Calycadenia (western rosinweed)
Calycoseris (tackstem)
Calyptocarpus (calyptocarpus)
Canadanthus (mountain aster)
Carduus (plumeless thistle)
Carlina (carline thistle)
Carminatia (carminatia)
Carphephorus (chaffhead)
Carphochaete (bristlehead)
Carthamus (distaff thistle)
Castalis (castalis)
Celmisia (celmisia)
Centaurea (knapweed)
Centipeda (centipeda)
Centratherum (centratherum)
Chaenactis (pin cushion)
Chaetadelpha (skeletonweed)
Chaetopappa (least daisy)
Chamaemelum (dog fennel)
Chamaechaenactis (chamaechaenactis)
Chaptalia (sun bonnets)
Chloracantha (chloracantha)
Chondrilla (chondrilla)
Chromolaena (thoroughwort)
Chrysactinia (chrysactinia)
Chrysoma (chrysoma)
Chrysanthemum (daisy)
Chrysogonum (chrysogonum)
Chrysopsis (golden aster)
Chrysothamnus (rabbit brush)
Cichorium (chicory)
Cineraria (cineraria)
Cirsium (thistle)
Clappia (clapdaisy)
Clibadium (clibadium)
Cnicus (cnicus)
Columbiadoria (columbiadoria)
Condylidium (villalba)
Conoclinium (thoroughwort)
Conyza (horseweed)
Coreocarpus (coreocarpus)
Coreopsis (tickseed)
Corethrogyne (sand aster)
Cosmos (cosmos)
Cotula (waterbuttons)
Crassocephalum (ragleaf)
Crepis (hawksbeard)
Critonia (thoroughwort)
Crocidium (spring gold)
Croptilon (scratch daisy)
Crupina (crupina)
Cyanopsis (knapweed)
Cyanthillium (ironweed)
Cymophora (cymophora)

Cynara (cynara)
Dahlia (dahlia)
Delairea (capeivy)
Dendranthema (arctic daisy)
Dichaetophora (dichaetophora)
Dicoria (twin bugs)
Dicranocarpus (dicranocarpus)
Dimeresia (dimeresia)
Dimorphotheca (cape marigold)
Dittrichia (dittrichia)
Doellingeria (whitetop)
Doronicum (false leopardbane)
Dracopis (coneflower)
Dubautia (dubautia)
Dysodiopsis (dog fennel)
Dyssodia (dyssodia)
Eastwoodia (eastwoodia)
Eatonella (eatonella)
Echinacea (purple coneflower)
Echinops (globe thistle)
Eclipta (eclipta)
Egletes (tropic daisy)
Elephantopus (elephant's foot)
Eleutheranthera (eleutheranthera)
Emilia (tasselflower)
Encelia (brittlebush)
Enceliopsis (sunray)
Engelmannia (Engelmann's daisy)
Enydra (swampwort)
Erechtites (burnweed)
Ericameria (goldenbush)
Erigeron (fleabane)
Eriophyllum (woolly sunflower)
Erlangea (erlangea)
Eucephalus (aster)
Euchiton (euchiton)
Eupatorium (thoroughwort)
Eurybia (aster)
Euryops (euryops)
Euthamia (goldentop)
Evax (pygmy cudweed)
Facelis (trampweed)
Filago (cottonrose)
Fitchia (fitchia)
Flaveria (yellowtops)
Fleischmannia (thoroughwort)
Florestina (florestina)
Flourensia (tarwort)
Flyriella (brickell bush)
Gaillardia (blanket flower)
Galinsoga (gallant-soldier)
Gamochaeta (everlasting)
Garberia (garberia)
Gazania (gazania)
Geraea (desertsunflower)
Gerbera (Transvaal daisy)
Glyptopleura (glyptopleura)
Gnaphalium (cudweed)
Gochnatia (gochnatia)
Grindelia (gumweed)
Guardiola (guardiola)
Guizotia (guizotia)
Gundelia (gundelia)
Gundlachia (gundlachia)
Gutierrezia (snakeweed)

(Continued)

(Continued)

Table 13 (Continued)

Gymnosperma (gymnosperma)
Gymnostyles (burrweed)
Gynura (gynura)
Haploesthes (false broomweed)
Haplocarpha (onefruit)
Haplopappus (haplopappus)
Hartwrightia (hartwrightia)
Hasteola (false Indian plaintain)
Hazardia (bristleweed)
Hebeclinium (thoroughwort)
Hecastocleis (hecastocleis)
Hedypnois (hedypnois)
Helenium (sneezeweed)
Helianthell (helianthella)
Helianthus (sunflower)
Helichrysum (strawflower)
Heliopsis (heliopsis)
Heliomeris (false goldeneye)
Hemizonia (tarweed)
Hesperevax (dwarf-cudweed)
Hesperomannia (island aster)
Hesperodoria (glowweed)
Heteranthemis (ox eye)
Heterosperma (heterosperma)
Heterotheca (false golden aster)
Hieracium (hawkweed)
Holocarpha (tarweed)
Holozonia (holozonia)
Hulsea (alpinegold)
Hymenoclea (burrobrush)
Hymenopappus (hymenopappus)
Hymenothrix (thimblehead)
Hymenoxys (rubberweed)
Hypochaeris (cat's ear)
Hypochoeris (cat's ear)
Inula (yellowhead)
Ionactis (aster)
Isocarpha (pearlhead)
Isocoma (golden bush)
Iva (marsh elder)
Ixeris (ixeris)
Jamesianthus (jamesianthus)
Jaumea (jaumea)
Jefea (jefea)
Kalimeris (aster)
Koanophyllon (thoroughwort)
Krigia (dwarf dandelion)
Lactuca (lettuce)
Laennecia (laennicia)
Lagascea (lagascea)
Lagenifera (island daisy)
Lagophylla (hareleaf)
Lapsana (nipplewort)
Lapsanastrum (nipplewort)
Lasianthaea (lasianthaea)
Lasiospermum (cocoonhead)
Lasthenia (goldfields)
Launaea (launaea)
Layia (tidy tips)
Leibnitzia (sun bonnets)
Lembertia (lembertia)
Leontodon (hawkbit)
Lepidospartum (broom sage)
Lessingia (lessingia)

(Continued)

Leucanthemum (daisy)
Leucanthemella (leucanthemella)
Liatris (blazing star)
Lindheimera (lindheimera)
Lipochaeta (nehe)
Logfia (cotton rose)
Luina (silverback)
Lygodesmia (skeleton plant)
Machaeranthera (tansy aster)
Madia (tarweed)
Malacothrix (desert dandelion)
Malperia (malperia)
Mantisalca (mantisalca)
Marshallia (Barbara's buttons)
Matricaria (mayweed)
Megalodonta (water marigold)
Melampodium (blackfoot)
Melanthera (squarestem)
Micropus (cottonseed)
Microseris (silver puffs)
Mikania (hemp vine)
Monolopia (monolopia)
Monoptilon (desert star)
Montanoa (montanoa)
Mycelis (mycelis)
Neurolaena (neurolaena)
Nicolletia (hole-in-the-sand plant)
Nothocalais (prairie dandelion)
Oclemena (aster)
Olearia (daisy bush)
Oligoneuron (goldenrod)
Omalotheca (arctic cudweed)
Oncosiphon (oncosiphon)
Onopordum (cotton thistle)
Oonopsis (false goldenweed)
Oreochrysum (goldenrod)
Oreostemma (aster)
Orochaenactis (orochaenactis)
Osmadenia (osmadenia)
Osteospermum (daisy bush)
Packera (ragwort)
Palafoxia (palafox)
Parasenecio (Indian plantain)
Parthenice (parthenice)
Parthenium (feverfew)
Pascalia (Pascalia)
Pectis (cinchweed)
Pentachaeta (pygmy daisy)
Pentzia (pentzia)
Pericome (pericome)
Pericallis (ragwort)
Perityle (rock daisy)
Petasites (butterbur)
Petradoria (rock goldenrod)
Peucephyllum (pygmy cedar)
Phalacroseris (mock dandelion)
Phoebanthus (false sunflower)
Picradeniopsis (bahia)
Picris (ox tongue)
Picrothamnus (bud sagebrush)
Pinaropappus (rock lettuce)
Piptocarpha (ash daisy)
Piptocoma (velvet shrub)
Pityopsis (silk grass)
Platyschkuhria (basin daisy)

(Continued)

Table 13 (Continued)

Pleurocoronis (pleurocoronis)
Pluchea (camphorweed)
Polymnia (polymnia)
Porophyllum (poreleaf)
Prenanthes (rattlesnake root)
Prenanthella (prenanthella)
Proustia (proustia)
Psacalium (Indianbush)
Psathyrotes (turtleback)
Pseudogynoxys (pseudogynoxys)
Pseudelephantopus (dog's tongue)
Pseudobahia (sunburst)
Pseudoclappia (false clap daisy)
Pseudognaphalium (cudweed)
Psilactis (tansy aster)
Psilocarphus (woolly heads)
Psilostrophe (paper flower)
Pterocaulon (blackroot)
Pulicaria (false fleabane)
Pyrrhopappus (desert chicory)
Pyrrocoma (goldenweed)
Rafinesquia (California chicory)
Raillardella (raillardella)
Raillardiopsis (raillardiopsis)
Rainiera (rainiera)
Ratibida (prairie coneflower)
Rayjacksonia (tansy aster)
Reichardia (bright eye)
Remya (remya)
Rhagadiolus (rhagadiolus)
Rigiopappus (rigiopappus)
Rolandra (yerba de plata)
Roldana (groundsel)
Rudbeckia (coneflower)
Rugelia (Rugel's Indian plantain)
Sachsia (sachsia)
Salmea (bejuco de miel)
Santolina (lavender cotton)
Sanvitalia (creeping zinnia)
Sartwellia (glowwort)
Saussurea (sawwort)
Schkuhria (false threadleaf)
Sclerocarpus (bone bract)
Sclerolepis (bog button)
Scolymus (golden thistle)
Scorzonera (scorzonera)
Senecio (ragwort)
Sericocarpus (whitetop aster)
Serratula (plumeless sawwort)
Shinnersoseris (beaked skeletonweed)
Sigesbeckia (St Paul's wort)
Silphium (rosinweed)
Silybum (milk thistle)
Simsia (bush sunflower)
Smallanthus (smallanthus)
Solidago (goldenrod)
Soliva (burrweed)
Sonchus (sow thistle)
Sphaeromeria (chicken sage)
Sphagneticola (creeping ox eye)
Spilanthes (spilanthes)
Spiracantha (dogwood leaf)
Stebbinsoseris (silver puffs)
Stenotus (mock goldenweed)

Stephanomeria (wire lettuce)
Stevia (candyleaf)
Stokesia (stokesia)
Struchium (struchium)
Stylocline (nest straw)
Symphyotrichum (aster)
Synedrella (synedrella)
Syntrichopappus (Fremont's gold)
Tagetes (marigold)
Tamaulipa (boneset)
Tanacetum (tansy)
Taraxacum (dandelion)
Tephroseris (groundsel)
Tetramolopium (tetramolopium)
Tetraneuris (four-nerve daisy)
Tetradymia (horsebrush)
Tetragonotheca (nerve ray)
Thelesperma (green thread)
Thurovia (thurovia)
Thymophylla (pricklyleaf)
Tithonia (tithonia)
Tolpis (umbrella milkwort)
Tonestus (serpentweed)
Townsendia (Townsend daisy)
Tracyina (Indian headdress)
Tragopogon (goat's beard)
Trichocoronis (bugheal)
Trichoptilium (trichoptilium)
Tridax (tridax)
Tripleurospermum (mayweed)
Tripolium (sea aster)
Trixis (threefold)
Tussilago (coltsfoot)
Uropappus (silver puffs)
Urospermum (urospermum)
Vanclevea (vanclevea)
Varilla (varilla)
Venegasia (venegasia)
Venidium (venidium)
Verbesina (crownbeard)
Vernonia (ironweed)
Viguiera (goldeneye)
Wedelia (creeping ox eye)
Whitneya (whitneya)
Wilkesia (iliau)
Wollastonia (watermeal)
Wyethia (mule-ears)
Xanthisma (sleepy daisy)
Xanthium (cocklebur)
Xanthocephalum (xanthocephalum)
Xylorhiza (woody aster)
Xylothamia (desert goldenrod)
Yermo (desert yellowhead)
Youngia (youngia)
Zinnia (zinnia)

redness of the pharynx and stomachache after drinking tea prepared from yarrow and camomile. Skin tests were positive to chrysanthemum with cross-reactions to sunflower, arnica, camomile, yarrow, tansy, mugwort, and frullania (a lichen that does not occur in the Northern part of Germany). Patch-testing with primin showed high-grade hypersensitivity to Primula.

(Continued)

A. millefolium contains sesquiterpene lactones, polyacetylenes, coumarins, and flavonoids. Extracts have often been used in cosmetics in concentrations of 0.5–10%. *A. millefolium* was weakly genotoxic in *Drosophila melanogaster*. In provocative testing, patients reacted to a mix of *Compositae* that contained yarrow, as well as to yarrow itself. In clinical use, a formulation containing a 0.1% extract was not a sensitizer and alcoholic extracts of the dried leaves and stalks of the flower were not phototoxic (6). However, positive patch tests to *A. millefolium* have been reported (7).

Anthemis species and *Matricaria recutita* (chamomile)

Chamomile is the vernacular name of *Anthemis* genus and *Matricaria recutita* (German chamomile, pinhead). The former are more potent skin sensitizers (delayed-type) than the latter, presumably because they can contain a higher concentration of the sesquiterpene lactone, anthecotullid. Cross-sensitivity with related allergenic sesquiterpene lactones in other plants is possible.

Adverse effects
Internal use of chamomile tea has been associated with rare cases of anaphylactic reactions (8) and its use in eyewashes can cause allergic conjunctivitis (9).

Arnica montana

Arnica montana (arnica) contains a variety of terpenoids and has mostly been used in the treatment of sprains and bruises but is also used in cosmetics.

Ingestion of tea prepared from *Arnica montana* flowers can result in gastroenteritis.

- A 27-year-old woman presented with a rapidly enlarging necrotic lesion on her face and left leg together with malaise and high fever (10). She reported that she had applied a 1.5% arnica cream to her face before these symptoms had occurred. The diagnosis was Sweet's syndrome elicited by pathergy to arnica. She was treated with prednisolone and her skin lesions disappeared within 3 weeks.

Of 443 individuals who were tested for contact sensitization, 5 had a positive reaction to *A. montana* and 9 to *Calendula officinalis* (marigold); a mixture of the two was positive in 18 cases (3). Sensitization was often accompanied by reactions to nickel, *Myroxylon pereirae* resin, fragrance mix, propolis, and colophon.

Artemisia species

There are about 60 different species of *Artemisia*, of which the principal are *Artemisia absinthium*, *Artemisia annua*, *Artemisia cina*, and *Artemisia vulgaris*.

Artemisia absinthium
The volatile oil of *A. absinthium* (wormwood), which gives the alcoholic liqueur absinthe its flavor, can damage the nervous system and cause mental deterioration. This toxicity is attributed to thujones (alpha-thujone and beta-thujone), which constitute 0.25–1.32% in the whole herb and 3–12% of the oil. Alcoholic extracts and the essential oil are forbidden in most countries.

Artemisia annua
Artemisia annua, known in China as Qinghaosu, contains artemisinin, which has antimalarial activity. Several derivatives of the original compound have proved effective in the treatment of *Plasmodium falciparum* malaria and are currently available in a variety of formulations: artesunate (intravenous, rectal, oral), artelinate (oral), artemisinin (intravenous, rectal, oral), dihydroartemisinin (oral), artemether (intravenous, oral, rectal), and artemotil (intravenous). Artemisinic acid (qinghao acid), the precursor of artemisin, is present in the plant in a concentration up to 10 times that of artemisinin. Several semisynthetic derivatives have been developed from dihydroartemisinin (11).

Artemisia cina
Artemisia cina (wormseed) contains the toxic lactone, santonin, which was formerly used as an antihelminthic drug, but has now been superseded by other less toxic compounds.

Artemisia vulgaris
Artemisia vulgaris (common wormwood) contains the toxic lactone, santonin, which was formerly used as an antihelminthic drug, but has now been superseded by other less toxic compounds. Depending on the origin of the plant, 1,8-cineole, camphor, linalool, and thujone may all be major components. Allergic skin reactions (12) and abortive activity have been described.

Calendula officinalis

Calendula officinalis (marigold) contains a variety of carotenoids, saponins, steroids, sesquiterpenoids, and triterpenoids.

Of 443 individuals who were tested for contact sensitization, five had a positive reaction to *A. montana* and nine to *C. officinalis*; a mixture of the two was positive in 18 cases (3). Sensitization was often accompanied by reactions to nickel, *Myroxylon pereirae* resin, fragrance mix, propolis, and colophon.

Callilepis laureola

Callilepis laureola (impila, ox-eye daisy) contains the toxic compound atractyloside and related compounds. The plant is responsible for the deaths of many Zulu people in Natal, who use its roots as a herbal medicine.

Adverse effects
Necropsy records of 50 children who had taken herbal medicines made from *C. laureola* showed typical hepatic and renal tubular necrosis (13). In young Black children the plant causes hypoglycaemia, altered consciousness, and hepatic and renal dysfunction. This syndrome can be hard to distinguish from Reye's syndrome.

Acute renal insufficiency has been attributed to *C. laureola* (14).

Chrysanthemum vulgaris

Chrysanthemum vulgaris (common tansy) contains essential oils and thujone in such amounts that even normal doses can be neurotoxic (15).

Cynara scolymus

Cynara scolymus (artichoke) contains a variety of flavonoids, phenols, and sesquiterpenoids, including cynarapicrin, cynaratriol, cynarolide, and isoamberboin. It has been used to lower serum cholesterol, with little evidence of efficacy (16).

Adverse effects

Two vegetable warehouse workers developed occupational rhinitis and bronchial asthma by sensitization to *C. scolymus* (17). Skin prick tests to artichoke were positive and IgE specific for artichoke was found. Nasal challenge with artichoke extract triggered a reduction in peak nasal inspiratory flow of 81 and 85%. One patient had a reduction in peak expiratory flow rate of up to 36% after exposure to artichoke in the workplace.

Allergic contact dermatitis (18) and occupational contact urticaria (19) have also been reported.

Echinacea species

The three most commonly used species *Echinacea* are *E. angustifolia*, *E. pallida*, and *E. purpurea*. *Echinacea* is recommended for the prevention and treatment of the common cold.

Echinacea species (coneflower, black Sampson hedgehog, Indian head, snakeroot, red sunflower, scurvy root) have become increasingly popular, particularly for the prophylaxis and treatment and prevention of cold and flu symptoms. However, the claimed efficacy of *Echinacea* in the common cold has not been confirmed in a randomized, double-blind, placebo-controlled trial (20) or a systematic review (21). *Echinacea* is claimed to have antiseptic and antiviral properties and is under investigation for its immunostimulant action. The active ingredients are glycosides (echinacoside), polysaccharides, alkamides, and flavonoids.

Adverse effects

Between July 1996 and November 1998, the Australian Adverse Drug Reactions Advisory Committee received 37 reports of suspected adverse drug reactions in association with *Echinacea* (22). Over half of these ($n = 21$) described allergic-like effects, including bronchospasm ($n = 9$), dyspnea ($n = 8$), urticaria ($n = 5$), chest pain ($n = 4$), and angioedema ($n = 3$). The 21 patients were aged 3–58 (median 31) years and 12 had a history of asthma ($n = 7$) and/or allergic rhinitis/conjuctivitis/hayfever ($n = 5$). *Echinacea* was the only suspected cause in 19 of the 21 cases. The symptoms began at variable times, within 10 minutes of the first dose to a few months, and all but two cases occurred within 3 days of starting treatment. At the time of reporting 17 of the patients had recovered, 2 had not yet recovered, and the outcome was unknown in the other two cases.

A systematic review of all clinical reports of adverse events in clinical trials, post-marketing surveillance studies, surveys, spontaneous reporting schemes, and to manufacturers, the WHO, and national drug safety bodies has suggested that short-term use of *Echinacea* is associated with a relatively good safety profile, with a slight risk of transient, reversible, adverse events, of which gastrointestinal upsets and rashes occur most often (23). In rare cases, *Echinacea* is associated with allergic reactions, which can be severe.

Hematologic

Possible leukopenia has been associated with long-term use of *Echinacea* (24).

Skin

Recurrent erythema nodosum has been attributed to *Echinacea*.

- A 41-year-old man, who had taken *Echinacea* intermittently for the previous 18 months, had four episodes of erythema nodosum, preceded by myalgia and arthralgia, fever, headache, and malaise (25). The skin lesions resolved within 2–5 weeks and responded to oral prednisolone. He was advised to discontinue *Echinacea* and 1 year later remained free from further recurrence.

Echinacea has been reported to have caused a flare up of pemphigus vulgaris (26).

- A 55-year-old man with pemphigus vulgaris in remission self-administered *Echinacea* for an upper respiratory tract infection. Within 1 week he developed an acute exacerbation of the pemphigus vulgaris. Withdrawal of *Echinacea* resulted in improvement of his symptoms, but he had to be treated with prednisolone, azathioprine, and dapsone to achieve a partial remission.

The authors suggested that the immunostimulatory properties of *Echinacea* may have caused this flare up.

Immunologic

Intravenous administration of *Echinacea* has been associated with severe allergic reactions. Oral ingestion can cause allergic skin and respiratory responses (27).

Five cases of adverse drug reactions have been attributed to oral *Echinacea* extracts (28). Two of the patients had anaphylaxis and one had an acute attack of asthma. The authors also tested 100 atopic subjects and found that 20 of them, who had never before taken *Echinacea*, had positive reactions to skin prick tests.

An anaphylactic reaction to *Echinacea angustifolia* has been reported (29).

- A 37-year-old woman who took various food supplements on an irregular basis self-medicated with 5 ml of an extract of *E. angustifolia*. She had immediate burning of the mouth and throat followed by tightness of the chest, generalized urticaria, and diarrhea. She made a full recovery within 2 hours.

The basis for this anaphylactic reaction was hypersensitivity to *Echinacea*, confirmed by skin prick and RAST testing. However, others have challenged the notion of a causal relation in this case (30). Nevertheless, the author affirmed his belief that *Echinacea* was the causal agent and reported that at that time *Echinacea* accounted for 22 of 266 suspected adverse reactions to complementary medicines reported to the Australian Adverse Drug Reaction Advisory Committee (30).

Sjögren's syndrome has been attributed to *Echinacea* (31).

- A 36-year-old woman developed generalized muscle weakness (31). She was found to have hypokalemia, which was treated with electrolyte replacement. Her muscular complaints disappeared but she then complained of joint stiffness, dry mouth, and dry eyes. The diagnosis of Sjögren's syndrome was confirmed by laboratory tests.

She had been taking a herbal mixture that included *Echinacea*, which is known to stimulate the immune system, and the authors speculated that *Echinacea* had aggravated a pre-existing autoimmune disease.

Teratogenicity

Of 412 pregnant Canadian women who contacted a specialized information service between 1996 and 1998 with concerns about the use of *Echinacea* during pregnancy, 206 had already taken the remedy and the other 206 eventually decided not to use it (32). In the *Echinacea* group, 54% had taken it during the first trimester of pregnancy; 12 babies had malformations, six major and six minor. The figures in the control group were seven and seven respectively. Thus, there was no difference in the incidence of birth defects. However, the study lacked sufficient power to generate reliable data.

Eupatorium species

Several *Eupatorium* species, such as *Eupatorium cannabinum* (hemp agrimony) and *Eupatorium purpureum* (gravel root), have hepatotoxic potential due to the presence of pyrrolizidine alkaloids, which are covered in a separate monograph.

There is no evidence of pyrrolizidine alkaloids in *Eupatorium rugosum* (white snakeroot) but this plant also has poisonous properties, which are attributed to an unstable toxin called tremetol. Transfer from cow's milk to humans can produce a condition known as milk sickness, including trembles, weakness, nausea and vomiting, prostration, delirium, and even death.

Inula helenium

Large doses of the root of *Inula helenium* (elecampane) can cause vomiting, diarrhea, cramps, and paralytic symptoms.

Petasites species

Petasites species have hepatotoxic potential, owing to the presence of pyrrolizidine alkaloids, which are covered in a separate monograph. Extracts of *Petasites hybridus* (blatterdock, bog rhubarb, butterbur, butterdock) contain little in the way of these alkaloids (33). Butterbur has been used to treat allergic rhinitis and asthma and in the prevention of migraine.

Senecio species

Many species of *Senecio*, such as *Senecio jacobaea* (ragwort) and *Senecio longilobus* (thread leaf groundsel), contain hepatotoxic amounts of pyrrolizidine alkaloids (which are covered in a separate monograph). Honey made from *Senecio* plants also contains pyrrolizidine alkaloids (34).

Adverse effects

Veno-occlusive disease has been attributed to *Senecio* after chronic use (35,36).

- Hepatic veno-occlusive disease occurred in a 38-year-old woman who had occasionally consumed "Huamanrripa" (*Senecio tephrosioides*) as a cough remedy for many years (37). She had abdominal pain, jaundice, and anasarca. A hepatic biopsy showed pronounced congestion with a centrilobular predominance, foci of necrosis, and in some areas a reversed lobulation pattern. During the next 13 months she was hospitalized four times with complications of portal hypertension.
- An infant developed hepatic veno-occlusive disease after having been fed a herbal tea known as gordolobo yerba, commonly used as a folk remedy among Mexican-Americans; there was acute hepatocellular disease and portal hypertension, which progressed over 2 months to extensive hepatic fibrosis (38).

In one case hepatic damage due to *Senecio* mimicked Reye's syndrome (39).

Silybum marianum

Silybum marianum (holy thistle, lady's thistle, milk thistle, St. Mary's thistle) has been used to treat liver problems, such as hepatitis, and prostatic cancer. It contains a variety of lignans, including silandrin, silybin, silychristin, silydianin, silymarin, and silymonin.

Adverse effects

- A 57-year-old Australian woman presented with a 2-month history of intermittent episodes of sweating, nausea, colicky abdominal pain, fluid diarrhea, vomiting, weakness, and collapse (40). She was taking ethinylestradiol and amitriptyline and had taken milk thistle for 2 months. A thorough check-up showed no abnormalities. On reflection she realized that all her attacks had invariably occurred after taking the milk thistle. She stopped taking it and had no symptoms until a few weeks later, when she tried another capsule and had the same symptoms.

This idiosyncratic reaction to milk thistle seems to be a rarity. The Australian authorities knew of only two other adverse drug reactions associated with milk thistle.

Immunologic
Anaphylactic shock has been reported after the use of a herbal tea containing an extract of the fruit of the milk thistle (41).

Drug-drug interactions
Indinavir
Milk thistle inhibits CYP3A4 and uridine diphosphoglucuronosyl transferase in human hepatocyte cultures (42).

In 10 healthy subjects silymarin 160 mg tds had no effect on the pharmacokinetics of indinavir 800 mg tds (43). In a similar study silymarin 175 mg tds had no effect on the pharmacokinetics of indinavir 800 mg tds (44).

Tanacetum parthenium

Tanacetum parthenium (feverfew, bachelor's buttons, motherherb) has been used in the prevention of migraine, with some benefit (45), and for rheumatoid arthritis, without (46).

Adverse effects
As *Tanacetum parthenium* is rich in allergenic sesquiterpene lactones, such as parthenolide, it is not surprising that contact dermatitis has been observed (SEDA-11, 426). The most common adverse effect of oral feverfew is mouth ulceration. A more widespread inflammation of the oral mucosa and tongue, swelling of the lips, and loss of taste have also been reported.

Feverfew inhibits platelet aggregation (47), and its concomitant use with anticoagulants such as warfarin is therefore not advised.

Tussilago farfara

Tussilago farfara (coltsfoot) has hepatotoxic potential owing to the presence of pyrrolizidine alkaloids (see separate monograph).

- An 18-month-old boy who had regularly consumed a herbal tea mixture since the 3rd month of life developed veno-occlusive disease with portal hypertension and severe ascites (48). Histology of the liver showed centrilobular sinusoidal congestion with perivenular bleeding and parenchymal necrosis without cirrhosis. The child was given conservative treatment only and recovered completely within 2 months.

The tea contained peppermint and what the mother thought was coltsfoot (*T. farfara*), analysis of which revealed high amounts of pyrrolizidine alkaloids. Seneciphylline and the corresponding *N*-oxide were identified as the major components, and the child had consumed at least 60 µg/kg/day of the toxic pyrrolizidine alkaloid mixture over 15 months. Macroscopic and microscopic analysis of the leaf material indicated that *Adenostyles alliariae*

(Alpendost) had been erroneously gathered by the parents in place of coltsfoot. The two plants can easily be confused especially after the flowering period.

References

1. Kanerva L, Estlander T, Alanko K, Jolanki R. Patch test sensitization to Compositae mix, sesquiterpene–lactone mix, Compositae extracts, laurel leaf, chlorophorin, mansonone A, and dimethoxydalbergione. Am J Contact Dermat 2001;12(1):18–24.
2. Guin JD, Skidmore G. Compositae dermatitis in childhood. Arch Dermatol 1987;123(4):500–2.
3. Reider N, Komericki P, Hausen BM, Fritsch P, Aberer W. The seamy side of natural medicines: contact sensitization to arnica (*Arnica montana* L.) and marigold (*Calendula officinalis* L.) Contact Dermatitis 2001;45(5):269–72.
4. Jovanovic M, Poljacki M, Duran V, Vujanovic L, Sente R, Stojanovic S. Contact allergy to Compositae plants in patients with atopic dermatitis. Med Pregl 2004;57(5-6):209–18.
5. Hausen BM, Schulz KH. Polyvalente Kontaktallergie bei einer Floristin. [Polyvalent contact allergy in a florist.] Derm Beruf Umwelt 1978;26(5):175–6.
6. Anonymous. Final report on the safety assessment of yarrow (*Achillea millefolium*) Extract. Int J Toxicol 2001;20(Suppl 2):79–84.
7. Stingeni L, Agea E, Lisi P, Spinozzi F. T lymphocyte cytokine profiles in compositae airborne dermatitis. Br J Dermatol 1999;141(4):689–93.
8. Subiza J, Subiza JL, Hinojosa M, Garcia R, Jerez M, Valdivieso R, Subiza E. Anaphylactic reaction after the ingestion of chamomile tea: a study of cross-reactivity with other composite pollens. J Allergy Clin Immunol 1989;84(3):353–8.
9. Subiza J, Subiza JL, Alonso M, Hinojosa M, Garcia R, Jerez M, Subiza E. Allergic conjunctivitis to chamomile tea. Ann Allergy 1990;65(2):127–32.
10. Delmonte S, Brusati C, Parodi A, Rebora A. Leukemia-related Sweet's syndrome elicited by pathergy to *Arnica*. Dermatology 1998;197(2):195–6.
11. Ridley RG, Hudson AT. Chemotherapy of malaria. Curr Opin Infect Dis 1998;11:691–705.
12. Kurz G, Rapaport MJ. External/internal allergy to plants (*Artemesia*). Contact Dermatitis 1979;5(6):407–8.
13. Watson AR, Coovadia HM, Bhoola KD. The clinical syndrome of impila (*Callilepis laureola*) poisoning in children. S Afr Med J 1979;55(8):290–2.
14. Seedat YK, Hitchcock PJ. Acute renal failure from *Callilepsis laureola*. S Afr Med J 1971;45(30):832–3.
15. Holstege CP, Baylor MR, Rusyniak DE. Absinthe: return of the Green Fairy. Semin Neurol 2002;22(1):89–93.
16. Pittler MH, Thompson CO, Ernst E. Artichoke leaf extract for treating hypercholesterolaemia. Cochrane Database Syst Rev 2002;(3):CD003335.
17. Miralles JC, Garcia-Sells J, Bartolome B, Negro JM. Occupational rhinitis and bronchial asthma due to artichoke (*Cynara scolymus*). Ann Allergy Asthma Immunol 2003;91(1):92–5.
18. Meding B. Allergic contact dermatitis from artichoke, *Cynara scolymus*. Contact Dermatitis 1983;9(4):314.
19. Quirce S, Tabar AI, Olaguibel JM, Cuevas M. Occupational contact urticaria syndrome caused by globe artichoke (*Cynara scolymus*). J Allergy Clin Immunol 1996;97(2):710–1.
20. Yale SH, Liu K. Echinacea purpurea therapy for the treatment of the common cold: a randomized, double-blind,

placebo-controlled clinical trial. Arch Intern Med 2004;164(11):1237–41.

21. Melchart D, Linde K, Fischer P, Kaesmayr J. *Echinacea* for preventing and treating the common cold. Cochrane Database Syst Rev 2000;(2):CD000530.

22. Anonymous. *Echinacea*-allergic reactions. WHO Pharm Newslett 1999;5/6:7.

23. Huntley AL, Thompson Coon J, Ernst E. The safety of herbal medicinal products derived from Echinacea species: a systematic review. Drug Saf 2005;28(5):387–400.

24. Kemp DE, Franco KN. Possible leukopenia associated with long-term use of *Echinacea*. J Am Board Fam Pract 2002;15(5):417–9.

25. Soon SL, Crawford RI. Recurrent erythema nodosum associated with *Echinacea* herbal therapy. J Am Acad Dermatol 2001;44(2):298–9.

26. Lee AN, Werth VP. Activation of autoimmunity following use of immunostimulatory herbal supplements. Arch Dermatol 2004;140:723–27.

27. Anonymous. Wie verträglich sind *Echinacea*-haltige Präparate? Dtsch Arzteblatt 1996;93:2723.

28. Mullins RJ, Heddle R. Adverse reactions associated with echinacea: the Australian experience. Ann Allergy Asthma Immunol 2002;88(1):42–51.

29. Mullins RJ. *Echinacea*-associated anaphylaxis. Med J Aust 1998;168(4):170–1.

30. Myers SP, Wohlmuth H. *Echinacea*-associated anaphylaxis. Med J Aust 1998;168(11):583–4.

31. Logan JL, Ahmed J. Critical hypokalemic renal tubular acidosis due to Sjögren's syndrome: association with the purported immune stimulant *Echinacea*. Clin Rheumatol 2003;22(2):158–9.

32. Gallo M, Sarkar M, Au W, Pietrzak K, Comas B, Smith M, Jaeger TV, Einarson A, Koren G. Pregnancy outcome following gestational exposure to *Echinacea*: a prospective controlled study. Arch Intern Med 2000;160(20):3141–3.

33. Kalin P. Gemeine Pestwurz (*Petasites hybridus*)—Portrait einer Arzneipflanze. [The common butterbur (*Petasites hybridus*)—portrait of a medicinal herb.] Forsch Komplementarmed Klass Naturheilkd 2003;10(Suppl 1): 41–4.

34. Deinzer ML, Thomson PA, Burgett DM, Isaacson DL. Pyrrolizidine alkaloids: their occurrence in honey from tansy ragwort (*Senecio jacobaea* L.) Science 1977;195(4277):497–9.

35. Ortiz Cansado A, Crespo Valades E, Morales Blanco P, Saenz de Santamaria J, Gonzalez Campillejo JM, Ruiz Tellez T. Enfermedad venooclusiva hepatica por ingestion de infusiones de *Senecio vulgaris*. [Veno-occlusive liver disease due to intake of *Senecio vulgaris* tea.] Gastroenterol Hepatol 1995;18(8):413–6.

36. Radal M, Bensaude RJ, Jonville-Bera AP, Monegier Du Sorbier C, Ouhaya F, Metman EH, Autret-Leca E.

Maladie veino-occlusive apres ingestion chronique d'une specialite a base de senecon. [Veno-occlusive disease following chronic ingestion of drugs containing senecio.] Therapie 1998;53(5):509–11.

37. Tomioka M, Calvo F, Siguas A, Sanchez L, Nava E, Garcia U, Valdivia M, Reategui E. Enfermedad hepatica veno-oclusiva asociada a la ingestion de huamanrripa (*Senecio tephrosioides*). [Hepatic veno-occlusive disease associated with ingestion of *Senecio tephrosioides*.] Rev Gastroenterol Peru 1995;15(3):299–302.

38. Stillman AS, Huxtable R, Consroe P, Kohnen P, Smith S. Hepatic veno-occlusive disease due to pyrrolizidine (*Senecio*) poisoning in Arizona. Gastroenterology 1977;73(2):349–52.

39. Fox DW, Hart MC, Bergeson PS, Jarrett PB, Stillman AE, Huxtable RJ. Pyrrolizidine (*Senecio*) intoxication mimicking Reye syndrome. J Pediatr 1978;93(6):980–2.

40. Adverse Drug Reactions Advisory Committee. An adverse reaction to the herbal medication milk thistle (*Silybum marianum*). Med J Aust 1999;170(5):218–9.

41. Geier J, Fuchs T, Wahl R. Anaphylaktischer Schock durch einen Mariendistel-Extrakt bei Soforttyp-Allergie auf Kiwi. Allergologie 1990;13:387–8.

42. Venkataramanan R, Ramachandran V, Komoroski BJ, Zhang S, Schiff PL, Strom SC. Milk thistle, a herbal supplement, decreases the activity of CYP3A4 and uridine diphosphoglucuronosyl transferase in human hepatocyte cultures. Drug Metab Dispos 2000;28(11):1270–3.

43. DiCenzo R, Shelton M, Jordan K, Koval C, Forrest A, Reichman R, Morse G. Coadministration of milk thistle and indinavir in healthy subjects. Pharmacotherapy 2003;23(7):866–70.

44. Piscitelli SC, Formentini E, Burstein AH, Alfaro R, Jagannatha S, Falloon J. Effect of milk thistle on the pharmacokinetics of indinavir in healthy volunteers. Pharmacotherapy 2002;22(5):551–6.

45. Pittler MH, Vogler BK, Ernst E. Feverfew for preventing migraine. Cochrane Database Syst Rev 2000;(3):CD002286.

46. Pattrick M, Heptinstall S, Doherty M. Feverfew in rheumatoid arthritis: a double blind, placebo controlled study. Ann Rheum Dis 1989;48(7):547–9.

47. Groenewegen WA, Heptinstall S. A comparison of the effects of an extract of feverfew and parthenolide, a component of feverfew, on human platelet activity in-vitro. J Pharm Pharmacol 1990;42(8):553–7.

48. Sperl W, Stuppner H, Gassner I, Judmaier W, Dietze O, Vogel W. Reversible hepatic veno-occlusive disease in an infant after consumption of pyrrolizidine-containing herbal tea. Eur J Pediatr 1995;154(2):112–6.

Basidiomycetes

General Information

Basidiomycetes are fungi that include mushrooms, puffballs, and bracket fungi.

Lentinus edodes

Lentinus edodes (shiitake) is an edible mushroom that contains a polysaccharide, lentinan. Its use is occasionally associated with skin reactions (1,2).

Adverse effects

Most adverse effects of *Lentinus edodes* occur in shiitake workers.

Respiratory

Lentinus edodes can cause an interstitial hypersensitivity pneumonitis called mushroom worker's lung, associated with IgG antibodies against shiitake spore antigens; those who cultivate white button mushrooms (*Agaricus bisporus*) or the oyster mushroom (*Pleurotus* species) have only low titers (3).

Workers at a shiitake farm developed cough and sputum production after a variable period of exposure to shiitake mushrooms (4). All four had abnormal diffusing capacity and three had abnormal spirometry. Chest X-rays showed an interstitial pattern in one case. Pulmonary function tests fell significantly during several days of work, with a more than 20% fall in forced vital capacity and/or maximal mid-expiratory flow. Antigens to shiitake spore antigens, in common with antigens from other cultivated mushrooms (*Agaricus* and *Pleurotus*), were demonstrated by ELISA.

Bronchial asthma has been attributed to shiitake (5).

Skin

Skin reactions due to *Lentinus edodes* are not uncommon (1,2).

- A 42-year-old female shiitake grower developed skin lesions while planting shiitake hyphae into bed logs (6). She complained of repeated eczematous skin lesions during the planting season, from March to July, for 10 years. Each day she handled 7000 pieces of small conic blocks made of beech, with shiitake hyphae attached to their surface, and altogether 300 000 pieces each season. Patch tests with extracts of shiitake hyphae were positive. In contrast, female shiitake growers with skin lesions associated with work other than planting, and without skin lesions, were negative on patch-testing.

Drug interactions

Lentinan inhibits CYP1A (7), but the relevance of this to drug interactions in man is not known.

References

1. Nakamura T, Kobayashi A. Toxikodermie durch den Speisepilz Shiitake (Lentinus edodes). [Toxicodermia cause by the edible mushroom shiitake (Lentinus edodes).] Hautarzt 1985;36(10):591–3.
2. Nakamura T. Shiitake (Lentinus edodes) dermatitis. Contact Dermatitis 1992;27(2):65–70.
3. Van Loon PC, Cox AL, Wuisman OP, Burgers SL, Van Griensven LJ. Mushroom worker's lung. Detection of antibodies against shii-take (Lentinus edodes) spore antigens in shii-take workers. J Occup Med 1992;34(11):1097–101.
4. Sastre J, Ibanez MD, Lopez M, Lehrer SB. Respiratory and immunological reactions among shiitake (Lentinus edodes) mushroom workers. Clin Exp Allergy 1990;20(1):13–9.
5. Kondo T. [Case of bronchial asthma caused by the spores of Lentinus edodes (Berk) Sing.]Arerugi 1969;18(1):81–5.
6. Ueda A, Obama K, Aoyama K, Ueda T, Xu BH, Li Q, Huang J, Kitano T, Inaoka T. Allergic contact dermatitis in shiitake (Lentinus edodes (Berk) Sing) growers. Contact Dermatitis 1992;26(4):228–33.
7. Okamoto T, Kodoi R, Nonaka Y, Fukuda I, Hashimoto T, Kanazawa K, Mizuno M, Ashida H. Lentinan from shiitake mushroom (Lentinus edodes) suppresses expression of cytochrome P450 1A subfamily in the mouse liver. Biofactors 2004;21(1–4):407–9.

Berberidaceae

General Information

The genera in the family of Berberidaceae (Table 14) include lychee and soapberry.

Berberis vulgaris

Barberry (pipperidge bush) is a vernacular name for *Berberis vulgaris* (the European barberry), but it can also refer to *Mahonia aquifolium* and *Mahonia nervosa*. In the USA only the *Mahonia* species have had official status as a source of barberry, but *Berberis vulgaris* is said to serve similar medicinal purposes and to contain similar principles. Its root bark yields the quaternary isoquinoline alkaloid berberine and several other tertiary and quaternary alkaloids. Berberine is also found in *Hydrastis canadensis* (goldenseal) and *Coptis chinensis* (goldenthread).

Adverse effects

In man berberine has positive inotropic, negative chronotropic, antidysrhythmic, and vasodilator properties (1) and there is experimental evidence that it can cause arterial hypotension (2,3).

Berberine displaces bilirubin from albumin and there is therefore a risk of kernicterus in jaundiced neonates (4).

In a study of the effect of berberine in acute watery diarrhea, oral doses of 400 mg were well tolerated, except for complaints about its bitter taste and a few instances of transient nausea and abdominal discomfort. However, patients with cholera given tetracycline plus berberine were more ill, suffered longer from diarrhea, and required larger volumes of intravenous fluid than those given tetracycline alone (5).

Caulophyllum thalictroides

Caulophyllum thalictroides (blue cohosh) contains vasoactive glycosides and quinolizidine alkaloids that produce toxic effects on the myocardium in animals.

Table 14 The genera of Berberidaceae

Achlys (achlys)
Berberis L (barberry)
Caulophyllum (cohosh)
Diphylleia (umbrellaleaf)
Epimedium (epimedium)
Jeffersonia (jeffersonia)
Mahonia (barberry)
Nandina (nandina)
Podophyllum (may apple)
Vancouveria (insideout flower)

Adverse effects
Cardiovascular

Heart failure occurred in the fetus of a mother who used blue cohosh.

- A 41-week-old boy weighing 3.66 kg developed respiratory distress, acidosis, and shock shortly after a spontaneous vaginal delivery (6). His 36-year-old mother had a history of adequately controlled hypothyroidism and had taken tablets of blue cohosh for 1 month to induce uterine contractions. Subsequently she felt more contractions and less fetal activity. After delivery, the baby continued to be critically ill for several weeks and required treatment for respiratory failure and cardiogenic shock. He gradually improved and was extubated after 21 days. There were no congenital abnormalities or other reasons to explain the infant's problems. He remained in hospital for 31 days and an electrocardiogram at discharge was consistent with a resolving anterolateral myocardial infection. Two years later he had fully recovered, but cardiomegaly and impaired left ventricular function persisted.

The authors believed that the consumption of blue cohosh by the mother had caused heart failure in the child.

Drug contamination

Caulophyllum thalictroides has been reportedly adulterated with cocaine (7).

- A 24-year-old woman who had taken blue cohosh for induction of labor was delivered by cesarean section of an apparently healthy female infant weighing 3860 g, but 26 hours later the infant had focal motor seizures of the right arm, which turned out to be due to an infarct in the distribution of the left middle cerebral artery. The seizures were managed with phenobarbital and phenytoin. There were no clotting abnormalities and the family history was negative. The infant's urine tested positive for the cocaine metabolic benzoylecgonine and so did the mother's blue cohosh tablets.

The authors pointed out that maternal cocaine use is a recognized cause of perinatal stroke and speculated that either benzoylecgonine is a metabolite of both cocaine and blue cohosh or the tablets had been contaminated with cocaine.

Dysosma pleianthum

Dysosma pleianthum (bajiaolian), a species of May apple, is a traditional Chinese herbal medicine rich in podophyllotoxin. It has been widely used in China for thousands of years as a general remedy and for the treatment of snake bite, weakness, condyloma accuminata, lymphadenopathy, and tumors.

Adverse effects

Five people developed nausea, vomiting, diarrhea, abdominal pain, thrombocytopenia, leukopenia, abnormal liver function tests, sensory ataxia, altered consciousness, and persistent peripheral tingling or numbness after drinking infusions of bajiaolian (8). These effects were consistent with podophyllum intoxication.

Mahonia species

Barberry is a vernacular name for members of the *Berberis* species, such as *Berberis vulgaris* (European barberry), but is also used to refer to members of the *Mahonia* species, such as *Mahonia aquifolium* and *Mahonia nervosa*. In the USA only the latter species have had official status as a source of barberry, but *Berberis vulgaris* is said to serve similar medicinal purposes and to contain similar principles. Its root bark yields the quaternary isoquinoline alkaloid berberine and several other tertiary and quaternary alkaloids. It has been used to treat a variety of skin conditions (9,10). The literature sometimes cautions that barberry alkaloids can cause arterial hypotension.

References

1. Lau CW, Yao XQ, Chen ZY, Ko WH, Huang Y. Cardiovascular actions of berberine. Cardiovasc Drug Rev 2001;19(3):234–44.
2. Sabir M, Bhide NK. Study of some pharmacological actions of berberine. Indian J Physiol Pharmacol 1971;15(3):111–32.
3. Chun YT, Yip TT, Lau KL, Kong YC, Sankawa U. A biochemical study on the hypotensive effect of berberine in rats. Gen Pharmacol 1979;10(3):177–82.
4. Chan E. Displacement of bilirubin from albumin by berberine. Biol Neonate 1993;63(4):201–8.
5. Khin-Maung-U, Myo-Khin, Nyunt-Nyunt-Wai, Aye-Kyaw, Tin-U. Clinical trial of berberine in acute watery diarrhoea. BMJ (Clin Res Ed) 1985;291(6509):1601–5.
6. Jones TK, Lawson BM. Profound neonatal congestive heart failure caused by maternal consumption of blue cohosh herbal medication. J Pediatr 1998;132(3 Pt 1):550–2.
7. Finkel RS, Zarlengo KM. Blue cohosh and perinatal stroke. N Engl Med J 2004;351:302–3.
8. Kao WF, Hung DZ, Tsai WJ, Lin KP, Deng JF. Podophyllotoxin intoxication: toxic effect of Bajiaolian in herbal therapeutics. Hum Exp Toxicol 1992;11(6):480–7.
9. Turner NJ, Hebda RJ. Contemporary use of bark for medicine by two Salishan native elders of southeast Vancouver Island, Canada. J Ethnopharmacol 1990;29(1):59–72.
10. Grimme H, Augustin M. Phytotherapie bei chronischen Dermatosen und Wunden: was ist gesichert?. [Phytotherapy in chronic dermatoses and wounds: what is the evidence?.] Forsch Komplementarmed 1999;6(Suppl 2):5–8.

Boraginaceae

General Information

The genera in the family of Boraginaceae (Table 15) include bluebells, borage, comfrey, and forget-me-nots.

Cynoglossum officinale

Cynoglossum officinale (hound's tongue) contains alkaloids with curare-like activity as well as hepatotoxic and carcinogenic pyrrolizidine alkaloids, which are covered in a separate monograph.

Symphytum officinale

Symphytum officinale (black wort, boneset, bruise wort, comfrey, knitback, knitbone, slippery root) contains pyrrolizidine alkaloids, such as lasiocarpine and symphytine,

Table 15 The genera of Boraginaceae

Alkanna (alkanna)
Amsinckia (fiddleneck)
Anchusa (bugloss)
Antiphytum (saucerflower)
Argusia (sea rosemary)
Asperugo (German-madwort)
Borago (borage)
Bothriospermum (bothriospermum)
Bourreria (strongbark)
Brunnera (brunnera)
Buglossoides (buglossoides)
Carmona (scorpionbush)
Cordia (cordia)
Cryptantha (cryptantha)
Cynoglossum (hound's tongue)
Dasynotus (whitethroat)
Echium (vipersbugloss)
Ehretia (ehretia)
Eritrichium (alpine forget-me-not)
Hackelia (stickseed)
Harpagonella (grapplinghook)
Heliotropium (heliotrope)
Lappula (stickseed)
Lithospermum (stoneseed)
Macromeria (giant-trumpets)
Mertensia (bluebells)
Myosotis (forget-me-not)
Myosotidium (giant forget-me-not)
Nonea (monkswort)
Omphalodes (navelwort)
Onosmodium (marbleseed)
Onosma (onosma)
Pectocarya (combseed)
Pentaglottis (pentaglottis)
Plagiobothrys (popcorn flower)
Pulmonaria (lungwort)
Rochefortia (rochefortia)
Symphytum (comfrey)
Tiquilia (crinklemat)
Tournefortia (soldierbush)

and their N-oxides, and has repeatedly been associated with hepatotoxicity.

Comfrey products have been withdrawn from the market in several countries, including the USA and the UK. The German Federal Health Office has restricted the availability of botanical medicines containing unsaturated pyrrolizidine alkaloids (1,2). Herbal medicines that provide more than 1 mg internally or more than 100 mg externally per day, when used as directed, are not permitted; herbal medicines that provide 0.1–1 mg internally or 10–100 mg externally per day, when used as directed, may be applied only for a maximum of 6 weeks per year, and they should not be used during pregnancy or lactation.

Adverse effects

Symphytum officinale contains pyrrolizidine alkaloids, which are the subject of a separate monograph. Certain representatives of this class and the plants in which they occur are hepatotoxic, as well as mutagenic and carcinogenic. They can produce veno-occlusive disease of the liver with clinical features like abdominal pain with ascites, hepatomegaly, splenomegaly, anorexia, nausea, vomiting, and diarrhea. Sometimes there is also damage to the lungs.

Liver

The main type of liver damage caused by *S. officinale* is veno-occlusive disease, a non-thrombotic obliteration of small hepatic veins leading to cirrhosis and eventually liver failure (3). Patients can present with acute or chronic signs; portal hypertension, hepatomegaly, and abdominal pain are the main features.

- A 23-year-old man who had taken comfrey leaves presented with hepatic veno-occlusive disease and severe portal hypertension and subsequently died from liver failure (4). Light microscopy and hepatic angiography showed occlusion of sublobular veins and small venous radicles of the liver, associated with widespread hemorrhagic necrosis of hepatocytes.

Other reports of hepatotoxicity due to *S. officinale* have appeared (5–7).

However, the author of a review of the toxicity of *S. officinale* pointed out that since 1990 no cases of adverse events have been reported and stated that comfrey has a history of effective therapeutic use in humans and that it "might not be as dangerous to humans as current restrictions indicate" (8).

Fetotoxicity

It is prudent to avoid exposing unborn or suckling children to herbal remedies containing pyrrolizidine alkaloids. Animal studies have shown that transplacental passage and transfer to breast milk are possible, and there is a human case on record of fatal neonatal liver injury, in which the mother had used a herbal cough tea containing pyrrolizidine alkaloids throughout her pregnancy.

Heliotropium species

Heliotropium species contain various pyrrolizidine alkaloids (which are covered in a separate monograph).

Adverse effects

It is prudent to avoid exposure of unborn or suckling children to herbal remedies containing pyrrolizidine alkaloids. Animal studies have shown that transplacental passage and transfer to breast milk can occur, and fatal neonatal liver damage has been reported, when the mother used an herbal cough tea containing pyrrolizidine alkaloids throughout her pregnancy.

The German Federal Health Office has restricted the availability of botanical medicines containing unsaturated pyrrolizidine alkaloids (1,2). Herbal medicines that provide over 1 mg/day internally or over 100 mg/day externally are not permitted; herbal medicines that provide 0.1–1 mg/day internally or 10–100 mg/day may be applied only for a maximum of 6 weeks per year, and they should not be used during pregnancy or lactation.

References

1. Anonymous. Vorinformation Pyrrolizidinalkaloidhaltige Human Arzneimittel. Pharm Ztg 1990;135:2532–31990;135:2623–4.
2. Anonymous. Aufbereitungsmonographien Kommission E. Pharm Ztg 1990;135:2081–2.
3. Stickel F, Seitz HK. The efficacy and safety of comfrey. Public Health Nutr 2000;3(4A):501–8.
4. Yeong ML, Swinburn B, Kennedy M, Nicholson G. Hepatic veno-occlusive disease associated with comfrey ingestion. J Gastroenterol Hepatol 1990;5(2):211–4.
5. Ridker PN, McDermont WV. Hepatotoxicity due to comfrey herb tea. Am J Med 1989;87(6):701.
6. Bach N, Thung SN, Schaffner F. Comfrey herb tea-induced hepatic veno-occlusive disease. Am J Med 1989;87(1):97–9.
7. Weston CF, Cooper BT, Davies JD, Levine DF. Veno-occlusive disease of the liver secondary to ingestion of comfrey. BMJ (Clin Res Ed) 1987;295(6591):183.
8. Rode D. Comfrey toxicity revisited. Trends Pharmacol Sci 2002;23(11):497–9.

Brassicaceae

General Information

The genera in the family of Brassicaceae (Table 16) include various types of brassica (cabbage, broccoli, brussels sprouts, kale, kohlrabi, pak choi, rape, turnip), mustard, and cress.

Table 16 The genera of Brassicaceae

Alliaria (alliaria)
Alyssum (madwort)
Anelsonia (anelsonia)
Aphragmus (aphragmus)
Arabidopsis (rock cress)
Arabis (rock cress)
Armoracia (armoracia)
Athysanus (sandweed)
Aubrieta (lilac bush)
Aurinia (aurinia)
Barbarea (yellow rocket)
Berteroa (false madwort)
Brassica (broccoli, cabbage, mustard, rape)
Braya (northern-rock cress)
Bunias (warty cabbage)
Cakile (searocket)
Calepina (ballmustard)
Camelina (false flax)
Capsella (capsella)
Cardamine (bittercress)
Cardaria (whitetop)
Caulanthus (wild cabbage)
Caulostramina (caulostramina)
Chlorocrambe (chlorocrambe)
Chorispora (chorispora)
Cochlearia (scurvy grass)
Coincya (star mustard)
Conringia (hare's ear mustard)
Coronopus (swine cress)
Crambe (crambe)
Cusickiella (cusickiella)
Descurainia (tansy mustard)
Dimorphocarpa (spectacle pod)
Diplotaxis (wallrocket)
Dithyrea (shield pod)
Draba (draba)
Dryopetalon (dryopetalon)
Eruca (rocket salad)
Erucastrum (dog mustard)
Erysimum (wallflower)
Euclidium (mustard)
Eutrema (eutrema)
Glaucocarpum (waxfruit mustard)
Guillenia (mustard)
Halimolobos (fissurewort)
Hesperis (rocket)
Heterodraba (heterodraba)
Hirschfeldia (hirschfeldia)
Hutchinsia (hutchinsia)
Iberis (candytuft)

(Continued)

Idahoa (idahoa)
Iodanthus (iodanthus)
Ionopsidium (ionopsidium)
Isatis (woad)
Leavenworthia (gladecress)
Lepidium (pepperweed)
Lesquerella (bladderpod)
Lobularia (lobularia)
Lunaria (lunaria)
Lyrocarpa (lyrepod)
Malcolmia (malcolmia)
Mancoa (mancoa)
Matthiola (stock)
Microthlaspi (penny cress)
Moricandia (moricandia)
Myagrum (myagrum)
Neobeckia (lake cress)
Nerisyrenia (fan mustard)
Neslia (neslia)
Parrya (parrya)
Pennellia (mock thelypody)
Phoenicaulis (phoenicaulis)
Physaria (twinpod)
Polyctenium (combleaf)
Raphanus (radish)
Rapistrum (bastard cabbage)
Rorippa (yellowcress)
Schoenocrambe (plains mustard)
Selenia (selenia)
Sibara (winged rock cress)
Sibaropsis (sibaropsis)
Sinapis (mustard)
Sisymbrium (hedge mustard)
Smelowskia (candytuft)
Stanfordia (stanfordia)
Stanleya (prince's plume)
Streptanthella (streptanthella)
Streptanthus (twist flower)
Stroganowia (stroganowia)
Subularia (awlwort)
Synthlipsis (synthlipsis)
Teesdalia (shepard's cress)
Thelypodium (thelypody)
Thelypodiopsis (tumble mustard)
Thlaspi (penny cress)
Thysanocarpus (fringepod)
Tropidocarpum (tropidocarpum)
Warea (pineland cress)

Several of the Brassicaceae contain allyl isothiocyanate, which is a potent irritant and has mutagenic activity in bacteria and fetotoxic and carcinogenic effects in rats. However, as allyl isothiocyanate also occurs in ordinary mustard, it would not be realistic to ban all botanical drugs that contain it, since they commonly provide no more than a normal daily dose of mustard (for example 5 mg of allyl isothiocyanate per 5 g of mustard).

Armoracia rusticana

Armoracia rusticana (horseradish) contains 0.05–0.2% of essential oils of which 85% is allyl isothiocyanate.

Adverse effects

Abdominal discomfort and convulsive syncope occurred after the ingestion of raw horseradish that had not been properly aired before use (1).

Brassica nigra

Brassica nigra (black mustard) contains allyl isothiocyanate. External application of preparations from black mustard has declined because of skin irritation (see *Sinapis* species).

Raphanus sativus

The root of *Raphanus sativus* var. niger (black radish) contains 0.0025% of essential oil with glycosides yielding allyl isothiocyanate and butyl isothiocyanate.

Adverse effects

The consumption of several roots of *Raphanus sativus* can produce miosis, pain, vomiting, slowed respiration, stupor, and albuminuria.

Skin

Allergic contact dermatitis has been attributed to radish (2).

Sinapis **species**

The *Sinapis* genus contains several different types of mustard species, including *Sinapis alba* (white mustard) and *Sinapis arvensis* (charlock mustard). Mustard has traditionally been used in heated compresses (sinapisms) to draw blood away from underlying infections and to act as a counterirritant. This can cause direct skin damage.

- A 50-year-old woman experienced a second-degree burn after applying a heated mustard compress to her chest to relieve pulmonary congestion associated with a recent episode of pneumonia (3). The injury resulted in permanent hyperpigmentation and hypertrophic scarring.

References

1. Rubin HR, Wu AW. The bitter herbs of Seder: more on horseradish horrors. JAMA 1988;259(13):1943.
2. Mitchell JC, Jordan WP. Allergic contact dermatitis from the radish, *Raphanus sativus*. Br J Dermatol 1974;91(2):183–9.
3. Linder SA, Mele JA 3rd, Harries T. Chronic hyperpigmentation from a heated mustard compress burn: a case report. J Burn Care Rehabil 1996;17(4):351–2.

Burseraceae

General information

The genera in the family Burseraceae (Table 17) include olive and myrrh.

Boswellia serrata

Preparations from the gum resin of *Boswellia serrata* have been used as traditional remedies in Ayurvedic medicine in India for inflammatory diseases (1). The gum contains substances that have anti-inflammatory properties, pentacyclic triterpenes related to boswellic acid, which inhibit leukotriene biosynthesis in neutrophilic granulocytes by inhibiting 5-lipoxygenase. Certain boswellic acids also inhibit elastase in leukocytes, inhibit proliferation, induce apoptosis, and inhibit topoisomerases in cancer cell lines.

In a double-blind, placebo-controlled study in 40 patients, aged 18-75 years, with bronchial asthma, who were given *Boswellia serrata* 300 mg tds for 6 weeks, 28 had improved dyspnea, a reduced number of attacks, increases in FEV_1, FVC, and peak flow, and reduced eosinophil counts and ESR (2). Of 40 patients, aged 14-58 years, given lactose 300 mg tds for 6 weeks, only 11 improved.

In patients with ulcerative colitis *Boswellia serrata* 350 mg tds for 6 weeks produced improvements in stool properties, histopathology of rectal biopsies, haemoglobin, serum iron, calcium, phosphorus, proteins, and total leukocyte and eosinophil counts, with remission in 82% of patients (3). The corresponding figure with sulfasalazine 1 g tds was 75%.

In 30 patients, aged 18-48 years, with chronic colitis, characterized by vague lower abdominal pain, rectal bleeding, diarrhea, and palpable tender descending and sigmoid colons *Boswellia serrata* (900 mg/day for 6 weeks) produced remission in 14 of 20 patients, and sulfasalazine (3 g/day for 6 weeks) in four of 10 patients (4). There were few adverse effects.

Commiphora species

Myrrh is an oleo gum resin obtained from the stem of *Commiphora molmol*, a tree that grows in north-east Africa and the Arabian Peninsula. In mice, myrrh showed no mutagenic effects and was a potent cytotoxic drug against solid tumor cells; the antitumor potential of *Commiphora molmol* was comparable with that of cyclophosphamide (5).

Commiphora mukul (also called *Commiphora wightii*, the mukul myrrh tree) is found from northern Africa to central Asia, but is most common in northern India. It is

Table 17 The genera of Burseraceae

Boswellia (boswellia)
Bursera (bursera)
Canarium (olive)
Commiphora (myrrh)
Dacryodes (dacryodes)
Tetragastris (tetragastris)

part of the Ayurvedic system of medicine as an extract called gugulipid, guggulipid or guglipid; the active ingredient is called guggulsterone.

Adverse effects

The efficacy and adverse effects of myrrh and the most effective dosage schedule have been studied in 204 (169 men and 35 women) patients with schistosomiasis aged 12–68 years and 20 healthy non-infected age- and sex-matched volunteers (6). The patients were divided into two groups: 86 patients with schistosomal colitis and 118 with hepatosplenic schistosomiasis; the latter were further divided into two subgroups—77 patients with compensated disease and 41 with decompensated disease. All but 12 had received one or more courses of praziquantel. The dosage of myrrh was 10 mg/kg/day for 3 days on an empty stomach 1 hour before breakfast. A second course of 10 mg/kg/day for 6 days was given to patients who still had living ova in rectal or colonic biopsy specimens. The response rate to a single course of myrrh was 92% in 187 patients. The cure rates were 91%, 94%, and 90% in patients with schistosomal colitis, compensated hepatosplenic schistosomiasis and decompensated hepatosplenic schistosomiasis respectively. The cure rate was less in patients who had previously taken praziquantel and in patients with impaired liver function. *Schistosoma hematobium* infection was the most responsive (n=4, cure rate 100%), followed by mixed infections (n=29, cure rate 93%). Those infected with *Schistosoma mansoni* had the lowest cure rate (n=171, cure rate 91%). There was no impairment of liver function after treatment with myrrh. In contrast, liver function tests significantly improved in patients with impaired liver function. There were no significant effects of myrrh on the electrocardiogram. Adverse effects of myrrh were reported in 24 of the 204 patients. Giddiness, somnolence, or mild fatigue were the most common (2.5%); all other adverse effects were minor and less frequent. None of the healthy volunteers reported any adverse effects, nor were there any significant changes in liver or kidney function. A second course of myrrh resulted in a cure in 13 of the 17 patients who did not respond to a single course.

The efficacy of myrrh has been studied in seven patients aged 10–41 years (five men, two women) with fascioliasis and 10 age- and sex-matched healthy volunteers (7). Myrrh was given orally in the morning on an empty stomach in a dosage of 12 mg/kg/day for 6 days. All the patients were passing *Fasciola* eggs in their stools (mean 36 eggs per gram of stool). The symptoms and signs of fascioliasis resolved during treatment with myrrh, and *Fasciola* eggs could not be demonstrated in the stools 3 weeks and 3 months after treatment. Antifasciola antibody titers became negative in six of the seven patients. There were no adverse effects.

Metabolism

The results of several studies have suggested that gugulipid lowers cholesterol concentrations (8). However, in a placebo-controlled study it increased LDL cholesterol by 4% (300 mg/day of guggulipid) and 5% (600 mg/day of guggulipid) (9).

References

1. Ammon HP. Boswelliasauren (Inhaltsstoffe des Weihrauchs) als wirksame Prinzipen zur Behandlung chronisch entzundlicher Erkrankungen. [Boswellic acids (components of frankincense) as the active principle in treatment of chronic inflammatory diseases.]. Wien Med Wochenschr 2002;152(15–16):373–8.

2. Gupta I, Gupta V, Parihar A, Gupta S, Lüdtke R, Safayhi H, Ammon HP. Effects of *Boswellia serrata* gum resin in patients with bronchial asthma: results of a double-blind, placebo-controlled, 6-week clinical study. Eur J Med Res 1998;3(11):511–4.

3. Gupta I, Parihar A, Malhotra P, Singh GB, Lüdtke R, Safayhi H, Ammon HP. Effects of *Boswellia serrata* gum resin in patients with ulcerative colitis. Eur J Med Res 1997;2(1):37–43.

4. Gupta I, Parihar A, Malhotra P, Gupta S, Lüdtke R, Safayhi H, Ammon HP. Effects of gum resin of *Boswellia serrata* in patients with chronic colitis. Planta Med 2001;67(5):391–5.

5. Al Harbi MM, Qureshi S, Ahmed MM, Rafatulla S, Shah AH. Effect of Commiphora molmol (oleogum-resin) on the cytological and biochemical changes induced by cyclophosphamide in mice. Am J Chin Med 1994;22:77–82.

6. Sheir Z, Nasr AA, Massoud A, Salama O, Badra GA, El Shennawy H, Hassan N, Hammad SM. A safe, effective, herbal antischistosomal therapy derived from myrrh. Am J Trop Med Hyg 2001;65:700–4.

7. Massoud A, El Sisi S, Salama O, Massoud A. Preliminary study of therapeutic efficacy of a new fasciolicidal drug derived from *Commiphora molmol* (myrrh). Am J Trop Med Hyg 2001;65:96–9.

8. Deng R. Therapeutic effects of guggul and its constituent guggulsterone: cardiovascular benefits. Cardiovasc Drug Rev 2007;25(4):375–90.

9. Szapary PO, Wolfe ML, Bloedon LT, Cucchiara AJ, DerMarderosian AH, Cirigliano MD, Rader DJ. Gugulipid for the treatment of hypercholesterolemia: a randomised controlled trial. J Am Med Assoc 2003;290:765–72.

Campanulaceae

General Information

The genera in the family of Campanulaceae (Table 18) include lobelia and Venus' looking glass.

Table 18 The genera of Campanulaceae

Asyneuma (harebell)
Brighamia (brighamia)
Campanula (bell flower)
Campanulastrum (bell flower)
Canarina (canarina)
Clermontia (clermontia)
Cyanea (cyanea)
Delissea (delissea)
Downingia (calico flower)
Gadellia (gadellia)
Githopsis (bluecup)
Heterocodon (heterocodon)
Hippobroma (hippobroma)
Howellia (howellia)
Jasione (jasione)
Legenere (false Venus' looking glass)
Legousia (legousia)
Lobelia (lobelia)
Nemacladus (threadplant)
Parishella (parishella)
Platycodon (platycodon)
Porterella (porterella)
Rollandia (rollandia)
Trematolobelia (false lobelia)
Triodanis (Venus' looking-glass)
Wahlenbergia (wahlenbergia)

Lobelia inflata

Lobelia inflata (Indian tobacco) contains lobeline and other pyridine alkaloids. It has been used as an emetic, antidepressant, respiratory stimulant, an aid to smoking cessation, and a treatment for metamfetamine abuse (1).

Adverse effects

Lobeline has peripheral effects similar to those of nicotine, whereas its central activity may be different. It has been associated with nausea, vomiting, headache, tremors, and dizziness. Symptoms caused by overdosage include profuse sweating, paresis, tachycardia, hypertension, Cheyne-Stokes respiration, hypothermia, coma, and death. Large doses are convulsant.

Reference

1. Dwoskin LP, Crooks PA. A novel mechanism of action and potential use for lobeline as a treatment for psychostimulant abuse. Biochem Pharmacol 2002;63(2):89–98.

Cannabaceae

General Information

The family of Cannabaceae contains two genera:

1. *Cannabis* (hemp)
2. *Humulus* (hop).

The cannabinoids are covered in a separate monograph.

Humulus lupulus

Humulus lupulus (hop) contains a variety of sesquiterpenoids, diterpenoids, and triterpenoids, phytoestrogens, and the flavonoid xanthohumol, which has some in vitro anti-HIV activity (1). Apart from their use in brewing beer, hops have been used for sedative purposes, but evidence of efficacy is poor.

Adverse effects

Hop farmers are exposed to air that can contain dust, endotoxin, and micro-organisms. In one study of 19 farms in Poland Gram-positive bacteria formed 22–96% of the total count; among them, corynebacteria and endospore-forming bacilli were prevalent (2). Fungi constituted 3.7–65% of the total count; the dominant species were *Penicillium citrinum*, *Alternaria alternata*, and *Cladosporium epiphyllum*. Thermophilic actinomycetes and Gram-negative bacteria were detected in the air of only 10 and six farms respectively. The concentrations of endotoxin were 313–6250 µg/g. The hop growers seem to be exposed to lower concentrations of dust, micro-organisms, and endotoxin than other branches of agriculture, which the authors partly attributed to antimicrobial properties of *H. lupulus*.

Eight of twenty-three hops farmers, who had been exposed to organic dust from *H. lupulus*, reported symptoms of chronic bronchitis and five reported work-related symptoms, including dry cough and dyspnea (3).

After 30 years of working with hop without any health problems a 46-year-old farmer developed erythema of the face, neck, and upper chest, edema of the eyelids, conjunctivitis, and dermatitis of the hands (4). Her symptoms were provoked by both fresh and dried hops, appeared after half-an-hour of working, and persisted over 1–2 days. Skin tests yielded the following results: hop leaves (saline extract)—prick positive, patch negative; hop leaves (glycerol extract)—prick positive, patch negative; hop cones (saline extract)—prick positive, patch negative; hop cones (glycerol extract)—prick negative, patch positive after 48 and 72 hours. Despite discontinuing work, the patient had several relapses attributed to other sources of hop allergens: a beauty cream, a herbal sedative, and her husband, also a hop farmer.

Systemic urticaria has been reported in patients who had been exposed to *H. lupulus* (5), in one case with arthralgia and fever (6).

Flavonoids in hops that persist in beer can inhibit certain isoforms of cytochrome P450 in vitro (7). At a concentration of 10 µmol/l xanthohumol almost completely inhibited CYP1A1 and other hop flavonoids inhibited it by 27–91%. At a concentration of 10 µmol/l xanthohumol completely inhibited CYP1B1 and other hop flavonoids inhibited it by 2–99%. The most effective inhibitors of CYP1A2 were the two prenylated flavonoids, 8-prenylnaringenin and isoxanthohumol, which produced over 90% inhibition in concentrations of 10 µmol/l. However, the flavonoids were poor inhibitors of CYP2E1 and CYP3A4 and so in vivo drug interactions are unlikely to be of importance.

References

1. Wang Q, Ding ZH, Liu JK, Zheng YT. Xanthohumol, a novel anti-HIV-1 agent purified from Hops Humulus lupulus. Antiviral Res 2004;64(3):189–94.
2. Gora A, Skorska C, Sitkowska J, Prazmo Z, Krysinska-Traczyk E, Urbanowicz B, Dutkiewicz J. Exposure of hop growers to bioaerosols. Ann Agric Environ Med 2004;11(1): 129–38.
3. Skorska C, Mackiewicz B, Gora A, Golec M, Dutkiewicz J. Health effects of inhalation exposure to organic dust in hops farmers. Ann Univ Mariae Curie Sklodowska [Med] 2003;58(1):459–65.
4. Spiewak R, Dutkiewicz J. Occupational airborne and hand dermatitis to hop (Humulus lupulus) with non-occupational relapses. Ann Agric Environ Med 2002;9(2):249–52.
5. Estrada JL, Gozalo F, Cecchini C, Casquete E. Contact urticaria from hops (Humulus lupulus) in a patient with previous urticaria-angioedema from peanut, chestnut and banana. Contact Dermatitis 2002;46(2):127.
6. Pradalier A, Campinos C, Trinh C. Urticaire systemique induite par le houblon. [Systemic urticaria induced by hops.] Allerg Immunol (Paris) 2002;34(9):330–2.
7. Henderson MC, Miranda CL, Stevens JF, Deinzer ML, Buhler DR. In vitro inhibition of human P450 enzymes by prenylated flavonoids from hops, Humulus lupulus. Xenobiotica 2000;30(3):235–51.

Capparaceae

General Information

The genera in the family of Capparaceae (Table 19) include caper and stinkweed.

Table 19 The genera of Capparaceae

Atamisquea (atamisquea)
Capparis (caper)
Cleome (spiderflower)
Cleomella (stinkweed)
Forchhammeria (forchhammeria)
Koeberlinia (allthorn)
Morisonia (morisonia)
Oxystylis (oxystylis)
Polanisia (clammyweed)
Wislizenia (wislizenia)

Capparis spinosa

The leaf and fruit of *Capparis spinosa* (caper plant) both contain isothiocyanates, in addition to the flavonoid rutinoside and the quaternary alkaloid stachydrine.

Adverse effects
Allergic contact dermatitis can follow the application of *C. spinosa* in the form of wet compresses (1).

Reference

1. Angelini G, Vena GA, Filotico R, Foti C, Grandolfo M. Allergic contact dermatitis from *Capparis spinosa* L. applied as wet compresses. Contact Dermatitis 1991;24(5):382–3.

Celastraceae

General Information

The genera in the family of Celastraceae (Table 20) include bittersweet and khat.

Catha edulis

Khat, or qat, is a stimulant commonly used in East Africa, Yemen, and Southern Saudi Arabia. Khat leaves from the evergreen bush Catha edulis are typically chewed while fresh, but can also be smoked, brewed in tea, or sprinkled on food. Its use is culturally based.

Chewing the leaves of *Catha edulis* (khat, qat) results in subjective mental stimulation, increased physical endurance, and increased self-esteem and social interaction. Until recently, this habit was confined to Arabian and East African countries, because only fresh leaves are active, but because of increased air transportation, khat is now also chewed in other parts of the world. Although cathine (norpseudoephedrine) is quantitatively the main alkaloid, the amphetamine-like euphorigenic and sympathomimetic cardiovascular effects of khat are primarily attributed to cathinone (1). Cathinone, a phenylalkylamine, is structurally similar to amfetamine. It degrades to norpseudoephedrine and norephedrine within days of leaf picking. Cathinone increases dopamine release and reduces dopamine reuptake (2). In Yemen chewers of khat produced in fields where chemical pesticides are used regularly have more symptoms than chewers of khat produced in fields where chemical pesticides are rarely or never used (3).

The toxicologist Louis Lewin described the effects of chewing khat in his monograph *Phantastica* (1924): "The khat eater is happy when he hears everyone talk in turn and tries to contribute to this social entertainment. In this way the hours pass in a rapid and agreeable manner. Khat produces joyous excitation and gaiety. Desire for sleep is banished, energy is revived during the hot hours of the day, and the feeling of hunger on long marches is dispersed. Messengers and warriors use khat because it makes the ingestion of food unnecessary for several days."

Khat is often used in social gatherings called "sessions", which can last 3–4 hours. They are generally attended by men, although khat use among women is growing. Men are also more likely to be daily users. Users pick leaves from the khat branch, chew them on one side of the mouth, swallowing only the juice, and adding fresh leaves periodically. The khat chewer may experience increased alertness and euphoria. About 100–300 grams of khat may be chewed during each session, and 100 grams of khat typically contains 36 mg of cathinone.

Khat has been recognized as a substance of abuse with increasing popularity. It is estimated that 10 million people chew khat worldwide, and it is used by up to 80% of adults in Somalia and Yemen. It now extends to immigrant African communities in the UK and USA. It is banned in Saudi Arabia, Egypt, Morocco, Sudan, and Kuwait. It is also banned in the USA and European countries. However, in Australia, its importation is controlled by a licence issued by the Therapeutic Goods Administration, which allows up to 5 kg of khat per month per individual for personal use.

There have been several review articles describing khat and its growth (2). The World Health Organization Advisory Group's 1980 report reviewed the pharmacological effects of khat in animals and humans (4). The societal context of khat use has also been reviewed (5).

Adverse effects

Tachycardia and increased blood pressure, irritability, psychosis, and psychic dependence have been described as acute adverse effects of khat.

The long term adverse effects were well described by Louis Lewin: "Those organs functions which are incessantly subjected to the influence of the drug finally flag or are diverted into another channel of activity··· The khat eater is seized with a restlessness which robs him of sleep. The excited cerebral hemispheres do not return to their normal state of repose, and in consequence the functions of the peripheral organs, especially those of the heart, suffer to such a degree that serious cardiac affectations have been ascertained in a great number of khat eaters. The disorders of the nervous system in many cases also give rise to troubles of general metabolism partly due to the chronic loss of appetite from the consumption of khat··· In Yemen it was openly stated that inveterate eaters of khat were indifferent to sexual excitation and desire, and did not marry at all, or for economic reasons waited until they had saved enough money. The lost of libido sexualis has been also observed in other inhabitants of these countries."

Cardiovascular

Khat is a sympathomimetic amine and increases blood pressure and heart rate. Limited evidence suggests that khat increases the risk of acute myocardial infarction. In Yemen 100 patients admitted to an intensive care unit

Table 20 The genera of Celastraceae

Canotia (canotia)
Cassine (cassine)
Catha (khat)
Celastrus (bittersweet)
Crossopetalum (crossopetalum)
Euonymus (spindle tree)
Gyminda (false box)
Lophopyxis
Maytenus (mayten)
Mortonia (saddlebush)
Pachystima (pachystima)
Paxistima (paxistima)
Perrottetia (perrottetia)
Pristimera (pristimera)
Salacia
Schaefferia (schaefferia)
Torralbasia (torralbasia)
Tripterygium

with an acute myocardial infarction were compared with 100 sex- and age-matched controls recruited from an ambulatory clinic (6). They completed a questionnaire on personal habits, such as khat use and cigarette smoking, past medical history, and a family history of myocardial infarction. Use of khat was an independent risk factor for acute myocardial infarction, with an odds ratio of 5.0 (95% CI = 1.9, 13). The relation was dose-related: "heavy" khat users were at higher risk than "moderate" users, although the extent of use and the potency of khat used were estimated, being hard to quantify. To explain the increased risk of acute myocardial infarction, the authors suggested that it may have been related to increased blood pressure and heart rate, with a resultant increase in myocardial oxygen demand. They also suggested that khat could have acted via the mechanisms proposed to explain acute myocardial infarction after the use of amphetamines, such as catecholamine-induced platelet aggregation and coronary vasospasm.

When 80 healthy volunteers chewed fresh khat leaves for 3 hours there were significant progressive rises in systolic and diastolic blood pressures and heart rate, without return to baseline 1 hour after chewing had ceased (7).

Of 247 chronic khat chewers 169 (62%) had hemorrhoids and 124 (45%) underwent hemorrhoidectomy; by comparison, of 200 non-khat chewers 8 (4%) had hemorrhoids and one underwent hemorrhoidectomy (8).

Nervous system

Of 19 khat users suspected of driving under the influence of drugs, three had impaired driving and 10 had marked impairment of psychophysical functions with effects on the nervous system (slow pupil reaction to light, dry mouth, increased heart rate), trembling, restlessness/nervousness, daze/apathy/dullness, and impaired attention, walking, and standing on one leg; however, the concentrations of the khat alkaloids assayed in blood did not correlate with the symptoms of impairment (9).

The prevalence and health effects of headache in Africa have been reviewed (10). In 66 khat users 25% reported headaches (11) and in people with migraine 12% reported using khat (12).

A leukoencephalopathy has been associated with khat (13).

Psychiatric

Khat has amphetamine-like effects and can cause psychoses (14–20), including mania (21) and hypnagogic hallucinations (22). Two men developed relapsing short-lasting psychotic episodes after chewing khat leaves; the psychotic symptoms disappeared without any treatment within 1 week (23).

In addition to the acute stimulant effects of euphoria and alertness caused by khat, there is the question of whether continued khat use alters mood, behavior, and mental health.

- A 33-year-old unemployed Somali man with a 10-year history of khat chewing, who had lived in Western Australia for 4 years, wand who was socially isolated, started to sleep badly, and had weight loss and

persecutory delusions (24). His mental state deteriorated over 2–3 months and he thought that his relatives were poisoning him and that he was being followed by criminals. He had taken rifampicin and ethambutol for pulmonary tuberculosis for 1 year but became noncompliant for 2 months before presentation. He had reportedly chewed increasing amounts of khat daily from his backyard for last 2 years. There was no history of other drug use and his urine drug screen was negative. He responded well to olanzapine 20 mg/day and was discharged after 4 weeks, as his psychosis was gradually improving.

In Hargeisa, Somalia, trained local interviewers screened 4854 individuals for disability due to severe psychiatric problems and identified 169 cases (137 men and 32 women) (25). A subset of 52 positive screening cases was randomly selected for interview and were matched for age, sex, and education with controls. In all, 8.4% of men screened positive and 83% of those who screened positive had severe psychotic symptoms. Khat chewing and the use of greater amounts of khat were more common in this group. Khat users were also more likely to have had active war experience. Only 1.9% of women had positive screening. Khat use starting at an earlier age and in larger amounts (in "bundles" per day) correlated positively with psychotic symptoms.

In 800 Yemeni adults (aged 15–76 years) symptoms that might have been caused by the use of khat were elicited by face-to-face interviews; 90 items covered nine scales of the following domains: somatization, depression, anxiety, phobia, hostility, interpersonal sensitivity, obsessive-compulsive, hostility, interpersonal sensitivity, paranoia, and psychoticism (26). At least one life-time episode of khat use was reported in 82% of men and 43% of women. The incidence of adverse psychological symptoms was not greater in khat users, and there was a negative association between the use of khat and the incidence of phobic symptoms.

Psychological

In 25 daily khat-chewing flight attendants, 39 occasional khat-chewing flight attendants, and 24 non-khat-chewing aircrew members, memory function test scores were significantly lower in khat chewers than non-chewers and in regular chewers than occasional chewers (27).

The impact of khat use on psychological symptoms was one of several factors considered in a study in which 180 Somali refugees were interviewed about psychological symptoms and about migration-related experiences and traumas (28). Suicidal thinking was more common among those who used khat (41 of 180) after migration compared with those who did not (21 of 180). However, a causal relationship cannot be deduced from these data. The authors raised the concern that khat psychosis could be increasing in Australia because of a growing number of African refugees. Furthermore, factors related to immigration, such as social displacement and unemployment, may predispose to abuse, especially as khat is easily available in Australia.

In a cross-sectional survey of Yemeni adults the self-reported frequency of khat use and psychological symptoms was assessed using face-to-face interviews with members from a random sample of urban and rural households (26). Of 800 adults surveyed, 82% of men and 43% of women had used khat at least once. There was no association between khat and negative adverse psychological symptoms, and khat users had less phobic anxiety (56%) than non-users (38%). The authors were surprised by these results and offered several explanations: that the form of khat used in Yemen is less potent than in other locations; that prior reports of khat-related psychosis occurred in users in unfamiliar environments; that the sampling procedure may have under-represented heavier khat users; and that their measurement tool was not sensitive enough to detect psychological symptoms.

Sensory systems
Bilateral optic atrophy occurred in two patients who were long-standing users of khat leaves and had chewed larger quantities than usual (29).

Metabolism
Chronic khat chewing increased plasma glucose and C-peptide concentrations in people with type 2 diabetes mellitus (30).

Mouth
In 20 volunteers who chewed khat regularly (10–160 g/day), there was an eight-fold increase in micronucleated buccal mucosa cells compared with 10 controls (31). Among heavy khat chewers, 81% of nuclei had a centromere signal, suggesting that khat is aneuploidogenic. The effects of khat, tobacco, and alcohol were additive. The highest frequency of abnormality occurred during the fourth week after consumption.

Of 2500 Yemeni citizens (mean age 27 years) 1528 (61%) were khat chewers; of them, 342 cases (22%) had oral keratotic white lesions at the site of khat chewing, while only 6 (0.6%) non-chewers had such lesions; the prevalence and severity of these lesions increased as duration and frequency of use increased (32).

However, in a case-control study in 85 khat-chewing Kenyans and 141 matched controls, smoking unprocessed tobacco (Kiraiku) and smoking cigarettes were the most significant factors for oral leukoplakia; traditional beer, khat, and chili peppers were not significantly associated with oral leukoplakia (33).

A 30-year-old immigrant from Somalia developed a plasma-cell gingivitis in the mandibular gums, probably caused by chewing khat (34).

Chewing khat can cause a generalized mousy brown pigmentation of the gums (35).

Gastrointestinal
In 12 healthy volunteers who chewed Khat leaves or lettuce for 2 hours gastric emptying was significantly prolonged by khat (36).

Chewing khat is said to be a risk factor of duodenal ulceration. (37).

In a case-control study 175 patients with duodenal ulceration (all diagnosed by endoscopy) and 150 controls completed a questionnaire about their health habits (38). Khat use, defined as chewing khat at least 14 hours/week, was significantly more common among the cases (76 versus 35%). Potential confounding variables, including smoking, use of alcohol or NSAIDs, a family history, and chronic hepatic and renal disease, were not significantly different between the two groups. The authors postulated several mechanisms, including a physiological reaction to the stress response to cathine or exposure to chemicals, such as pesticides.

Anorexia due to khat is probably due to norephedrine (4).

The tannins in khat have been associated with delayed intestinal absorption, stomatitis, gastritis, and esophagitis observed with khat use (4).

Constipation is a common physical complaint among khat users (5). The World Health Organization Advisory Group attributed this adverse effect to the tannins and norpseudoephedrine found in khat (4).

Urinary tract
In 11 healthy men khat chewing produced a fall in average and maximum urine flow rate; this effect was inhibited by the alpha$_1$-adrenoceptor antagonist indoramin (39).

Sexual function
Chewing khat lowers libido and can also lead to erectile impotence after long-term use (40).

Drug dependence
Khat use can lead to dependence, which is more important because of its social consequences than because of the effects of physical withdrawal. Khat users may devote significant amounts of time to acquiring and using khat, to the detriment of work and social responsibilities. The physical effects of early khat withdrawal are generally mild. Chronic users may experience craving, lethargy, and a feeling of warmth during early khat abstinence.

Tumorigenicity
Three studies of the association of khat with head and neck cancers have been reviewed (41) The studies showed a trend towards an increased risk of oral cancer and head and neck cancer with the use of khat, but there were too few data for a definitive conclusion. Tobacco use, which is common among khat users, and alcohol use were confounding factors.

A possible association of cancer of the oral cavity with khat has been studied in exfoliated buccal and bladder cells from healthy male khat users; the cells were examined for micronuclei, a marker of genotoxic effects (42). Of 30 individuals who did not use cigarettes or alcohol, 10 were non-users of khat and the other 20 used 10–60 g/day. The 10 individuals who used more than 100 g/day had an eight-fold increase in the frequency of micronuclei compared with non-users. There was a statistically significant dose-response relation between khat use and the number of micronuclei in oral mucosal cells but not urothelial cells. In a separate set of samples taken from khat users and non-users who also used cigarettes and alcohol,

alcohol and cigarette use caused a 4.5-fold increase in the frequency of micronuclei and use of khat in addition to alcohol and cigarettes further doubled the frequency. In buccal mucosal cells from four individuals who ingested 100 g/day for 3 days, the maximum frequency of micronuclei occurred at 27 days after chewing and returned to baseline after 54 days. These data together suggest that khat, especially in combination with alcohol and smoking, may contribute to or cause oral malignancy.

However, in a small study of biopsies taken from the oral mucosa of 40 Yemeni khat users and 10 non-users there were no histopathological changes consistent with malignancy (43). In the khat users, there were changes such as acanthosis, orthokeratosis, epithelial dysplasia, and intracellular edema on both the chewing and non-chewing sides of the oral mucosa. However, none of these lesions was malignant or pre-malignant. The authors thought that these changes were most probably due to mechanical friction or possibly due to chemical components of the khat or pesticides that had been used on the plants.

Fertility

Reduced sperm count, semen volume, and sperm motility have been associated with khat dependence (5).

In 65 Yemeni khat addicts (mean duration of addiction 25 years; mean age 40 years), semen volume, sperm count, sperm motility and motility index, and percentage of normal spermatozoa were lower than in 50 controls (44). There were significant negative correlations between the duration of khat consumption and all semen parameters. On electron microscopy, about 65% of the spermatozoa were deformed, with different patterns of deformation, including both the head and flagella in complete spermatozoa, aflagellate heads, headless flagella, and multiple heads and flagella. Deformed heads had aberrant nuclei with immature nuclear chromatin and polymorphic intranuclear inclusions; these were associated with acrosomal defects.

Fetotoxicity

In pregnant women, consumption of khat affects fetal growth by inhibiting placental blood flow (41), and birth weights are reduced (45, 2). In 1141 consecutive deliveries in Yemen, non-users of khat (n = 427) had significantly fewer low birth-weight babies (less than 2500 g) than occasional users (n = 223) and regular users (n = 391) (46). Khat-chewing mothers were older, of greater parity, and had more surviving children than the non-chewers. Significantly more khat-chewers had concomitant diseases. There was no difference in rates of stillbirth or congenital malformations.

Lactation

Khat chewing by a breastfeeding mother can lead to the presence of cathine in the urine of the suckling child.

Reduced lactation has been reported in khat-using mothers (2).

Drug contamination

Some khat leaves are grown with chemical pesticides. In 114 male khat users in two different mountainous areas of Yemen, users of khat that had been produced in fields in which chemical pesticides were used regularly had more acute gastrointestinal adverse effects (nausea and abdominal pain) and chronic body weakness and nasal problems (47). The authors suggested that organic chemical pesticides such as dimethoatecide can cause such adverse effects.

Drug overdose

Acute toxicity requiring emergency medical treatment is rare. When it occurs there is a typical sympathomimetic syndrome, which should be treated with fluids, control of hyperthermia, bed rest, and, if necessary, sedation with benzodiazepines (2).

Drug interactions

Ampicillin

The speed and extent of ampicillin systemic availability were reduced significantly by khat chewing in eight healthy adult Yemeni men, except when they took it 2 hours after the khat (48).

Euonymus species

The fruit of *Euonymus europaeus* (the European spindle tree) and the bark of *Euonymus atropurpureus* (Wahoo bark) have cathartic and emetic activity, due to sesquiterpenoids and cardiac glycosides that they contain.

Adverse effects

IgE-mediated type I allergy to *E. europaeus* wood has been described in a 44-year-old goldsmith who developed rhinitis and conjunctivitis after having worked with dust from the wood for 15 years (49).

Tripterygium wilfordii

Extracts from the root of *Tripterygium wilfordii* (Lei gong teng) are used in China for the treatment of various disorders, such as rheumatoid arthritis, ankylosing spondylitis, systemic lupus erythematosus, and glomerulonephritis. The potential benefits in such serious diseases should be carefully weighed against a substantial risk of adverse reactions, including gastrointestinal disturbances, skin rashes, amenorrhea, leukopenia, and thrombocytopenia (50).

Adverse effects
Cardiovascular

- A previously healthy young man developed profuse vomiting and diarrhea, leukopenia, renal insufficiency, profound hypotension, and shock after taking an extract of *T. wilfordii* (51). Serial electrocardiograms, cardiac enzymes, and echocardiography showed evidence of coexisting cardiac damage. He died of intractable shock 3 days later.

Endocrine

Tripterygium wilfordii (thundergod vine) is often used in traditional Chinese medicine, for instance to treat arthritis.

- A 36-year-old woman developed vaginal dryness, reduced libido, and hot flushes after taking *Tripterygium* for 3 months to treat psoriasis (52). Her follicle stimulating hormone and luteinizing hormone concentrations were abnormally high, while her 17-beta-estriadol concentration was abnormally low. Her signs and symptoms normalized after the herbal remedy was withdrawn.

The authors suggested that thundergod vine has suppressive effects on both male and female gonads.

Gastrointestinal
In a double-blind, placebo-controlled study in 35 patients with long-standing rheumatoid arthritis in whom conventional therapy had failed, an extract of *T. wilfordii* was used in either a low-dose (180 mg/day) or a high-dose (360 mg/day) for 20 weeks, followed by an open extension period (53). Only 21 patients completed the study, of whom one in each group withdrew because of adverse events. The most common adverse effect of *T. wilfordii* was diarrhea, which caused one patient in the high-dose group to withdraw.

Reproductive system
In men, prolonged use of *T. wilfordii* can cause oligospermia and azoospermia, and a reduction in testicular size (54–56).

In 14 women with rheumatoid arthritis *T. wilfordii* caused amenorrhea associated with increased FSH and LH concentrations, which began to rise after 2–3 months and reached menopausal values after 4–5 months; estradiol concentrations began to fall after 3–4 months and reached very low concentrations at 5 months, suggesting an effect on the ovary (57).

Immunologic
The immunosuppressive properties of Lei gong teng can promote the development of infectious diseases (58).

Teratogenicity
A boy whose mother had taken *T. wilfordii* for rheumatoid arthritis early in her pregnancy was born with an occipital meningoencephalocele and cerebellar agenesis, which the authors attributed to the herb (59).

References

1. Widler P, Mathys K, Brenneisen R, Kalix P, Fisch HU. Pharmacodynamics and pharmacokinetics of khat: a controlled study. Clin Pharmacol Ther 1994;55(5):556–62.
2. Haroz R, Greenberg MI. Emerging drugs of abuse. Med Clin N Am 2005;89:1259–76.
3. Date J, Tanida N, Hobara T. Qat chewing and pesticides: a study of adverse health effects in people of the mountainous areas of Yemen. Int J Environ Health Res 2004;14(6):405–14.
4. World Health Organization Advisory Group. Review of the pharmacology of khat. Bull Narcotics 1980;32:83–93.
5. Al-Motarreb A, Baker K, Broadley KJ. Khat: pharmacological and medical aspects and its social use in Yemen. Phytother Res 2002;16:403–13.
6. Al-Motarreb A, Briancon S, Al-Jaber N, Al-Adhi B, Al-Jailani F, Salek MS, Broadley KJ. Khat chewing is a risk factor for acute myocardial infarction: a case-control study. Br J Clin Pharmacol 2005;59:574–81.
7. Hassan NA, Gunaid AA, Abdo-Rabbo AA, Abdel-Kader ZY, al-Mansoob MA, Awad AY, Murray-Lyon IM. The effect of qat chewing on blood pressure and heart rate in healthy volunteers. Trop Doct 2000;30(2):107–8.
8. Al-Hadrani AM. Khat induced hemorrhoidal disease in Yemen. Saudi Med J 2000;21(5):475–7.
9. Toennes SW, Kauert GF. Driving under the influence of khat—alkaloid concentrations and observations in forensic cases. Forensic Sci Int 2004;140(1):85–90.
10. Tekle Haimanot R. Burden of headache in Africa. J Headache Pain 2003;4:S47–54.
11. Mekasha A. The clinical effects of khat. In: The International Symposium on Khat, Ethiopia 1984;77–81.
12. Tekle Haimanot R, Seraw B, Forsgren L, Ekbom K, Ekstedt J. Migraine, chronic tension-type headache, and cluster headache in an Ethiopian rural community. Cephalagia 1995;15:482–8.
13. Morrish PK, Nicolaou N, Brakkenberg P, Smith PE. Leukoencephalopathy associated with khat misuse. J Neurol Neurosurg Psychiatry 1999;67(4):556.
14. Alem A, Shibre T. Khat induced psychosis and its medicolegal implication: a case report. Ethiop Med J 1997;35(2):137–9.
15. Jager AD, Sireling L. Natural history of khat psychosis. Aust NZ J Psychiatry 1994;28(2):331–2.
16. Pantelis C, Hindler CG, Taylor JC. Use and abuse of khat (*Catha edulis*): a review of the distribution, pharmacology, side effects and a description of psychosis attributed to khat chewing. Psychol Med 1989;19(3):657–68.
17. Maitai CK, Dhadphale M. khat-induced paranoid psychosis. Br J Psychiatry 1988;152:294.
18. McLaren P. Khat psychosis. Br J Psychiatry 1987;150:712–3.
19. Kalix P. Amphetamine psychosis due to khat leaves. Lancet 1984;1(8367):46.
20. Dhadphale M, Mengech A, Chege SW. Miraa (*Catha edulis*) as a cause of psychosis. East Afr Med J 1981;58(2):130–5.
21. Giannini AJ, Castellani S. A manic-like psychosis due to khat (*Catha edulis* Forsk.) J Toxicol Clin Toxicol 1982;19(5):455–9.
22. Granek M, Shalev A, Weingarten AM. Khat-induced hypnagogic hallucinations. Acta Psychiatr Scand 1988;78(4):458–61.
23. Nielen RJ, van der Heijden FM, Tuinier S, Verhoeven WM. Khat and mushrooms associated with psychosis. World J Biol Psychiatry 2004;5(1):49–53.
24. Stefan J, Mathew B. Khat chewing: an emerging drug concern in Australia? Aust N Z J Psychiatry 2005;39:842–3.
25. Odenwald M, Neuner F, Schauer M. Khat use as risk factor for psychotic disorders: a cross-sectional and case-control study in Somalia. BMC Med 2005;3:5.
26. Numan N. Exploration of adverse psychological symptoms in Yemeni khat users by the Symptoms Checklist-90 (SCL-90). Addiction 2004;99(1):61–5.
27. Khattab NY, Amer G. Undetected neuropsychophysiological sequelae of khat chewing in standard aviation medical examination. Aviat Space Environ Med 1995;66(8):739–44.
28. Bhui K, Abdi A, Abdi M, Pereira S, Dualeh M. Traumatic events, migration characteristics and psychiatric symptoms among Somali refugees—preliminary communication. Soc Psychiatry Psychiatr Epidemiol 2003;38:35–43.
29. Roper JP. The presumed neurotoxic effects of *Catha edulis*—an exotic plant now available in the United Kingdom. Br J Ophthalmol 1986;70(10):779–81.
30. Saif-Ali R, Al-Qirbi A, Al-Geiry A, AL-Habori M. Effect of *Catha edulis* on plasma glucose and C-peptide in both type 2 diabetics and non-diabetics. J Ethnopharmacol 2003;86(1):45–9.

31. Kassie F, Darroudi F, Kundi M, Schulte-Hermann R, Knasmuller S. Khat (*Catha edulis*) consumption causes genotoxic effects in humans. Int J Cancer 2001;92(3):329–32.

32. Ali AA, Al-Sharabi AK, Aguirre JM, Nahas R. A study of 342 oral keratotic white lesions induced by qat chewing among 2500 Yemeni. J Oral Pathol Med 2004;33(6):368–72.

33. Macigo FG, Mwaniki DL, Guthua SW. The association between oral leukoplakia and use of tobacco, alcohol and khat based on relative risks assessment in Kenya. Eur J Oral Sci 1995;103(5):268–73.

34. Marker P, Krogdahl A. Plasma cell gingivitis apparently related to the use of khat: report of a case. Br Dent J 2002;192(6):311–3.

35. Ashri N, Gazi M. More unusual pigmentations of the gingiva. Oral Surg Oral Med Oral Pathol 1990;70(4):445–9.

36. Heymann TD, Bhupulan A, Zureikat NE, Bomanji J, Drinkwater C, Giles P, Murray-Lyon IM. Khat chewing delays gastric emptying of a semi-solid meal. Aliment Pharmacol Ther 1995;9(1):81–3.

37. Raja'a YA, Noman TA, al Warafi AK, al Mashraki NA, al Yosofi AM. Khat chewing is a risk factor of duodenal ulcer. East Mediterr Health J 2001;7(3):568–70.

38. Raja'a YA, Noma T, Warafi TA. Khat chewing is a risk factor for duodenal ulcer. Saudi Med J 2000;21:887–8.

39. Nasher AA, Qirbi AA, Ghafoor MA, Catterall A, Thompson A, Ramsay JW, Murray-Lyon IM. Khat chewing and bladder neck dysfunction. A randomized controlled trial of alpha 1-adrenergic blockade. Br J Urol 1995;75(5):597–8.

40. Mwenda JM, Arimi MM, Kyama MC, Langat DK. Effects of khat (*Catha edulis*) consumption on reproductive functions: a review. East Afr Med J 2003;80(6):318–23.

41. Goldenberg D, Lee J, Koch WM, Kim MM, Trink B, Sidransky D, Moon CS. Habitual risk factors for head and neck cancer. Otolaryngol Head Neck Surg 2004;131:986–93.

42. Kassie F, Darroudi F, Kundi M, Schulte-Hermann R, Knasmuller S. Khat (Catha edulis) consumption causes genotoxic effects in humans. Int J Cancer 2001;92:329–32.

43. Ali AA, Al-Sharabi AK, Aguirre JM. Histopathological changes in oral mucosa due to takhzeen al-qat: a study of 70 biopsies. J Oral Pathol Med 2006;35:81–5.

44. el-Shoura SM, Abdel Aziz M, Ali ME, el-Said MM, Ali KZ, Kemeir MA, Raoof AM, Allam M, Elmalik EM. Deleterious effects of khat addiction on semen parameters and sperm ultrastructure. Hum Reprod 1995;10(9):2295–300.

45. Abdul Ghani N, Eriksson M, Kristiansson B, Qirbi A. The influence of khat-chewing on birth-weight in full-term infants. Soc Sci Med 1987;24(7):625–7.

46. Eriksson M, Ghani NA, Kristiansson B. Khat-chewing during pregnancy-effect upon the off-spring and some characteristics of the chewers. East Afr Med J 1991;68(2):106–11.

47. Date J, Tanida N, Hobara T. Qat chewing and pesticides: a study of adverse health effects in people in the mountainous areas of Yemen. Int J Environ Health Res 2004;6:405–14.

48. Attef OA, Ali AA, Ali HM. Effect of khat chewing on the bioavailability of ampicillin and amoxycillin. J Antimicrob Chemother 1997;39(4):523–5.

49. Herold DA, Wahl R, Maasch HJ, Hausen BM, Kunkel G. Occupational wood-dust sensitivity from *Euonymus europaeus* (spindle tree) and investigation of cross reactivity between E.e. wood and *Artemisia vulgaris* pollen (mugwort) Allergy 1991;46(3):186–90.

50. Pyatt DW, Yang Y, Mehos B, Le A, Stillman W, Irons RD. Hematotoxicity of the chinese herbal medicine *Tripterygium wilfordii* Hook F in CD34-positive human bone marrow cells. Mol Pharmacol 2000;57(3):512–8.

51. Chou WC, Wu CC, Yang PC, Lee YT. Hypovolemic shock and mortality after ingestion of *Tripterygium wilfordii* Hook F.: a case report Int J Cardiol 1995;49(2):173–7.

52. Edmonds SE, Montgomery JC. Reversible ovarian failure induced by a Chinese herbal medicine: lei gong teng. Br J Obstet Gynaecol 2003;110:77–8.

53. Tao X, Younger J, Fan FZ, Wang B, Lipsky PE. Benefit of an extract of *Tripterygium wilfordii* Hook F in patients with rheumatoid arthritis: a double-blind, placebo-controlled study. Arthritis Rheum 2002;46(7):1735–43.

54. Yu DY. Clinical observation of 144 cases of rheumatoid arthritis treated with glycoside of radix *Tripterygium wilfordii*. J Tradit Chin Med 1983;3(2):125–9.

55. Tao XL, Sun Y, Dong Y, Xiao YL, Hu DW, Shi YP, Zhu QL, Dai H, Zhang NZ. A prospective, controlled, double-blind, cross-over study of *Tripterygium wilfodii* Hook F in treatment of rheumatoid arthritis. Chin Med J (Engl) 1989;102(5):327–32.

56. Qian SZ. *Tripterygium wilfordii*, a Chinese herb effective in male fertility regulation. Contraception 1987;36(3):335–45.

57. Gu CX. [Cause of amenorrhea after treatment with *Tripterygium wilfordii* F.]Zhongguo Yi Xue Ke Xue Yuan Xue Bao 1989;11(2):151–3.

58. Guo JL, Yuan SX, Wang XC, Xu SX, Li DD. *Tripterygium wilfordii* Hook F in rheumatoid arthritis and ankylosing spondylitis. Preliminary report. Chin Med J (Engl) 1981;94(7):405–12.

59. Takei A, Nagashima G, Suzuki R, Hokaku H, Takahashi M, Miyo T, Asai J, Sanada Y, Fujimoto T. Meningoencephalocele associated with *Tripterygium wilfordii* treatment. Pediatr Neurosurg 1997;27(1):45–8.

Chenopodiaceae

General Information

The genera in the family of Chenopodiaceae (Table 21) include beet and spinach.

Table 21 The genera of Chenopodiaceae

Allenrolfea (allenrolfea)
Aphanisma (coastalcreeper)
Arthrocnemum (arthrocnemum)
Atriplex (saltbush)
Axyris (Russian pigweed)
Bassia (smotherweed)
Beta (beet)
Ceratoides (winterfat)
Chenopodium (goosefoot)
Corispermum (bugseed)
Cycloloma (cycloloma)
Dysphania (dysphania)
Endolepis (endolepis)
Grayia (hopsage)
Halogeton (saltlover)

(Continued)

Kochia (molly)
Krascheninnikovia (winterfat)
Microtea (jumby pepper)
Monolepis (povertyweed)
Nitrophila (niterwort)
Polycnemum (polycnemum)
Proatriplex (proatriplex)
Salicornia (pickleweed)
Salsola (Russian thistle)
Sarcobatus (greasewood)
Sarcocornia (swampfire)
Spinacia (spinach)
Suaeda (seepweed)
Suckleya (suckleya)
Zuckia (zuckia)

Chenopodium ambrosioides

Chenopodium ambrosioides (American wormseed) contains the diterpenoid aritasone and the toxic principle ascaridole, which was formerly used as an antihelminthic drug, but has now been superseded.

Clusiaceae

General Information

The genera in the family of Clusiaceae (Table 22) include St. John's wort.

Hypericum perforatum

Hypericum perforatum (devil's scourge, goat weed, rosin rose, St. John's wort, Tipton weed, witch's herb) contains the naphthodianthrones hypericin and pseudohypericin, flavonoids, such as hyperoside, isoquercitin, and rutin, and phloroglucinols, such as adhyperforin and hyperforin. It is effective in mild to moderate depression (1).

Adverse effects
A meta-analysis of 23 randomized, controlled trials of St. John's wort showed that the herbal extract is more effective than placebo for mild to moderate depression, but that current evidence was inadequate to establish whether St. John's wort is as effective as standard antidepressants (2). In clinical trials, St. John's wort appeared to have fewer short-term adverse effects than some conventional antidepressants, but information on long-term adverse effects is lacking.

In a meta-analysis of three randomized, placebo-controlled studies with very similar methods in 594 patients with mild to moderate depression, who took 900 mg/day of a standardized hypericum extract for 6 weeks, the numbers of patients who had adverse effects were similar with hypericum and placebo (3). Specifically, hypericum was devoid of sedative or anticholinergic effects and did not cause gastrointestinal or sexual problems, which can be a problem with conventional antidepressants.

Two reviews have systematically addressed the safety of St. John's wort. One showed that the most common adverse events were gastrointestinal symptoms, dizziness, confusion, tiredness, sedation, and dry mouth (4). A second review compared the adverse effect profile of St. John's wort with those of conventional antidepressants (5). The adverse effects of St. John's wort were fewer and less serious than those associated with conventional antidepressant drugs.

A review of all reported adverse effects associated with St. John's wort has shown that it has an encouraging safety profile (4). Adverse effects reported in clinical trials were invariably mild and transient: gastrointestinal

Table 22 The genera of Clusiaceae

Calophyllum (calophyllum)
Clusia (attorney)
Garcinia (sap tree)
Hypericum (St. John's wort)
Mammea (mammea)
Pentadesma (pentadesma)
Platonia (platonia)
Triadenum (marsh St. John's wort)

Table 23 Reports of adverse effects of formulations of St. John's wort (up to May 1998)

System	Number of cases
Cardiovascular (edema)	2
Cardiovascular (bradycardia)	1
Respiratory	4
Nervous system	5
Nervous system (stroke)	1
Psychiatric	15
Hematological (coagulation)	4
Gastrointestinal	2
Liver	4
Urinary tract (interstitial nephritis)	1
Skin (allergic reactions)	16
Skin (conjunctivitis)	1
Reduced therapeutic response	1

symptoms (8.5%), dizziness/confusion (4.5%), tiredness/sedation (4.5%), and dry mouth (4.0%). Synthetic drugs used in comparative trials were burdened with significantly higher rates of adverse effects. Data obtained from the WHO Collaborating Center for International Drug Monitoring are summarized in Table 23.

Cardiovascular
A 41-year-old man who had taken St. John's wort had a hypertensive crisis after taking cheese with red wine (6). St. John's wort is a monoamine oxidase inhibitor, and the authors believed that this explained how the concomitant use of a tyramine-rich food with St. John's wort had caused this problem.

Nervous system
The mechanism of action of St. John's wort in depression is not understood, but serotonin re-uptake inhibition is one possibility, for which evidence is increasing. In one case this may have led to the serotonin syndrome in a 33-year-old woman who developed extreme acute anxiety after taking only three doses of extracts of St. John's wort and recovered after withdrawal (7).

- A 40-year-old man with a history of anxiety disorder and depression presented with flushing, sweating, agitation, weakness of the legs, dry mouth, tightness in the chest, and inability to focus (8). He was taking clonazepam (0.5 mg bd) and had started to take St. John's wort 10 days before. He had previously had two similar episodes after having taken sertraline.

The authors concluded that self-medication with St. John's wort, which has SSRI activity, had caused the serotonin syndrome.

Psychiatric
Delirium has been attributed to St. John's wort.

- A 76-year-old woman began taking an extract of St. John's wort (75 mg/day) and developed delirium and psychosis 3 weeks later (9). She had no relevant medical history and did not take any other medications.

She was given risperidone and donepezil hydrochloride, and her paranoid delusions and visual hallucinations improved.

The final diagnosis was acute psychotic delirium associated with St. John's wort in a woman with underlying Alzheimer's dementia.

Hypomania has been reported with St. John's wort (10).

- A 47-year-old woman with an 8-year history of nocturnal panic attacks and a recent history of major depression had a poor response to SSRIs and instead took a 0.1% tincture of St. John's wort. After 10 days she noted racing and distorted thoughts, increased irritability, hostility, aggressive behavior, and a reduced need for sleep. After discontinuing the herbal treatment, her symptoms resolved within 2 days.

The author suggested that St. John's wort had caused this episode of hypomania.

Two cases of mania have been associated with the use of St. John's wort (11). The authors pointed out that St. John's wort, like all antidepressants, can precipitate hypomania, mania, or increased cycling of mood states, particularly in patients with occult bipolar disorder. Alternatively, the mania experienced by these patients could simply be the expression of the natural cause of their psychiatric illness.

The clinical evidence associating *Hypericum* with psychotic events has been summarized in a systematic review of 17 case reports (12). In 12 instances the diagnosis was mania or hypomania. In most of these cases, causality between the herbal remedy and the adverse effect was rated as possible; in no instance was there a positive re-challenge.

Endocrine

In a retrospective case-control study, 37 patients with raised TSH concentrations were compared with 37 individuals with normal TSH concentrations (13). Exposure to St. John's wort during the previous 3–6 months increased the odds of a raised TSH concentration by a factor of 2.12 (95% CI = 0.36, 12). The authors concluded that an association between St. John's wort and raised TSH concentrations is probable.

Skin

Photosensitivity has been attributed to St. John's wort (14).

- A 35-year-old woman took ground whole St. John's wort (500 mg/day) for mild depression. After 4 weeks she developed stinging pain on her face and the backs of both hands, which worsened with sun exposure. She was seen after the area of pain had spread following exposure to the sun. After withdrawal of St. John's wort, the symptoms gradually disappeared during the next 2 months.

The authors thought that photoactive hypericins had caused demyelination of cutaneous nerve axons. If they are correct, then this would be the first human case of photosensitivity after St. John's wort, a condition previously only reported in animals.

Of 43 users of St. John's wort surveyed by telephone, 47% reported adverse effects that they related to the remedy (15). The only potentially serious complaint was photosensitivity, which was noted by four individuals.

An Australian dermatologist has reported three cases of phototoxic reactions to St. John's wort (16). The patients were fair-skinned and had had significant exposure to ultraviolet light. In two cases St. John's wort was applied topically. In all cases complete recovery occurred after withdrawal of St. John's wort and cessation of exposure to ultraviolet light.

Hair

Alopecia has been attributed to St. John's wort.

- A 24-year-old woman with schizophrenia who self-medicated with St. John's wort while also taking olanzapine (5–10 mg/day) developed hair loss on her scalp and eyebrows 5 months later; it persisted for 12 months (8). Her medical history was otherwise unremarkable.

The authors speculated that, like SSRIs, St. John's wort can cause hair loss.

Drug withdrawal

Like most antidepressants *Hypericum perforatum* can cause withdrawal symptoms.

- A 58-year-old woman had taken a hypericum extract 1800 mg tds for 32 days when she decided to stop taking it because of a photosensitivity reaction (which is a rare adverse effect of hypericum) (17). Within 24 hours, she developed nausea, anorexia, dizziness, dry mouth, thirst, cold chills, and extreme fatigue. These symptoms peaked on day 3 and gradually improved until they abated on day 8.

The authors believed that these symptoms had been caused by the withdrawal of St John's wort, similar to withdrawal from conventional antidepressants.

Pregnancy

Like all herbal remedies, there is insufficient evidence that it is safe to take St. John's wort during pregnancy. Two women who took St. John's wort during pregnancy in order to avoid potential harmful effects of synthetic antidepressants to the fetus also discontinued their prescribed medications and did not discuss their decisions with their doctor (18). Although the effects of St. John's wort on the fetus are not known, the authors cautioned against using it under these circumstances and argued that tricyclic antidepressants or fluoxetine would be safer.

Drug-drug interactions

Interactions of St. John's wort with prescribed drugs are due to its ability to induce the activity of both CYP3A4 and the P glycoprotein transporter system (19). Prescribed drugs that interact with St. John's wort include ciclosporin, digoxin, HIV protease inhibitors, oral contraceptives, serotonin re-uptake inhibitors, theophylline, and warfarin (20).

All pharmacokinetic trials in which interactions between *Hypericum perforatum* and conventional drugs were examined have been evaluated in a systematic review (22 studies in all) (21). In 17 trials there was a reduction in the systemic availability of a range of conventional drugs. The drugs that were affected included amitriptyline, ciclosporin, digoxin, indinavir, irinotecan, mycophenolic acid, omeprazole, oral contraceptives, tacrolimus, theophylline, and warfarin.

The mechanism of hepatic enzyme induction by St. John's wort has been intensively researched (22,23). Treatment of human hepatocytes with extracts of St. John's wort or with hyperforin (one of its active constituents) resulted in induction of CYP3A4 expression (24).

Following reports about the potential for herb–drug interactions, many national regulatory agencies issued warnings about the use of St. John's wort and its safety was reviewed (25,26).

In 12 healthy volunteers the effects of St. John's wort on CYP3A4 using dextromethorphan 30 mg and alprazolam 2 mg while were taking 3 x 300 mg hypericum extract for 14 days there was a two-fold reduction in AUC and a two-fold increase in drug clearance, consistent with the known enzyme inducing effect of St John's wort (27).

Ciclosporin and other immunosuppressants
St. John's wort is a hepatic enzyme inducer (28) and can lower the plasma concentrations of various prescribed drugs, including oral anticoagulants and ciclosporin.

In 30 patients a fall in ciclosporin concentrations by 33–62% after self-medication with St. John's wort necessitated a gradual increase in the dose of ciclosporin of 187% (range 84–292%) (29). No patient suffered any permanent consequences as a result.

In 11 renal transplant patients taking stable doses of ciclosporin, all required a ciclosporin dosage adjustment 3 days after the introduction of hypericum 600 mg/day (30). After 10 days the median daily dose of ciclosporin was increased from 2.7 to 4.2 mg/kg/day to maintain ciclosporin blood concentrations in the target range. The pattern of ciclosporin metabolism was also altered by hypericum as the relative concentrations of the individual ciclosporin metabolites were also changed.

There have also been anecdotal reports of interactions of ciclosporin with St. John's wort.

- A 29-year-old woman, who had received a cadaveric kidney and pancreas transplant, had stable organ function with ciclosporin when she decided to take St. John's wort (31). Subsequently her ciclosporin concentrations became subtherapeutic and she developed signs of organ rejection. St. John's wort was withdrawn and her ciclosporin concentrations returned to the target range. However, she developed chronic kidney rejection and had to return to dialysis.
- A 63-year-old patient with a liver allograft developed severe acute rejection 14 months after transplantation (32). Two weeks before he had taken St. John's wort, which significantly reduced his ciclosporin concentration. The dose of ciclosporin was doubled, at the expense of adverse effects. He recovered fully after St. John's wort was withdrawn.
- A 55-year-old woman, who had received a kidney transplant and had stable organ function with ciclosporin, took St. John's wort and 4 weeks later her ciclosporin concentration fell sharply (33). The concentration rose again on withdrawal of St. John's wort and fell on rechallenge. As the problem was identified early enough, the patient incurred no serious consequences.
- Two patients who took ciclosporin after kidney transplantation self-medicated with St. John's wort, and their ciclosporin concentrations became subtherapeutic (34). One subsequently had an acute transplant rejection. Withdrawal of St. John's wort resulted in normalization of ciclosporin concentrations.
- After a kidney transplant for end-stage renal insufficiency a 58-year-old man was given ciclosporin, azathioprine, and prednisolone (35). Four years later he started to take St. John's wort (300 mg bd) for depression, and 2 weeks later his previously stable ciclosporin concentrations had halved. Withdrawal of the St. John's wort resulted in normalization of his ciclosporin concentrations.

In a systematic review of all reports of interactions of St. John's wort with ciclosporin, 11 case reports and two case series were found (36). In most cases there was little doubt about causality.

The effect of hypericum extract 600 mg/day for 14 days on the pharmacokinetics of tacrolimus and mycophenolate mofetil has been studied in 10 renal transplant recipients who had received their transplant at least 2 years before and who were taking stable immunosuppressive therapy (37). The plasma concentrations of both drugs were significantly reduced.

Digoxin
In a randomized, placebo-controlled, double-blind study volunteers with steady-state digoxin concentrations took either placebo, a standardized extract of St. John's wort, encapsulated St. John's wort powder, St. John's wort tea, or an encapsulated fatty oil formulation of St. John's wort (38). The extract and the powder caused marked reductions in digoxin concentrations but the tea and the fatty oil formulation did not. The mechanism was not discussed, and it is not clear why the different formulations had different effects.

Fexofenadine
Fexofenadine is transported by P glycoprotein in vitro (39) and there is evidence that St John's wort can alter P glycoprotein activity (40). The effects of St John's wort on the pharmacokinetics of fexofenadine have therefore been studied (41). Fexofenadine 60 mg was given orally twice—before a single dose of St John's wort 900 mg and again after 3 weeks treatment with 300 mg tds. A single dose of St John's wort significantly inhibited intestinal P glycoprotein activity and increased the maximum plasma concentration of fexofenadine by 45%. In contrast, long-term St John's wort caused a 35% reduction in the C_{max} of fexofenadine and increased its oral clearance. However,

the authors concluded that these pharmacokinetic changes alone are unlikely to be clinically significant.

Irinotecan

In a randomized, crossover study five patients with cancer taking irinotecan were given St. John's wort (42). The plasma concentrations of irinotecan significantly fell in the presence of the herbal remedy. The authors pointed out that this is likely to reduce the effectiveness of the anticancer drug.

Oral contraceptives

The effects of *Hypericum perforatum* on oral contraceptive therapy have been evaluated in 16 healthy women who took a low-dose oral contraceptive containing ethinylestradiol 20 micrograms and norethindrone acetate 1 mg or a placebo for two consecutive 28-day cycles in a single-blind sequential study (43). Treatment with *Hypericum perforatum* 900 mg/day was added for two additional 28-day cycles and was associated with a significant 13–15% reduction in the dose exposure from the contraceptive. Breakthrough bleeding increased in the treatment cycles, as did evidence of follicle growth and probable ovulation. The authors suggested that *Hypericum perforatum* is associated with increased metabolism of norethindrone and ethinylestradiol, breakthrough bleeding, follicle growth, and ovulation.

In a controlled study, 18 women (aged 18-35 years) took ethinylestradiol + desogestrel 0.02/0.15 mg/day alone or in combination with hypericum 300 mg bd (cycle A) or tds (cycle B) (44). There was no evidence of ovulation during concomitant hypericum and oral contraceptive use; however, significantly more patients reported mid-cyclic bleeding during cycles A (76%) and B (88%) than during the control cycle (35%), and the AUC of 3-ketodesogestrel fell significantly during cycles A and B compared with the control cycle.

St John's wort may lower the blood concentrations of oral estrogens (45, 46).

- A 36-year-old woman who had taken St John's wort 1700 mg/day for 3 months for depression became pregnant despite regularly taking the combined oral contraceptive Valette®. A therapeutic abortion was carried out which revealed a healthy 17-week old 144 g male fetus.

The authors believed that this pregnancy was related to an interaction of hypericum with ethinylestradiol. They also mentioned four similar cases that had been reported to the German regulatory authorities.

Selective serotonin re-uptake inhibitors

Inhibition of serotonin re-uptake by St. John's wort can lead to interactions with SSRIs (47).

- A 28-year-old man without a previous psychiatric history was given sertraline 50 mg/day for depression after bilateral orchidectomy. Against medical advice, he also took St. John's wort and subsequently became manic.

The authors suggested that inhibition of serotonin re-uptake by sertraline had been potentiated by the use of St. John's wort.

Warfarin

In an open, crossover, randomized study in 12 healthy men St. John's wort significantly induced the apparent clearance of both *S*-warfarin and *R*-warfarin (by 29 and 23% respectively), which in turn resulted in a significant reduction in the pharmacological effect of racemic warfarin (48).

References

1. Stevinson C, Ernst E. Hypericum for depression. An update of the clinical evidence. Eur Neuropsychopharmacol 1999;9(6):501–5.
2. Linde K, Ramirez G, Mulrow CD, Pauls A, Weidenhammer W, Melchart D. St. John's wort for depression—an overview and meta-analysis of randomised clinical trials. BMJ 1996;313(7052):253–8.
3. Trautmann-Sponsel RD, Dienel A. Safety of Hypericum extract in mildly to moderately depressed outpatients. A review based on data from three randomized, placebo-controlled trials. J Affect Disord 2004;82:303–7.
4. Ernst E, Rand JI, Barnes J, Stevinson C. Adverse effects profile of the herbal antidepressant St. John's wort (Hypericum perforatum L.) Eur J Clin Pharmacol 1998;54(8):589–94.
5. Stevinson C, Ernst E. Safety of Hypericum in patients with depression. A comparison with conventional antidepressants. CNS Drugs 1999;11(2):125–32.
6. Patel S, Robinson R, Burk M. Hypertensive crisis associated with St. John's wort. Am J Med 2002;112(6):507–8.
7. Brown TM. Acute St. John's wort toxicity. Am J Emerg Med 2000;18(2):231–2.
8. Parker V, Wong AH, Boon HS, Seeman MV. Adverse reactions to St. John's wort. Can J Psychiatry 2001;46(1):77–9.
9. Laird RD, Webb M. Psychotic episode during use of St. John's wort. J Herbal Pharmacother 2001;1:81–7.
10. Schneck C. St. John's wort and hypomania. J Clin Psychiatry 1998;59(12):689.
11. Nierenberg AA, Burt T, Matthews J, Weiss AP. Mania associated with St. John's wort. Biol Psychiatry 1999;46(12):1707–8.
12. Stevinson C, Ernst E. Can St John's wort trigger psychoses? Int J Clin Pharmacol Ther 2004;42:473–80.
13. Ferko N, Levine MA. Evaluation of the association between St. John's wort and elevated thyroid-stimulating hormone. Pharmacotherapy 2001;21(12):1574–8.
14. Bove GM. Acute neuropathy after exposure to sun in a patient treated with St. John's Wort. Lancet 1998;352(9134):1121–2.
15. Beckman SE, Sommi RW, Switzer J. Consumer use of St. John's wort: a survey on effectiveness, safety, and tolerability. Pharmacotherapy 2000;20(5):568–74.
16. Lane-Brown MM. Photosensitivity associated with herbal preparations of St. John's wort (Hypericum perforatum). Med J Aust 2000;172(6):302.
17. Dean AJ. Suspected withdrawal syndrome after cessation of St John's wort. Ann Pharmacother 2003;37:150.
18. Grush LR, Nierenberg A, Keefe B, Cohen LS. St. John's wort during pregnancy. JAMA 1998;280(18):1566.
19. Karyekar CS, Eddington ND, Dowling TC. Effect of St. John's wort extract on intestinal expression of cytochrome P4501A2: studies in LS180 cells. J Postgrad Med 2002;48(2):97–100.

20. Henderson L, Yue QY, Bergquist C, Gerden B, Arlett P. St. John's wort (Hypericum perforatum): drug interactions and clinical outcomes. Br J Clin Pharmacol 2002;54(4):349–56.

21. Mills E, Montori VM, Wu P, Gallicano K, Clarke M, Guyatt G. Interaction of St John's wort with conventional drugs: systematic review of clinical trials BMJ 2005;329:27–30.

22. Budzinski JW, Foster BC, Vandenhoek S, Arnason JT. An in vitro evaluation of human cytochrome P450 3A4 inhibition by selected commercial herbal extracts and tinctures. Phytomedicine 2000;7(4):273–82.

23. Obach RS. Inhibition of human cytochrome P450 enzymes by constituents of St. John's wort, an herbal preparation used in the treatment of depression. J Pharmacol Exp Ther 2000;294(1):88–95.

24. Moore LB, Goodwin B, Jones SA, Wisely GB, Serabjit-Singh CJ, Willson TM, Collins JL, Kliewer SA. St. John's wort induces hepatic drug metabolism through activation of the pregnane X receptor. Proc Natl Acad Sci USA 2000;97(13):7500–2.

25. Biffignandi PM, Bilia AR. The growing knowledge of St. John's wort (Hypericum perforatum L.) drug interactions and their clinical significance Curr Ther Res Clin Exp 2000;61:389–94.

26. Schulz V. Häufigkeit und klinische Relevanz der Interaktionen und Nebenwirkungen von Hypericum-Präparaten. Perfusion 2000;13:486.

27. Markowitz JS, Donovan JL, DeVane CL, Taylor RM, Ruan Y, Wang J-S, Chavin KD. Effect of St John's wort on drug metabolism by induction of cytochrome P450 3A4 enzyme. J Am Med Assoc 2003;290:1500–4.

28. Ernst E. Second thoughts about safety of St. John's wort. Lancet 1999;354(9195):2014–6.

29. Breidenbach T, Kliem V, Burg M, Radermacher J, Hoffmann MW, Klempnauer J. Profound drop of cyclosporin A whole blood trough levels caused by St. John's wort (Hypericum perforatum). Transplantation 2000; 69(10):2229–30.

30. Bauer S, Störmer E, Johne A, Kruger H, Budde K, Neumayer H-H, Roots I, Mai I. Alterations in cyclosporin A pharmacokinetics and metabolism during treatment with St John's wort in renal transplant patients. Br J Clin Pharmacol 2003;55:203–11.

31. Barone GW, Gurley BJ, Ketel BL, Lightfoot ML, Abul-Ezz SR. Drug interaction between St. John's wort and cyclosporine. Ann Pharmacother 2000;34(9):1013–6.

32. Karliova M, Treichel U, Malago M, Frilling A, Gerken G, Broelsch CE. Interaction of Hypericum perforatum (St. John's wort) with cyclosporin A metabolism in a patient after liver transplantation. J Hepatol 2000;33(5):853–5.

33. Mai I, Kruger H, Budde K, Johne A, Brockmoller J, Neumayer HH, Roots I. Hazardous pharmacokinetic interaction of Saint John's wort (Hypericum perforatum) with the immunosuppressant cyclosporin. Int J Clin Pharmacol Ther 2000;38(10):500–2.

34. Turton-Weeks SM, Barone GW, Gurley BJ, Ketel BL, Lightfoot ML, Abul-Ezz SR. St. John's wort: a hidden risk for transplant patients. Prog Transplant 2001;11(2):116–20.

35. Moschella C, Jaber BL. Interaction between cyclosporine and Hypericum perforatum (St. John's wort) after organ transplantation. Am J Kidney Dis 2001;38(5):1105–7.

36. Ernst E. St. John's wort supplements endanger the success of organ transplantation. Arch Surg 2002;137(3):316–9.

37. Mai I, Störmer E, Bauer S, Krüger H, Budde K, Roots I. Impact of St John's wort treatment on the pharmacokinetics of tacrolimus and mycophenolic acid in renal transplant patients. Nephrol Dial Transplant 2003;18:819–22.

38. Uehleke B, Mueller SC, Woehling H, Petzsch M, Riethling AK, Drewelow B. Interaction of St. John's wort with digoxin in relation to dosage and formulation. Phytomedicine 2000;SII:20.

39. Cvetkovic M, Leake B, Fromm MF, Wilkinson GR, Kim RB. OATP and P-glycoprotein transporters mediate the cellular uptake and excretion of fexofenadine. Drug Metab Dispos 1999;27(8):866–71.

40. Johne A, Brockmoller J, Bauer S, Maurer A, Langheinrich M, Roots I. Pharmacokinetic interaction of digoxin with an herbal extract from St John's wort (Hypericum perforatum). Clin Pharmacol Ther 1999;66(4):338–45.

41. Wang Z, Hamman MA, Huang SM, Lesko LJ, Hall SD. Effect of St John's wort on the pharmacokinetics of fexofenadine. Clin Pharmacol Ther 2002;71(6):414–20.

42. Mathijssen RH, Verweij J, de Bruijn P, Loos WJ, Sparreboom A. Effects of St. John's wort on irinotecan metabolism. J Natl Cancer Inst 2002;94(16):1247–9.

43. Murphy PA, Kern SE, Stanczyk FZ, Westhoff CL. Interaction of St John's wort with oral contraceptives: effects on the pharmacokinetics of norethindrone and ethinyl estradiol, ovarian activity and breakthrough bleeding. Contraception 2005;71(6):402–8.

44. Pfrunder A, Schiesser M, Gerber S, Haschke M, Bitzer J, Drewe J. Interaction of St John's wort with low-dose oral contraceptive therapy: a randomized controlled trial. Br J Clin Pharmacol 2003;56:683–90.

45. Schwarz UI, Buschel B, Kirch W. Failure of oral contraceptive because of St. Johns' wort. Eur J Clin Pharamacol 2001;57:A25.

46. Schwarz UI, Büschel B, Kirch W. Unwanted pregnancy on self-medication with St John's wort despite hormonal contraception. Br J Clin Pharmacol 2003;55:112–3.

47. Barbenel DM, Yusufi B, O'Shea D, Bench CJ. Mania in a patient receiving testosterone replacement postorchidectomy taking St. John's wort and sertraline. J Psychopharmacol 2000;14(1):84–6.

48. Jiang X, Williams KM, Liauw WS, Ammit AJ, Roufogalis BD, Duke CC, Day RO, McLachlan AJ. Effect of St. John's wort and ginseng on the pharmacokinetics and pharmacodynamics of warfarin in healthy subjects. Br J Clin Pharmacol 2004;57(5):592–9.

Convolvulaceae

General Information

The genera in the family of Convolvulaceae (Table 24) include bindweed and morning glory.

Table 24   The genera of Convolvulaceae

Aniseia (aniseia)
Argyreia (argyreia)
Bonamia (lady's nightcap)
Calystegia (false bindweed)
Convolvulus (bindweed)
Cressa (alkaliweed)
Dichondra (pony's foot)
Evolvulus (dwarf morning glory)
Ipomoea (morning glory)
Jacquemontia (cluster vine)
Merremia (wood rose)
Operculina (lidpod)
Poranopsis (poranopsis)
Stictocardia (stictocardia)
Stylisma (dawn flower)
Turbina (turbina)
Xenostegia (morning vine)

Convolvulus scammonia

Convolvulus scammonia (Mexican scammony) contains alkaloids that are drastic purgatives with irritant properties. It has now been superseded.

Ipomoea purga

Ipomoea purga (jalap) is a drastic cathartic with irritant action, which has been superseded by less toxic laxatives.

Coriariaceae

General Information

The family of Coriariaceae contains the single genus *Coriaria*.

Coriaria species

Coriaria arborea (tutu) plant is a traditional Maori medicine. It contains various sesquiterpenoids, including a toxin, tutin, that is allied to picrotoxin (1).

Adverse effects

Ingestion of portions or concoctions of *Coriaria* plants can result in tonic-clonic seizures, respiratory arrest, and death.

Three sisters who ate berries of *Coriaria myrtifolia* (redoul) suffered from acute poisoning; the adverse effects affected the gastrointestinal tract (nausea, vomiting, abdominal pain), the nervous system (obnubilation, convulsions, and their complications), and respiratory function (hyperpnea, apnea, short and superficial respiration), together with myositis of the pupils; one died (2).

Of 25 children who ate the fruits of *C. myrtifolia*, 13 had gastrointestinal digestive symptoms and nine had nervous system toxicity, including seizures and coma (3).

References

1. Anonymous. In: Twenty-fifth Annual Report of the National Toxicoloy Group. Dunedin: New Zealand National Poisons and Hazardous Chemicals Information Centre, 1990:4.
2. Skalli S, David JM, Benkirane R, Zaid A, Soulaymani R. Intoxication aigue par le redoul (*Coriaria myrtifolia* L.). [Acute intoxication by redoul (*Coriaria myrtifolia* L.). Three observations.] Presse Méd 2002;31(33):1554–6.
3. Garcia Martin A, Masvidal Aliberch RM, Bofill Bernaldo AM, Rodriguez, Alsina S. Intoxicacion por ingesta de *Coriaria myrtifolia* Estudio de 25 casos. [Poisoning caused by ingestion of *Coriaria myrtifolia*. Study of 25 cases.] An Esp Pediatr 1983;19(5):366–70.

Cucurbitaceae

General Information

The genera in the family of Cucurbitaceae (Table 25) include cucumbers, gourds, and melons.

Bryonia alba

Bryonia alba (white bryony) contains toxic triterpenoids called cucurbitacins.

Adverse effects
Cucurbitacins are drastic laxatives and emetics and can cause the symptoms of food poisoning (1).

Citrullus colocynthis

The dried pulp of the fruit of *Citrullus colocynthis* (colocynth) is a drastic laxative, which contains toxic cucurbitacins.

Adverse effects
A man experienced vomiting, colicky pain, and bloody diarrhea after self-medication with *C. colocynthis* (2). Hemorrhagic colitis secondary to ingestion of colocynth has been reported (2). In three cases of toxic acute colitis 8–12 hours after ingestion of colocynth for ritual purposes, the prominent clinical feature was dysenteric diarrhea; colonoscopic changes included congestion and

Table 25 The genera of Cucurbitaceae

Apodanthera (apodanthera)
Benincasa (benincasa)
Brandegea (starvine)
Bryonia (bryony)
Cayaponia (melonleaf)
Citrullus (watermelon)
Coccinia (coccinia)
Ctenolepis (ctenolepis)
Cucumis (melon)
Cucumeropsis (cucumeropsis)
Cucurbita (gourd)
Cyclanthera (cyclanthera)
Doyerea (doyeria)
Ecballium (squirting cucumber)
Echinocystis (echinocystis)
Echinopepon (balsam apple)
Fevillea (fevillea)
Hodgsonia (hodgsonia)
Ibervillea (globeberry)
Lagenaria (lagenaria)
Luffa (luffa)
Marah (manroot)
Melothria (melothria)
Momordica (momordica)
Psiguria (pygmymelon)
Sechium (sechium)

(Continued)

Sicana (sicana)
Sicyos (burr cucumber)
Sicyosperma (sicyosperma)
Telfairia (telfairia)
Thladiantha (thladiantha)
Trichosanthes (trichosanthes)
Tumamoca (tumamoca)

hyperemia of the mucosa with abundant exudates but no ulceration or pseudopolyp formation; there was rapid recovery within 3–6 days, with normal endoscopy at day 14 (3).

Ecballium elaterium

Ecballium elaterium (squirting cucumber) contains toxic cucurbitacins, which are violent purgatives. It is used in the Mediterranean as a purgative and in treating sinusitis.

Adverse effects
Intranasal use of *E. elaterium* has been associated with Quincke's edema (4) and with fatal cardiac and renal failure (5).

A report from a poisons unit in Israel included 13 patients who had used the juice of the squirting cucumber, either orally or topically, for unreported reasons (6). They subsequently had edema of the pharynx, dyspnea, drooling, dysphagia, vomiting, and conjunctivitis. With symptomatic treatment they recovered within a few days.

In a retrospective chart analysis in a Greek ENT department 42 patients with allergic reactions to *E. elaterium*, including upper airway edema, were identified (7). Treatment with glucocorticoids and antihistamines resulted in full recovery in all cases.

Momordica charantia

Oral formulations of *Momordica charantia* (karela fruit, bitter melon) have hypoglycemic activity in non-insulin dependent diabetes mellitus (8,9), and can interfere with conventional treatment with diet and chlorpropamide (10). In 15 patients aged 52–65 years a soft extract of *M. charantia* plus half doses of metformin or glibenclamide or both in combination caused hypoglycemia greater than that caused by full doses during treatment for 7 days (11). Subcutaneous injection of a principle obtained from the fruit may lower blood glucose concentrations in juvenile diabetes.

Adverse effects
Metabolism

M. charantia can cause hypoglycemic coma and convulsions in children (12).

Drug interactions
P glycoprotein
M. charantia inhibits P glycoprotein in vitro and drug interactions can therefore be expected (13).

Sechium edule

The tuber of *Sechium edule* (chayote) is valued as a potent diuretic by Latin American populations. Its use as a decoction by a pregnant woman suffering from pedal edema may have been the cause of a severe case of hypokalemia (14).

References

1. Kirschman JC, Suber RL. Recent food poisonings from cucurbitacin in traditionally bred squash. Food Chem Toxicol 1989;27(8):555–6.
2. Al Faraj S. Haemorrhagic colitis induced by *Citrullus colocynthis*. Ann Trop Med Parasitol 1995;89(6):695–6.
3. Goldfain D, Lavergne A, Galian A, Chauveinc L, Prudhomme F. Peculiar acute toxic colitis after ingestion of colocynth: a clinicopathological study of three cases. Gut 1989;30(10):1412–8.
4. Plouvier B, Trotin F, Deram R, De Coninck P, Baclet JL. Concombre d'ane (*Ecbalium elaterium*) une cause peu banale d'oedème de Quincke. [Squirting cucumber (*Ecbalium elaterium*), an uncommon cause of Quincke's edema.] Nouv Presse Méd 1981;10(31):2590.
5. Vlachos P, Kanitsakis NN, Kokonas N. Fatal cardiac and renal failure due to *Ecbalium elaterium* (squirting cucumber). J Toxicol Clin Toxicol 1994;32(6):737–8.
6. Raikhlin-Eisenkraft B, Bentur Y. *Ecbalium elaterium* (squirting cucumber)—remedy or poison? J Toxicol Clin Toxicol 2000;38(3):305–8.
7. Kloutsos G, Balatsouras DG, Kaberos AC, Kandiloros D, Ferekidis E, Economou C. Upper airway edema resulting from use of *Ecballium elaterium*. Laryngoscope 2001;111(9):1652–5.
8. Welihinda J, Karunanayake EH, Sheriff MH, Jayasinghe KS. Effect of *Momordica charantia* on the glucose tolerance in maturity onset diabetes. J Ethnopharmacol 1986;17(3):277–82.
9. Leatherdale BA, Panesar RK, Singh G, Atkins TW, Bailey CJ, Bignell AH. Improvement in glucose tolerance due to *Momordica charantia* (karela). BMJ (Clin Res Ed) 1981;282(6279):1823–4.
10. Aslam M, Stockley IH. Interaction between curry ingredient (karela) and drug (chlorpropamide). Lancet 1979;1(8116):607.
11. Tongia A, Tongia SK, Dave M. Phytochemical determination and extraction of *Momordica charantia* fruit and its hypoglycemic potentiation of oral hypoglycemic drugs in diabetes mellitus (NIDDM). Indian J Physiol Pharmacol 2004;48(2):241–4.
12. Basch E, Gabardi S, Ulbricht C. Bitter melon (*Momordica charantia*): a review of efficacy and safety. Am J Health Syst Pharm 2003;60(4):356–9.
13. Limtrakul P, Khantamat O, Pintha K. Inhibition of P-glycoprotein activity and reversal of cancer multidrug resistance by *Momordica charantia* extract. Cancer Chemother Pharmacol 2004;54(6):525–30.
14. Jensen LP, Lai AR. Chayote (*Sechium edule*) causing hypokalemia in pregnancy. Am J Obstet Gynecol 1986;155(5):1048–9.

Cupressaceae

General Information

The genera in the family of Cupressaceae (Table 26) include cedar, cypress, and juniper.

Table 26 The genera of Cupressaceae

Callitris (cypress pine)
Calocedrus (incense cedar)
Chamaecyparis (cedar)
Cupressus (cypress)
Cupressocyparis
Juniperus (juniper)
Platycladus (platycladus)
Tetraclinis (tetraclinis)
Thuja (red cedar)

Juniperus communis

Extracts of *Juniperus communis* and other species are used in cosmetics, as hair conditioners, and in fragrances (1). The volatile oil distilled from the berries of *Juniperus communis* (juniper) can act as a gastrointestinal irritant. It is said that excessive doses can cause renal damage, and use during pregnancy is discouraged because of a fear that this might also stimulate the uterus.

Reference

1. Anonymous. Final report on the safety assessment of *Juniperus communis* extract, *Juniperus oxycedrus* extract, *Juniperus oxycedrus* tar, *Juniperus phoenicea* extract, and *Juniperus virginiana* extract. Int J Toxicol 2001;20(Suppl 2): 41–56.

Cycadaceae

General Information

The family of Cycadaceae contains the single genus *Cycas*.

Cycas circinalis

The seeds of *Cycas circinalis* (false sago palm, queen sago) contain the non-protein amino acid beta-*N*-methylamino-L-alanine, which is similar to the neurotoxic amino acid beta-*N*-oxalylamino-L-alanine. Monkeys fed this amino acid develop a syndrome that closely resembles the disease that is known by the Chamorros of Guam as lytico-bodig, a complex of amyotrophic lateral sclerosis parkinsonism dementia that occurs in Guam, where the seeds of *C. circinalis* are a traditional staple of the indigenous diet (1). Consumption of flying foxes may also generate sufficiently high cumulative doses of *Cycas* neurotoxins, since the flying foxes forage on cycad seeds (2). The risk may be increased by the use of poultices prepared from cycad seeds as a topical cure for skin lesions in Eastern Irian Jaya (New Guinea), Indonesia.

Adverse effects

In a retrospective chart review at the Poison Control Center in Taiwan from 1990 to 2001 there were 21 cases of *Cycas* seed poisoning (3). The patients had taken 1–30 seeds for cosmetic use (5%), as edible food (70%), or for health promotion (10%), cancer prevention (10%), and gastrointestinal discomfort (5%). All had eaten the seeds after washing and cooking them. The time from ingestion to the onset of symptoms ranged from 30 minutes to 7 hours (mean 2.8 hours). All the patients except one presented with gastrointestinal disturbances, and 90% sought medical care at the emergency department. Severe vomiting was the most striking symptom. There was no respiratory depression. Within 24 hours all had recovered. Six patients had blood cyanide or thiocyanate concentrations measured, and although they were higher than normal, they did not reach the toxic range.

References

1. Spencer PS, Nunn PB, Hugon J, Ludolph A, Roy DN. Motorneurone disease on Guam: possible role of a food neurotoxin. Lancet 1986;1(8487):965.
2. Cox PA, Sacks OW. Cycad neurotoxins, consumption of flying foxes, and ALS-PDC disease in Guam. Neurology 2002;58(6):956–9.
3. Chang SS, Chan YL, Wu ML, Deng JF, Chiu T, Chen JC, Wang FL, Tseng CP. Acute *Cycas* seed poisoning in Taiwan. J Toxicol Clin Toxicol 2004;42(1):49–54.

Droseraceae

General Information

The family of Droseraceae contains two genera:

1. *Dionaea* (Venus flytrap)
2. *Drosera* (sundew).

Dionaea muscipula

The expressed sap of *Dionaea muscipula*, a fly-catching plant (Venus flytrap), was once available in Germany as an herbal oncolytic in the form of ampoules and oral drops (1). However, when it became apparent that intramuscular administration could produce shivers, fever, and anaphylactic shock, the health authorities banned the ampoules. They also ruled that the product information on the oral drops should warn against use in pregnancy and should list reddening of the face, headache, dyspnea, nausea, and vomiting as adverse effects.

Reference

1. Anonymous. Carnivorsa–Phytothexapeutikum zur Behandlung maligner Erkrankungen. Dokumentation Nr. 15. [Carnivora–phytotherapeutic agent for the treatment of malignant diseases. Documentation No. 15.] Schweiz Rundsch Med Prax 1988;77(11):283–7.

Dryopteraceae

General Information

The genera in the family of Dryopteraceae (Table 27) include various types of fern.

Dryopteris filix-mas

The rhizome of *Dryopteris filix-mas* (male fern) was formerly used as an antihelminthic drug (1), but it is highly toxic and has been superseded by other less-dangerous

Table 27 The genera of Dryopteraceae

Arachniodes (holly fern)
Athyrium (ladyfern)
Bolbitis (creeping fern)
Ctenitis (lacefern)
Cyclopeltis (cyclopeltis)
Cyrtomium (netvein hollyfern)
Cystopteris (bladderfern)
Deparia (false spleenwort)
Diplazium (twinsorus fern)
Dryopteris (wood fern)
Elaphoglossum (tongue fern)
Fadyenia (dotted fern)
Gymnocarpium (oak fern)
Hemidictyum (hemidictyum)
Hypoderris (hypoderris)
Lastreopsis (shield fern)

(Continued)

Lomagramma (lomagramma)
Lomariopsis (fringed fern)
Matteuccia (ostrich fern)
Megalastrum (megalastrum)
Nephrolepis (sword fern)
Nothoperanema (island lace fern)
Oleandra (oleander fern)
Olfersia (island fern)
Onoclea (sensitive fern)
Phanerophlebia (phanerophlebia)
Polystichum (holly fern)
Rumohra (rumohra)
Tectaria (halberd fern)
Triplophyllum (triplophyllum)
Woodsia (cliff fern)

agents. In spite of poor absorption, serious poisoning can occur, for example when absorption is increased by the presence of fatty foods. Poisoning is characterized by vomiting, diarrhea, vertigo, headache, tremor, cold sweats, dyspnea, cyanosis, convulsions, mental disturbances, disturbed vision, and even blindness, which in a few instances is permanent.

Reference

1. Vinkenborg J. [The male fern as a medicinal plant.]Pharm Weekbl 1961;96:726–36.

Ephedraceae

General information

The family of *Ephedraceae* contains one genus, *Ephedra*, which has several species, including *Ephedra vulgaris*, also known as sea-grape. *Ephedra* species contain ephedrine, which, with its stereoisomer pseudoephedrine, is covered in a separate monograph (p. 000).

Adverse effects

The complexity of the adverse effects associated with *Ephedra* has been summarized; problems arise mostly through improper use (1). *Ephedra* is often used under the name ma huang in traditional Chinese medicine. In the West it has been promoted as an aid to weight loss but has been banned in several countries because of serious safety concerns. The relative risk of an adverse reaction to *Ephedra* was more than 100 times higher than with other popular herbal medicines (2). *Ephedra* self-medication has been associated with many adverse effects, including myocardial infarction (3), psychoses (4, 5), and liver damage (6).

Cardiovascular

Metabolife 356, which is marketed as an aid to weight loss, contains ma huang (ephedrine 12 mg), guaraná extract (caffeine 40 mg), chromium picolinate, and various herbs and vitamins.

In a double-blind, crossover, randomized, placebo-controlled study in 15 young healthy volunteers with normal BMI, Metabolife 356 increased the mean maximal QTc interval and systolic blood pressure (7). Those who took it were more likely to have a shortening of the QTc interval by at least 30 milliseconds compared with placebo (RR 2.67; CI = 1.40, 5.10). All those who took Metabolife 356 reported non-specific symptoms (8). There were no adverse effects while patients were taking placebo. One woman developed a sinus tachycardia of 120/minute with palpitation, 1 hour after taking Metabolife 356. A hand tremor developed in one subject and subsided after 5 hours.

Two young men took *Ephedra* supplements and developed severe cardiomyopathies and global cardiac hypokinesis (9). Both were treated with standard treatments for heart failure but one died nevertheless.

Ephedra and coronary dissection have been linked (10).

- A 50-year-old African American woman developed a myocardial infarction 2 days after taking a supplement containing *Ephedra* standardized to ephedrine 20 mg/day. Subsequent investigations showed a dissection in the mid-distal segment of the left anterior descending artery and a thrombosis occluding a large first obtuse marginal branch. She had bypass surgery, but later developed refractory heart failure.

Nervous system

Five patients had ischemic strokes after self-medicating with *Ephedra* supplements (11). They had taken *Ephedra* products for weight loss or to increase their energy.

The use of Metabolife 356 has been associated with a transient ischemic attack (12).

- A 20-year-old woman, otherwise healthy, had symptoms of a transient ischemic attack less than 30 minutes after taking four tablets of Metabolife 356. She had also taken 6–15 tablets/day during the 3 days before in an attempt to lose weight. Two-point discrimination was reduced over the whole of the left side, but there were no other neurological abnormalities and her blood pressure was 134/84 mmHg. Her symptoms resolved within 4 hours. Rechallenge was not attempted.

Cerebral vasospasm was presumably the main mechanism in this case, but the authors noted reports of hemorrhagic as well as ischemic stroke associated with ephedrine-containing formulations. The manufacturer of this product has removed all ephedra-containing products, including Metabolife 356, from the market. However, the authors commented that although many ephedra-containing formulations are no longer available over the counter, they can be easily bought on the Internet.

Liver

Of 12 patients who experienced liver damage through herbal weight loss products most were women (n = 9), most had taken *Ephedra*-containing products (n = 10), eight recovered after medical treatment, three required liver transplants, and one died (13). There has been a further report of fulminant hepatic failure requiring liver transplantation after the use of *Ephedra sinica* or ma huang (14).

References

1. Soni MG, Carabin IG, Griffiths JC, Burdock GA. Safety of *Ephedra*: lessons learned. Toxicol Lett 2004;150:97–110.
2. Bent S, Tiedt TN, Odden MC, Shlipak MG. The relative safety of *Ephedra* compared with other herbal products. Ann Intern Med 2003;138:468–71.
3. Enders JM, Dobesh PP, Ellison JN. Acute myocardial infarction induced by ephedrine alkaloids. Pharmacotherapy 2003;23:1645–51.
4. Boerth JM, Caley CF. Possible case of mania associated with Ma-Huang. Pharmacotherapy 2003;23:380–3.
5. Walton R, Manos GH. Psychosis related to *Ephedra*-containing herbal supplement use. South Med J 2003;96:718–20.
6. Bajaj J, Knox JF, Komorowski PAC, Saeian K. The irony of herbal hepatitis. Ma-Huang-induced hepatotoxicity associated with compound heterozygosity for hereditary hemochromatosis. Dig Dis Sci 2003;48:1925–8.
7. McBride BF, Karapanos AK, Krudysz A, Kluger J, Coleman CI, White CM. Electrocardiographic and hemodynamic effects of a multicomponent dietary supplement containing *Ephedra* and caffeine. A randomized controlled trial. JAMA 2004;291:216–21.
8. Gardner SF, Frank AM, Gurley BJ, Haller CA, Singh BK, Mehta JL. Effect of a multicomponent, *Ephedra*–containing dietary supplement (Metabolife 356) on Holter monitoring

and hemostatic parameters in healthy volunteers. Am J Cardiol 2003;91:1510–13.

9. Naik SD, Freudenberger RS. *Ephedra*-associated cardiomyopathy. Ann Pharmacother 2004;38:400–3.

10. Sola S, Helmy T, Kacharava A. Coronary dissection and thrombosis after ingestion of Ephedra. Am J Med 2004;116:645–6.

11. Chen C, Biller J, Willing SJ, Lopez AM. Ischemic stroke after using over the counter products containing *Ephedra*. J Neurol Sciences 2004;217:55–60.

12. Lo Vecchio F, Sawyers B, Eckholdt PA. Transient ischemic attack associated with Metabolife 356 use. Am J Emerg Med 2005;23:199–200.

13. Neff GW, Reddy KR, Durazo FA, Meyer D, Marrero R, Kaplowitz N. Severe hepatotoxicity associated with the use of weight loss diet supplements containing Ma huang or usnic acid. J Hepatol 2004;41:1061–7.

14. Skoulidis F, Alexander GJ, Davies SE. Ma huang associated acute liver failure requiring liver transplantation. Eur J Gastroenterol Hepatol 2005;17(5):581–4.

Ericaceae

General Information

The genera in the family of Ericaceae (Table 28) include heather, huckleberry, and rhododendron.

Arctostaphylos uva-ursi

Arctostaphylos uva-ursi (bearberry) contains a steroid, sitosterol, and triterpenoids, such as amyrin, betulinic acid, lupeol, oleanolic acid, taraxenol, ursolic acid, and uvaol. The main constituent is a glucoside called arbutin. Other constituents are methylarbutin, ericolin, ursone, gallic acid, and ellagic acid.

Its reputed antibacterial activity is ascribed to the urinary metabolite hydroquinone, which is excreted in the form of inactive conjugates and needs an alkaline urine to be liberated. As the urine of people who consume a Western non-vegetarian diet is usually acidic, it is sometimes suggested that one should alkalinize the urine of bearberry users with sodium bicarbonate. However, as the dosage recommended for this purpose is usually high, this carries well-known risks such as a high sodium load and interference with the renal clearance of certain other drugs.

Table 28 The genera of Ericaceae

Agarista (Florida hobblebush)
Andromeda (bog rosemary)
Arbutus (madrone)
Arctostaphylos (manzanita)
Befaria (befaria)
Calluna (heather)
Cassiope (mountain heather)
Chamaedaphne (leather leaf)
Comarostaphylis (summer holly)
Elliottia (elliottia)
Epigaea (trailing arbutus)
Erica (heath)
Gaultheria (snowberry)
Gaylussacia (huckleberry)
Gonocalyx (brittle leaf)
Harrimanella (harrimanella)
Kalmia (laurel)
Kalmiopsis (kalmiopsis)
Ledum (Labrador tea)
Leiophyllum (leiophyllum)
Leucothoe (dog hobble)
Loiseleuria (loiseleuria)
Lyonia (stagger bush)
Menziesia (menziesia)
Ornithostaphylos (ornithostaphylos)
Oxydendrum (swamp cranberry)
Phyllodoce (mountain heath)
Pieris (fetter bush)
Rhododendron (rhododendron)
Symphysia (symphysia)
Vaccinium (blueberry)
Xylococcus (mission manzanita)
Zenobia (honeycup)

Adverse effects

The urine of patients taking bearberry can darken on standing.

Toxic reactions to bearberry have not been reported, except for an anomalous case of dyspnea, cyanosis, and skin reactions after the consumption of an aqueous decoction (1).

- A 56-year-old woman who had taken *uva ursi* for 3 years developed reduced visual acuity (2). She had a typical bull's-eye maculopathy bilaterally.

The authors suggested that the effect was due to impaired melanin synthesis.

Reports of carcinogenicity of hydroquinone after prolonged administration of high doses to rats or mice raise a question about the long-term safety of *A. uva-ursi* and other medicinal herbs that contain substantial amounts of arbutin (3).

Gaultheria procumbens

The volatile oil of *Gaultheria procumbens* (wintergreen) leaves consists largely of methyl salicylate (more than 95%).

Adverse effects

Salicylism has been attributed to the use of oil of wintergreen as part of an herbal skin cream for the treatment of psoriasis (4).

- A 40-year-old man became acutely unwell after receiving oil of wintergreen from an unregistered naturopath. Transcutaneous absorption of the methyl salicylate was enhanced in this case owing to the abnormal areas of skin that were covered and the use of an occlusive dressing. The patient developed tinnitus, vomiting, tachypnea, and the typical acid/base disturbance of salicylate toxicity. Decontamination of the skin followed by rehydration and establishment of good urine flow was successful.

Laryngeal edema has been attributed to oil of wintergreen (5).

Methyl salicylate in topical analgesic preparations can cause irritant or allergic contact dermatitis and anaphylactic reactions (6).

In a retrospective study 80 subjects who had taken aspirin tablets ($n = 42$) or topical oil of wintergreen ($n = 38$) were compared (7). The admission plasma salicylate concentrations were generally higher in those who had taken aspirin tablets, but the two highest readings (4.3 and 3.5 mmol/l) belonged to two of the subjects who had taken oil of wintergreen.

Oil of wintergreen in the form of candy flavoring was ingested by a 21-month-old boy who developed vomiting, lethargy, and hyperpnea but recovered rapidly with parenteral fluids and sodium bicarbonate (8).

Methylsalicylate is also an important constituent of the Red Flower Oil formulations that are popular herbal analgesics for topical application in Southeast Asia. Some users take small amounts of the oil orally to enhance its analgesic effects. There are many different brands, which provide variable amounts of declared or undeclared methylsalicylate (up to 0.78 g/ml of oil). A suicide attempt by deliberate ingestion of about 100 ml resulted in severe salicylate poisoning (9).

Ledum palustre

The essential oil of *Ledum palustre* (marsh Labrador tea), which contains flavones, monoterpenoids, and sesquiterpenoids, is a potent irritant of the gastrointestinal tract, kidneys, and urinary tract; other toxic effects include abortion.

Vaccinium macrocarpon

Vaccinium macrocarpon (cranberry, marsh apple) has been used to prevent and treat urinary tract infections, although it is not useful in established infections (10,11). It is supposed to act by preventing adhesion of bacteria to the bladder wall. It may also reduce the risk of formation of some types of urinary stone (12,13).

Adverse effects
Hematologic

Cranberry juice may contain small amounts of quinine, which can cause immune thrombocytopenia.

- After transurethral resection of his prostate, a 68-year-old man developed immune thrombocytopenic purpura (platelet count $1 \times 10^9/l$) (14). He had self-medicated with cranberry juice for 10 days before the operation and had also taken amlodipine and aspirin. He had oral petechiae, bleeding gums, hematuria, and bruises. He recovered within 3 days of being given human immunoglobulin and oral prednisolone, and 18 months later his platelet count was still normal.

Drug interactions

Cranberry juice has been reported to enhance the action of warfarin (15)

- After a chest infection a man in his 70s had a poor appetite for 2 weeks and ate next to nothing, taking only cranberry juice as well as his regular drugs (digoxin, phenytoin, and warfarin) (16). Six weeks later his international normalized ratio was over 50, having previously been stable. He died of gastrointestinal and pericardial haemorrhage. He had not taken any over the counter preparations or herbal medicines, and he had been taking his drugs correctly.

The Committee on Safety of Medicines has received several other reports through the yellow card-reporting scheme about a possible interaction between warfarin and cranberry juice. In one case this effect was suggested to have been due to contamination with salicylic acid, which displaces warfarin from protein binding sites (17).

References

1. Meijers FS. Idio-synkrasie voor folia uvae ursi. Ned Tijdschr Geneeskd 1902;46:1226–8.
2. Wang L, Del Priore LV. Bull's-eye maculopathy secondary to herbal toxicity from uva ursi. Am J Ophthalmol 2004;137(6):1135–7.
3. De Smet PA. Health risks of herbal remedies. Drug Saf 1995;13(2):81–93.
4. Bell AJ, Duggin G. Acute methyl salicylate toxicity complicating herbal skin treatment for psoriasis. Emerg Med (Fremantle) 2002;14(2):188–90.
5. Botma M, Colquhoun-Flannery W, Leighton S. Laryngeal oedema caused by accidental ingestion of oil of wintergreen. Int J Pediatr Otorhinolaryngol 2001;58(3):229–32.
6. Chan TY. Potential dangers from topical preparations containing methyl salicylate. Hum Exp Toxicol 1996;15(9):747–50.
7. Chan TY. The risk of severe salicylate poisoning following the ingestion of topical medicaments or aspirin. Postgrad Med J 1996;72(844):109–12.
8. Howrie DL, Moriarty R, Breit R. Candy flavoring as a source of salicylate poisoning. Pediatrics 1985;75(5):869–71.
9. Chan TH, Wong KC, Chan JC. Severe salicylate poisoning associated with the intake of Chinese medicinal oil ("red flower oil"). Aust NZ J Med 1995;25(1):57.
10. Raz R, Chazan B, Dan M. Cranberry juice and urinary tract infection. Clin Infect Dis 2004;38(10):1413–9.
11. Jepson RG, Mihaljevic L, Craig J. Cranberries for preventing urinary tract infections. Cochrane Database Syst Rev 2004;(2):CD001321.
12. Kessler T, Jansen B, Hesse A. Effect of blackcurrant-, cranberry- and plum juice consumption on risk factors associated with kidney stone formation. Eur J Clin Nutr 2002;56(10):1020–3.
13. McHarg T, Rodgers A, Charlton K. Influence of cranberry juice on the urinary risk factors for calcium oxalate kidney stone formation. BJU Int 2003;92(7):765–8.
14. Davies JK, Ahktar N, Ranasinge E. A juicy problem. Lancet 2001;358(9299):2126.
15. Grant P. Warfarin and cranberry juice: an interaction? J Heart Valve Dis 2004;13(1):25–6.
16. Suvarna R, Pirmohamed M, Henderson L. Possible interaction between warfarin and cranberry juice. BMJ 2003;327(7429):1454.
17. Isele H. Todliche Blutung unter Warfarin plus Preiselbeersaft. Liegt's an der Salizylsäaure. [Fatal bleeding under warfarin plus cranberry juice. Is it due to salicylic acid?] MMW Fortschr Med 2004;146(11):13.

Euphorbiaceae

General Information

The genera in the family of Euphorbiaceae (Table 29) include various spurges, such as poinsettia and croton.

Breynia officinalis

Breynia officinalis (Chi R Yun) contains the saponin breynin, and terpenic and phenolic glycosides (1). Its Chinese name, Chi R Yun, means dizziness or vertigo for 7 days. It has been used to treat venereal diseases, contusions, heart failure, growth retardation, and conjunctivitis in combination with other traditional Chinese medicines.

Adverse effects

Taiwanese doctors have reported two cases of hepatocellular injury after oral administration of infusions of *B. officinalis* (2).

- One of the two patients had taken 1500 mg of the lower stem and root of *B. officinalis* boiled in water in a suicide attempt. She was admitted with vomiting and headache and later developed gastritis, hematuria, and liver damage. Symptomatic treatment resulted in full recovery.
- A 51-year-old woman consumed 20 pieces of the lower stem and root of *B. officinalis* to treat dermatitis. She developed nausea, vomiting, and dizziness, and had raised liver enzymes. Her liver function recovered 1 month after withdrawal of *B. officinalis*.

When *B. officinalis* was mistaken for a similar plant, *Securinega suffruticosa*, and was cooked in a soup used for muscle aches, lumbago, or as a tonic by 19 patients, 14 developed diarrhea, 10 had nausea and felt cold, nine had sensations of abdominal fullness, and seven vomited (3). Liver enzymes rose and the median times to median peak activities were 3 days for alanine transaminase, 2 days for aspartate transaminase, 5 days for alkaline phosphatase, and 12 days for gamma glutamyltranspeptidase. The liver damage was hepatocellular liver injury rather than cholestatic and marked jaundice did not develop.

Table 29 The genera of Euphorbiaceae

Acalypha (copperleaf)
Adelia (wild lime)
Alchornea (alchornea)
Alchorneopsis (alchorneopsis)
Aleurites (aleurites)
Antidesma (china laurel)
Argythamnia (silverbush)
Bernardia (myrtle croton)
Bischofia (bishopwood)
Breynia (breynia)
Caperonia (false croton)
Chamaesyce (sandmat)
Chrozophora (chrozophora)
Claoxylon (claoxylon)
Cnidoscolusohl (cnidoscolus)
Codiaeum (codiaeum)
Croton (croton)
Dalechampia (dalechampia)
Ditta (ditta)
Drypetes (drypetes)
Euphorbia (spurge)
Flueggea (bushweed)
Gymnanthes (gymnanthes)
Hippomane (hippomane)
Hura (sandbox tree)
Hyeronima (hyeronima)
Jatropha (nettle spurge)
Leptopus (maidenbush)
Macaranga (macaranga)
Mallotus (mallotus)
Manihot (manihot)
Margaritaria (margaritaria)
Mercurialis (mercurialis)
Pedilanthus (pedilanthus)
Pera (pera)
Phyllanthus (leaf flower)
Reverchonia (reverchonia)
Ricinus (castor)
Sapium (milk tree)
Savia (savia)
Sebastiania (Sebastian bush)
Stillingia (toothleaf)
Tetracoccus (shrubby spurge)
Tragia (noseburn)
Triadica (tallow tree)
Vernicia (vernicia)

Croton tiglium

The oil of *Croton tiglium* (croton) has a violent purgative action and contains tumor-promoting phorbol diesters and triesters.

Adverse effects

Croton oil has been used as a peeling agent in phenol as a carrier; its use in concentrations of 2% and over is almost always associated with depigmentation of the skin and delays in healing in areas other than the thick skin of the lower nose and around the mouth (4).

Croton is also the name given to *Codiaeum variegatum*, a highly decorative potted plant. Handling this plant over a period of 6 months produced contact eczema of the hands in a nursery gardener. Patch tests with croton leaves were positive (5).

Ricinus communis

Castor oil is the fixed oil obtained from the seeds of *Ricinus communis* (castor) by cold expression. The whole beans contain a variety of substances that are not expressed with the oil, including the toxin ricin (6,7).

Adverse effects

When taken by mouth, especially in large doses, castor oil can cause violent purgation with nausea, vomiting, colic, and a risk of miscarriage.

Transdermal exposure to ricin is not serious, since it is not well absorbed through the skin. Oral exposure, for example by ingestion of castor beans, can cause severe gastroenteritis, gastrointestinal hemorrhage, and death due to circulatory collapse. Parenteral injection of ricin is rapidly fatal, as is aerosol exposure; the lethal dose by these routes is 5–10 micrograms/kg (8).

Immunologic

Anaphylaxis has been attributed to consumption of whole castor beans (9).

- A 44-year-old woman chewed a castor bean seed and within minutes developed urticaria, drowsiness, Quincke's edema, and extreme hypotension. Her anaphylactic shock was treated with adrenaline, intravenous glucocorticoids, antihistamines, and intravenous fluids. She quickly recovered and a subsequent blood test demonstrated CAP-RAST to castor beans.

Fetotoxicity

In 498 women who had recently taken castor oil and possibly herbal substances called "sihlambezo" before artificial rupture of membranes, fetal passage of meconium was more common (10). However, in a study of 100 women castor oil had no effect on the rate of meconium-stained liquor (11).

References

1. Morikawa H, Kasai R, Otsuka H, Hirata E, Shinzato T, Aramoto M, Takeda Y. Terpenic and phenolic glycosides from leaves of *Breynia officinalis* HEMSL. Chem Pharm Bull (Tokyo) 2004;52(9):1086–90.

2. Lin TJ, Tsai MS, Chiou NM, Deng JF, Chiu NY. Hepatotoxicity caused by *Breynia officinalis*. Vet Hum Toxicol 2002;44(2):87–8.

3. Lin TJ, Su CC, Lan CK, Jiang DD, Tsai JL, Tsai MS. Acute poisonings with *Breynia officinalis*—an outbreak of hepatotoxicity. J Toxicol Clin Toxicol 2003;41(5):591–4.

4. Hetter GP. An examination of the phenol-croton oil peel: part IV. Face peel results with different concentrations of phenol and croton oil. Plast Reconstr Surg 2000; 105(3):1061–83.

5. Hausen BM, Schulz KH. Occupational contact dermatitis due to croton (*Codiaeum variegatum* (L.) *A. Juss* var. *pictum* (Lodd.) Muell. Arg.). Sensitization by plants of the Euphorbiaceae Contact Dermatitis 1977;3(6):289–92.

6. Marks JD. Medical aspects of biologic toxins. Anesthesiol Clin North America 2004;22(3):509–32.

7. Doan LG. Ricin: mechanism of toxicity, clinical manifestations, and vaccine development. A review. J Toxicol Clin Toxicol 2004;42(2):201–8.

8. Bradberry SM, Dickers KJ, Rice P, Griffiths GD, Vale JA. Ricin poisoning. Toxicol Rev 2003;22(1):65–70.

9. Navarro-Rouimi R, Charpin D. Anaphylactic reaction to castor bean seeds. Allergy 1999;54(10):1117.

10. Mitri F, Hofmeyr GJ, van Gelderen CJ. Meconium during labour—self-medication and other associations. S Afr Med J 1987;71(7):431–3.

11. Kelly AJ, Kavanagh J, Thomas J. Castor oil, bath and/or enema for cervical priming and induction of labour. Cochrane Database Syst Rev 2001;(2):CD003099.

Fabaceae

General Information

The genera in the family of Fabaceae (Table 30; formerly Leguminosae) include broom, liquorice, senna, tamarind, and a variety of pulses, such as fava beans, lentil, peas, and vetches.

Table 30 The genera of Fabaceae

Abrus (abrus)
Acacia (acacia)
Adenanthera (bead tree)
Aeschynomene (joint vetch)
Afzelia (mahogany)
Albizia (albizia)
Alhagi (alhagi)
Alysicarpus (moneywort)
Amorpha (false indigo)
Amphicarpaea (hog peanut)
Anadenanthera (anadenanthera)
Andira (andira)
Anthyllis (kidney vetch)
Apios (groundnut)
Arachis (peanut)
Aspalathus (aspalathus)
Astragalus (milk vetch)
Baphia (baphia)
Baptisia (wild indigo)
Barbieria (barbieria)
Bauhinia (bauhinia)
Bituminaria (bituminaria)
Brongniartia (green twig)
Brya (coccuswood)
Butea (butea)
Caesalpinia (nicker)
Cajanus (cajanus)
Calliandra (stick pea)
Calopogonium (calopogonium)
Canavalia (jackbean)
Caragana (peashrub)
Carmichaelia (carmichaelia)
Cassia (cassia)
Centrosema (butterfly pea)
Ceratonia (ceratonia)
Cercis (redbud)
Chamaecystis (chamaecystis)
Chamaecrista (sensitive pea)
Chapmannia (chapmannia)
Christia (island pea)
Cicer (cicer)
Cladrastis (yellowwood)
Clianthus (glory pea)
Clitoria (pigeon wings)
Codariocalyx. (tick trefoil)
Cojoba (cojoba)
Cologania (cologania)
Colutea (colutea)
Copaifera (copaifera)
Coronilla (crown vetch)
Coursetia (baby bonnets)
Crotalaria (rattlebox)

Crudia (bedstraw)
Cullen (scurf pea)
Cyamopsis (cyamopsis)
Cynometra (cynometra)
Cytisus (broom)
Dalbergia (Indian rosewood)
Dalea (prairie clover)
Daniellia (daniellia)
Delonix (delonix)
Derris (derris)
Desmanthus (bundle flower)
Desmodium (tick trefoil)
Dialium (dialium)
Dichrostachys (dichrostachys)
Dioclea (dioclea)
Diphysa (diphysa)
Dipogon (dipogon)
Dipteryx (dipteryx)
Ebenopsis (Texas ebony)
Entada (callingcard vine)
Enterolobium (enterolobium)
Eriosema (sand pea)
Errazurizia (dunebroom)
Erythrina (erythrina)
Erythrophleum (sasswood)
Eysenhardtia (kidneywood)
Faidherbia (acacia)
Falcataria (peacock's plume)
Flemingia (flemingia)
Galactia (milk pea)
Galega (professor weed)
Genista (broom)
Genistidium (brush pea)
Gleditsia (locust)
Gliricidia (quickstick)
Glottidium (glottidium)
Glycine (soybean)
Glycyrrhiza (licorice)
Gymnocladus (coffee tree)
Haematoxylum (haematoxylum)
Halimodendron (halimodendron)
Havardia (havardia)
Hedysarum (sweet vetch)
Hippocrepis (hippocrepis)
Hoffmannseggia (rush pea)
Hoita (leather root)
Hymenaea (hymenaea)
Indigofera (indigo)
Inga (inga)
Inocarpus (chestnut)
Kanaloa (kanaloa)
Kummerowia (kummerowia)
Lablab (lablab)
Laburnum (golden chain tree)
Lathyrus (pea)
Lens (lentil)
Lespedeza (lespedeza)
Leucaena (lead tree)
Lonchocarpus (lance pod)
Lotononis (lotononis)
Lotus (trefoil)
Lupinus (lupin)
Lysiloma (falsetamarind)
Maackia (maackia)
Machaerium (machaerium)

(Continued)

(Continued)

Table 30 (Continued)

Macroptilium (bush bean)
Macrotyloma (macrotyloma)
Marina (false prairie clover)
Medicago (alfalfa)
Melilotus (sweet clover)
Mimosa (sensitive plant)
Mucuna (mucuna)
Myrospermum (myrospermum)
Myroxylon (myroxylon)
Neonotonia (neonotonia)
Neorudolphia (neorudolphia)
Neptunia (puff)
Nissolia (yellowhood)
Olneya (olneya)
Onobrychis (sainfoin)
Ononis (restharrow)
Orbexilum (leather root)
Ormosia (ormosia)
Ornithopus (bird's foot)
Oxyrhynchus (oxyrhynchus)
Oxytropis (locoweed)
Pachyrhizus (pachyrhizus)
Paraserianthes (paraserianthes)
Parkinsonia (paloverde)
Parkia (parkia)
Parryella (parryella)
Pediomelum (Indian breadroot)
Peltophorum (peltophorum)
Pentaclethra (pentaclethra)
Pericopsis (peperomia)
Peteria (peteria)
Phaseolus (bean)
Physostigma (physostigma)
Pickeringia (chaparral pea)
Pictetia (pictetia)
Piscidia (piscidia)
Pisum (pea)
Pithecellobium (blackbead)
Poitea (wattapama)
Prosopis (mesquite)
Psophocarpus (psophocarpus)
Psoralidium (scurf pea)
Psorothamnus (dalea)
Pterocarpus (pterocarpus)
Pueraria (kudzu)
Retama (bridal broom)
Rhynchosia (snoutbean)
Robinia (locust)
Rupertia (rupertia)
Samanea (raintree)
Schizolobium (Brazilian firetree)
Schleinitzia (strand tangantangan)
Scorpiurus (scorpion's tail)
Senna (senna)
Serianthes (vaivai)
Sesbania (riverhemp)
Sophora (necklace pod)
Spartium (broom)
Sphaerophysa (sphaerophysa)
Sphenostylis (sphenostylis)
Sphinctospermum (sphinctospermum)
Stahlia (stahlia)
Strongylodon (strongylodon)

(Continued)

Strophostyles (fuzzy bean)
Stryphnodendron (stryphnodendron)
Stylosanthes (pencil flower)
Sutherlandia (sutherlandia)
Tamarindus (tamarind)
Taralea (taralea)
Tephrosia (hoarypea)
Teramnus (teramnus)
Tetragonolobus (tetragonolobus)
Thermopsis (golden banner)
Ticanto (gray nicker)
Trifolium (clover)
Trigonella (fenugreek)
Ulex (gorse)
Vicia (vetch)
Vigna (cowpea)
Virgilia (virgilia)
Wisteria (wisteria)
Zapoteca (white stick pea)
Zornia (zornia)

Cassia species

The leaves and fruits of *Cassia angustifolia* and *Cassia senna* (senna) contain laxative anthranoid derivatives. Mutagenicity testing of sennosides has produced negative results in several bacterial and mammalian systems, except for a weak effect in *Salmonella typhimurium* strain TA102 (1,2). No evidence of reproductive toxicity of sennosides has been found in rats and rabbits (3).

Senna is widely used in fairly low doses without serious problems; it has also been used in a very high dosage form to clear the colon before radiological examination. In this form it is generally well tolerated, but it should not be used if there is any predisposition to colonic rupture.

Adverse effects

The safety and efficacy of senna have been reviewed (4). Its rhein-anthrone-induced laxative effects occur through two distinct mechanisms, an increase in intestinal fluid transport, which causes accumulation of fluid intraluminally, and an increase in intestinal motility. Senna can cause mild abdominal complaints, such as cramps or pain. Other adverse effects are discoloration of the urine and hemorrhoidal congestion. Prolonged use and overdose can result in diarrhea, extreme loss of electrolytes, especially potassium, damage to the surface epithelium, and impairment of bowel function by damage to autonomic nerves. Abuse of senna has also been associated with melanosis coli, but resolution occurs 8–11 months after withdrawal. Tolerance and genotoxicity do not seem to be problems associated with senna, especially when used periodically in therapeutic doses.

Liver
Abuse of senna can cause hepatitis (5,6).

Skin
Occupational allergic contact dermatitis has been attributed to a species of *Cassia* (7,8), as has contact urticaria (9).

Musculoskeletal

Finger clubbing (hypertrophic osteoarthropathy) has been reported in patients who have abused senna (10–15).

Lactation

When a standardized preparation containing senna pods (providing 15 mg of sennosides per day) was given to breastfeeding mothers, the suckling infants were only exposed to a non-laxative amount of rhein, which remained a factor of 10^{-3} below the maternal intake of this active metabolite (16).

Tumorigenicity

A well-defined purified senna extract was not carcinogenic, when administered orally to rats in daily doses up to 25 mg/kg for 2 years (17).

Drug interactions

Metronidazole

An unusual disulfiran (Antabuse) type reaction reported on one occasion seems to have been due to an interaction of metronidazole with the alcohol present in the high-dosage form X-Prep (SEDA-15, 398).

Arachis (peanut) oil

Immunologic

In 2003 all practising doctors in the UK were alerted by the Chief Medical Officer to the risks of topical medicines that contain arachis oil (18). Products are clearly labelled as containing this refined ingredient. The alert highlighted a study in children that suggested sensitization to peanuts may be caused by the application of creams containing arachis oil to inflamed skin (19). It also mentioned an earlier study that showed that small amounts of allergenic protein persist in peanut oil despite refinement (20).

Although the UK's Committee on Safety of Medicines has determined that there is insufficient evidence to conclude that exposure to topical medicines containing arachis oil leads to sensitization to peanut protein, it has issued a precautionary recommendation that patients with known peanut allergy should avoid such medicines, and the labelling of these products is to be updated with new warnings.

All this is highly relevant to practising otorhinolaryngologists, because a widely prescribed product, Naseptin nasal cream, contains peanut oil.

Crotalaria species

Most of the members of the *Crotalaria* (rattlebox) species contain pyrrolizidine alkaloids, such as crotaline and monocrotaline, and therefore have hepatotoxic potential (21). These alkaloids are covered in a separate monograph.

Adverse effects

Pyrrolizidine alkaloids cause obstruction of the hepatic venous system and can lead to hepatic necrosis. Clinical manifestations include abdominal pain, ascites, hepatomegaly, and raised serum transaminases. The prognosis is often poor, death rates of 20–30% being reported. In an outbreak of veno-occlusive disease in the Sarguja district of India, probably caused by consumption of cereals mixed with *Crotalaria* seeds, 28 of 67 patients died (22).

It is prudent to avoid exposing unborn or suckling children to herbal remedies containing pyrrolizidine alkaloids. Animal studies have shown that transplacental passage and transfer to breast milk are possible, and there is a human case on record of fatal neonatal liver injury, in which the mother had used a herbal cough tea containing pyrrolizidine alkaloids throughout her pregnancy.

Cyamopsis tetragonoloba

Guar gum, which is covered in a separate monograph, comes from the endosperm of the seeds of *Cyamopsis tetragonoloba* (cluster bean). It has a small hypolipidemic effect (23) and a small blood glucose-lowering effect (24).

Adverse effects

Guar gum causes abdominal pain, flatulence, diarrhea, and cramps (25). It has been repeatedly associated with esophageal obstruction (26).

Guar gum can cause occupational rhinitis and asthma. Of 162 employees at a carpet-manufacturing plant, in which guar gum is used to adhere the dye to the fiber, 37 (23%) had a history suggestive of occupational asthma and 59 (36%) of occupational rhinitis (27). Eight (5%) had immediate skin reactivity to guar gum. Eleven (8.3%) had serum IgE antibodies to guar gum.

Guar gum reduces the speed but not the extent of absorption of digoxin (28), nitrofurantoin (29), and paracetamol (30). It reduces both the speed and extent of absorption of phenoxymethylpenicillin (28).

Cytisus scoparius

Cytisus scoparius (Scotch broom) contains the toxic alkaloid sparteine and related quinolizidine alkaloids, such as isosparteine and cytisine. Sparteine has been used as a marker of drug metabolism by CYP2D6 (31) and is covered in a separate monograph.

Adverse effects

Inexpert self-medication with a broom tea has resulted in fatal poisoning with clinical symptoms of ileus, heart failure, and circulatory weakness.

It is prudent to avoid broom formulations during pregnancy, not only because sparteine has abortifacient potential but also because of evidence that the plant produced malformed lambs in feeding trials.

Dipteryx species

The seeds of *Dipteryx odorata* (Dutch tonka bean) and *Dipteryx oppositofolia* (English tonka bean) are said to yield 1–3% of the non-anticoagulant coumarin, which is covered in a separate monograph.

Genista tinctoria

Genista tinctoria (dyer's broom) contains 0.3–0.8% of toxic quinolizidine alkaloids, such as anagyrin, cytisine, and *N*-methylcytisine. The last two constituents have peripheral effects similar to those of nicotine, whereas their central activity may be different. Anagyrine is a suspected animal teratogen and cytisine has teratogenic activity in rabbits.

Glycyrrhiza glabra

Glycyrrhiza glabra (licorice) (32) contains a wide variety of chalcones, coumarins, flavonoids, and triterpenoids, some of which mimic the mineralocorticoid actions of aldosterone. The saponin glycoside glycyrrhizinic acid (glycyrrhetinic acid, carbenoxolone) (33) and deglycyrrhizinized licorice (Caved-S) (34) have both been used to treat gastric ulceration, but have been displaced by better drugs.

Licorice is used widely and is found in a variety of sources (35):

- *Confectionery*
  Licorice sticks, bricks, cakes, toffee, pipes, bars, balls, tubes, Catherine wheels, pastilles, and Licorice Allsorts
  Torpedos
  Blackcurrant
  Pomfret (Pontefract) cakes
  Servez vous
  Sorbits chewing gum
  Stimorol chewing gum
- *Health products*
  Liquirizia naturale
  Licorice-flavoured diet gum
  Throat pearls
  Licorice-flavored cough mixtures
  Herbal cough mixtures
  Antibron tablets
  Licorice tea
- *All types of licorice root*
  Afghan, Chinese, Iranian, Russian, Turkish, and of unknown origin
- *Chewing tobacco*
- *Alcoholic drinks*
  Belgian beers
  Pastis brands
  Anisettes (raki, ouzo, Pernod).

Adverse effects

Most individuals can consume 400 mg of glycyrrhizin daily without adverse effects, but some individuals develop adverse effects following regular daily intake of as little as 100 mg of glycyrrhizin (36).

Cardiovascular

Liquorice can cause hyperaldosteronism and hence hypertension.

- A 52-year-old woman had high blood pressure that had previously been resistant to antihypertensive drugs (37). Her history was unremarkable and her blood pressure readings varied between 140/70 and 200/80 mmHg

without apparent reason. On detailed questioning she admitted to eating two bars of an unnamed liquorice sweet daily. Discontinuation of this habit normalized her blood pressure within 1 month without the need for any other medical intervention.

In healthy volunteers who took licorice corresponding to glycyrrhizinic acid 75–540 mg/day for periods of 2–4 weeks, there was an average increase in systolic blood pressure of 3.1–14.4 mmHg (38). The increase in blood pressure was dose-related and the authors concluded that as little as 50 g/day of licorice for 2 weeks would have caused a significant rise in blood pressure.

Sensory systems

Five patients who had consumed large amounts (0.1–1 kg) of licorice subsequently had transient visual loss/aberrations (39). Glycyrrhizinic acid in licorice causes vasoconstriction in vascular smooth muscle and the authors therefore speculated that vasospasm of the retinal or occipital artery had caused the problems.

Electrolyte balance

The active ingredients of licorice inhibit the breakdown of mineralocorticoids by inhibiting 11-beta-hydroxysteroid dehydrogenase type 2, and its adverse effects relate mainly to mineralocorticoid excess, with sodium retention, potassium loss, and inhibition of the renin–angiotensin–aldosterone system (40).

In two cases prolonged intake of relatively small amounts of licorice resulted in hypertension, encephalopathy, and pseudohyperaldosteronism (41). Both patients were highly susceptible to the adverse effects of glycyrrhizinic acid because of 11-beta-hydroxysteroid dehydrogenase deficiency.

- A 44-year-old woman developed an irregular heart rhythm and repeated episodes of life-threatening torsade de pointes (42). Her serum potassium was 2.3 mmol/l. Treatment with an infusion of potassium and magnesium promptly restored normal rhythm. The cause of the problem was identified as the patient's habit of consuming large (but not more closely defined) quantities of licorice daily. One year after this episode she was still abstaining from licorice and showed no signs of cardiac disease.
- A 39-year-old woman had a potassium concentration of 2.9 mmol/l at a routine checkup (43). She denied taking any medications except a "cleansing tea" purchased from a health food company. The tea was analysed and found to contain significant amounts of licorice. Her potassium concentration normalized after withdrawal of the product and potassium supplementation.
- A 67-year-old Chinese man developed progressive muscular weakness (44). His medical history was unremarkable, but his urinary potassium excretion was high and he had hypokalemia and low plasma renin activity. He admitted taking a powdered Chinese herbal formulation for about 4 months, which was shown to contain large amounts of glycyrrhizinic acid (336 mg/day). He

was treated with spironolactone, and 2 weeks later his potassium values had normalized.

Three patients who used licorice-flavored confectionery developed severe hypokalemia (35).

- A 21-year-old woman who took licorice about 100 g/day developed a headache, hypertension (190/120 mmHg), and hypokalemia (2.6 mmol/l). She stopped eating licorice and instead used a chewing gum that also contained licorice (daily intake of glycyrrhizinic acid about 120 mg). Three weeks after she stopped using the gum her blood pressure was 110/80 mmHg and plasma potassium concentration 5.3 mmol/l.
- A 35-year-old woman developed pretibial edema and hypokalemia (2.2 mmol/1) after using a licorice-flavored chewing gum (about 50 mg of glycyrrhizinic acid per day). She stopped using the gum and improved within 2 weeks.
- A 56-year-old woman developed severe hypokalemic myopathy and pseudoaldosteronism (45). Her hypokalemia was corrected with an infusion of potassium chloride and all her symptoms resolved swiftly. The cause of the problem remained elusive until it was noted that she had eaten large amounts of Pontefract cakes, which contained 15 g of pure licorice, per day to treat chronic constipation.

In six male volunteers who took glycyrrhizinic acid for 7 days, serum, urinary, and sweat electrolytes values were consistent with a mineralocorticoid-like effect (46). Plasma renin activity was suppressed, and plasma cortisol and aldosterone progressively fell during treatment. The authors proposed that glycyrrhizinic acid initially acts by increasing the effect of endogenous mineralocorticoids, and then when it or its metabolites accumulate it may also have a direct mineralocorticoid-like effect itself.

Endocrine

Of 34 Japanese patients with diabetes and chronic hepatitis, 18 were given glycyrrhizin 240-525 mg for over 1 year (47). This resulted in a significant lowering of total testosterone concentrations and increased arteriosclerotic plaque formation. The authors suggested that glycyrrhizin treatment was an independent risk factor for arteriosclerosis. The testosterone lowering effect of liquorice has been confirmed in another trial (48).

Drug-drug interactions

The potential of liquorice to interact with prescribed drugs has been reviewed (49). The list of implicated co-medications includes prednisolone, aspirin, antibiotics, diuretics, cardiac glycosides, NSAIDs, oral contraceptives, antidiabetic drugs, antithrombotic drugs, and antidepressants.

Lupinus species

Lupinus (lupin) seeds are commonly taken as an appetizer in Southern Europe and the Middle East. Lupin flour has been used as a source of energy and protein (50,51).

An extract of *Lupus termis* has been used to treat chronic eczema (52).

Adverse effects

Anticholinergic effects of lupin poisoning have been reported (53).

- A 72-year-old Portuguese woman presented to an emergency department with classic anticholinergic signs: sudden onset of nausea and vomiting, blurred vision, generalized weakness, and tachycardia (54). She had taken a herbal product containing lupin seeds in the belief that it would cure her recently diagnosed diabetes mellitus.

Analysis of the product identified the preponderant compound as oxosparteine, which has powerful anticholinergic effects.

Medicago species

Medicago sativa (alfalfa) sprouts are commonly used as a source of protein.

Adverse effects

Skin

Dermatitis has been recorded after the ingestion of infusions made from alfalfa seeds. Dermatitis has also been reported in horse handlers exposed to the straw itch mite, *Pyemotes tritici*, contaminating alfalfa seeds (55).

Immunologic

Prolonged ingestion of alfalfa seeds or alfalfa tablets has been associated with the induction or exacerbation of a lupus-like syndrome in humans, perhaps because of the canavanine alfalfa contains (56,57).

Infection risk

Alfalfa seeds are readily infected and there have been numerous reports of infections arising from the consumption of alfalfa sprouts grown from contaminated seeds. The most common infections are with *Salmonella* species, including *Salmonella enterica* (58), *Salmonella havana* (59), *Salmonella kottbus* (60,61), *Salmonella mbandaka* (62), *Salmonella muenchen* (63), *Salmonella paratyphi* (64), *Salmonella stanley* (65), and other serotypes (66,67). Occasionally, other organisms occur, such as *Listeria* (68) and *Escherichia coli* (69,70).

Melilotus officinalis

Melilotus officinalis (sweet clover) contains coumarin, 3,4-dihydrocoumarin (melilotine), ortho-coumaric acid, ortho-hydroxycoumaric acid, and the ortho-glucoside of ortho-coumaric acid (melilotoside). Withering of the plant leads to enzymatic glycoside hydrolysis and the resulting ortho-coumaric acid is spontaneously transformed to coumarin; the dried herb therefore smells strongly of coumarin (see separate monograph). *M. officinalis* has been used to treat lymphedema and the edema of chronic venous insufficiency.

Myroxylon species

Balsam of Peru is derived from species of *Myroxylon* such as *Myroxylon balsamum* and *Myroxylon pereirae* and has been used in medicinal and cosmetic ointments for centuries. It contains about 25 different substances, including triterpenoids, and can cause allergic reactions. Its sensitizing constituents have been determined (71).

Adverse effects

Balsam of Peru is a topical photosensitizer (SEDA-19, 162) and can cause contact urticaria (72,73). A systemic contact dermatitis has been reported after oral administration (74). In one case an allergic contact dermatitis in a patient sensitive to balsam of Peru caused a primary eruption on the face with secondary purpuric vasculitis-like eruptions on both legs (75).

In 60 patients with positive patch-test reactions to a fragrance mix or *M. pereirae* resin (balsam of Peru) there were positive immediate contact reactions to *M. pereirae* resin in 57% and to the fragrance mix in 12% (76). In a control group ($n = 50$) of eczematous, patch test-negative patients there were positive immediate reactions to *M. pereirae* resin in 58% and to the fragrance mix in 12%. The authors commented that the absence of a significant difference between the fragrance-allergic group and the control group was in keeping with a non-immunological basis for the immediate contact reactions.

Pithecollobium jiringa

Pithecollobium jiringa (jering fruit) is valued in Malaysia and Indonesia as a delicacy and for its antidiabetic properties.

Adverse effects

Acute renal insufficiency is a rare complication of the use of jering. In one case there was dysuria, hematuria, vomiting, abdominal pain, and blue urine (77). In two other cases there was bilateral loin pain, fever, nausea, vomiting, oliguria, hematuria, and passage of sandy particles in the urine (78). Blood urea and serum creatinine were markedly raised. With conservative therapy, which included rehydration with isotonic saline and alkalinization of the urine with sodium bicarbonate, the acute renal insufficiency resolved.

Sophora falvescens

Sophora falvescens (Ku shen) contains a variety of matrine alkaloids, such as aloperine, cytosine, lehmannine, matrine, oxymartine, oxysophocarpine, sophocarpine, sophoramine, and sophoridine, the flavonoid kushenol, and the saponin sophoraflavoside. Ku shen is a Chinese herbal remedy made from the root of *S. falvescens*. And Qing luo yin, a Chinese herbal remedy for rheumatoid arthritis, is a combination of extracts of *Dioscorea hypoglauca*, *Phellodendron amurense*, *Sinomenium acutum*, and *S. flavescens*.

Adverse effects

A herbal record lists the following adverse effects of Ku shen: salivation, abnormal gait, dyspnea, tachycardia (79). In larger doses, nervous system stimulation with muscle spasm and seizures can occur. There have been three reports of adverse reactions such as nausea, vomiting, dizziness, bradycardia, palpitation, ataxia, pallor, sweating, seizures, and dysphasia.

Trifolium pratense

Trifolium pratense (red clover) contains flavones (isorhamnetin, pratensein), phenolic coumarins, phytoestrogens, and demethylpterocarpin. It has been used to treat menopausal symptoms, without good evidence of efficacy (80).

Adverse effects

The coumarins in red clover may potentiate the effects of oral anticoagulants (81).

References

1. Mengs U. Toxic effects of sennosides in laboratory animals and in vitro. Pharmacology 1988;36(Suppl 1):180–7.
2. Sandnes D, Johansen T, Teien G, Ulsaker G. Mutagenicity of crude senna and senna glycosides in *Salmonella typhimurium*. Pharmacol Toxicol 1992;71(3 Pt 1):165–72.
3. Mengs U. Reproductive toxicological investigations with sennosides. Arzneimittelforschung 1986;36(9):1355–8.
4. Mascolo N, Capasso R, Capasso F. Senna. A safe and effective drug. Phytother Res 1998;12(Suppl 1):S143–5.
5. Beuers U, Spengler U, Pape GR. Hepatitis after chronic abuse of senna. Lancet 1991;337(8737):372–3.
6. Seybold U, Landauer N, Hillebrand S, Goebel FD. Senna-induced hepatitis in a poor metabolizer. Ann Intern Med 2004;141(8):650–1.
7. De Benito V, Alzaga R. Occupational allergic contact dermatitis from *Cassia* (Chinese cinnamon) as a flavouring agent in coffee. Contact Dermatitis 1999;40(3):165.
8. Rudzki E, Grzywa Z. Immediate reactions to balsam of Peru, cassia oil and ethyl vanillin. Contact Dermatitis 1976;2(6):360–1.
9. Rietschel RL. Contact urticaria from synthetic cassia oil and sorbic acid limited to the face. Contact Dermatitis 1978;4(6):347–9.
10. Silk DB, Gibson JA, Murray CR. Reversible finger clubbing in a case of purgative abuse. Gastroenterology 1975;68(4 Pt 1):790–4.
11. Prior J, White I. Tetany and clubbing in patient who ingested large quantities of senna. Lancet 1978;2(8096):947.
12. Malmquist J, Ericsson B, Hulten-Nosslin MB, Jeppsson JO, Ljungberg O. Finger clubbing and aspartylglucosamine excretion in a laxative-abusing patient. Postgrad Med J 1980;56(662):862–4.
13. Levine D, Goode AW, Wingate DL. Purgative abuse associated with reversible cachexia, hypogammaglobulinaemia, and finger clubbing. Lancet 1981;1(8226):919–20.
14. Armstrong RD, Crisp AJ, Grahame R, Woolf DL. Hypertrophic osteoarthropathy and purgative abuse. BMJ (Clin Res Ed) 1981;282(6279):1836.
15. Fichter M, Chlond C. Hypertrophe Osteoarthropathie bei *Bulimia nervosa* mit chronischer Intoxikation mit Laxantien.

[Hypertrophic osteoarthropathy in *Bulimia nervosa* with chronic poisoning by laxatives.] Nervenarzt 1988;59(4):244–7.

16. Faber P, Strenge-Hesse A. Relevance of rhein excretion into breast milk. Pharmacology 1988;36(Suppl 1):212–20.

17. Lyden-Sokolowski A, Nilsson A, Sjoberg P. Two-year carcinogenicity study with sennosides in the rat: emphasis on gastro-intestinal alterations. Pharmacology 1993;47 (Suppl 1):209–15.

18. Hannan SA. Naseptin nasal cream "contains peanut oil". J Laryngol Otol 2003;117:1009–10.

19. Lack G, Fox D, Northstone K, Golding J. Factors associated with the development of peanut allergy in childhood. New Engl J Med 2003;348:977–85.

20. Olszewski A, Pons L, Moutete F, AimoneGastin J, Kanny G, Moneret-Vautrin DA. Isolation and characterisation of protein allergens in refined peanut oil. Clin Exp Allergy 1998;28:850–9.

21. Larrey D. Accidents hépatiques de la phytothérapie. [Liver involvement in the course of phytotherapy.] Presse Méd 1994;23(15):691–3.

22. Tandon BN, Tandon HD, Tandon RK, Narndranathan M, Joshi YK. An epidemic of veno-occlusive disease of liver in central India. Lancet 1976;2(7980):271–2.

23. Brown L, Rosner B, Willett WW, Sacks FM. Cholesterol-lowering effects of dietary fiber: a meta-analysis. Am J Clin Nutr 1999;69(1):30–42.

24. Fuessl HS, Williams G, Adrian TE, Bloom SR. Guar sprinkled on food: effect on glycaemic control, plasma lipids and gut hormones in non-insulin dependent diabetic patients. Diabet Med 1987;4(5):463–8.

25. Pittler MH, Ernst E. Guar gum for body weight reduction: meta-analysis of randomized trials. Am J Med 2001;110(9): 724–30.

26. Lewis JH. Esophageal and small bowel obstruction from guar gum-containing "diet pills": analysis of 26 cases reported to the Food and Drug Administration. Am J Gastroenterol 1992;87(10):1424–8.

27. Malo JL, Cartier A, L'Archeveque J, Ghezzo H, Soucy F, Somers J, Dolovich J. Prevalence of occupational asthma and immunologic sensitization to guar gum among employees at a carpet-manufacturing plant. J Allergy Clin Immunol 1990;86(4 Pt 1):562–9.

28. Huupponen R, Seppala P, Iisalo E. Effect of guar gum, a fibre preparation, on digoxin and penicillin absorption in man. Eur J Clin Pharmacol 1984;26(2):279–81.

29. Soci MM, Parrott EL. Influence of viscosity on absorption from nitrofurantoin suspensions. J Pharm Sci 1980;69(4):403–6.

30. Holt S, Heading RC, Carter DC, Prescott LF, Tothill P. Effect of gel fibre on gastric emptying and absorption of glucose and paracetamol. Lancet 1979;1(8117):636–9.

31. Meyer UA, Skoda RC, Zanger UM. The genetic polymorphism of debrisoquine/sparteine metabolism—molecular mechanisms. Pharmacol Ther 1990;46(2):297–308.

32. Shibata S. A drug over the millennia: pharmacognosy, chemistry, and pharmacology of licorice. Yakugaku Zasshi 2000;120(10):849–62.

33. Bianchi Porro G, Petrillo M, Lazzaroni M, Mazzacca G, Sabbatini F, Piai G, Dobrilla G, De Pretis G, Daniotti S. Comparison of pirenzepine and carbenoxolone in the treatment of chronic gastric ulcer. A double-blind endoscopic trial. Hepatogastroenterology 1985;32(6):293–5.

34. Morgan AG, McAdam WA, Pacsoo C, Darnborough A. Comparison between cimetidine and Caved-S in the treatment of gastric ulceration, and subsequent maintenance therapy. Gut 1982;23(6):545–51.

35. de Klerk GJ, Nieuwenhuis MG, Beutler JJ. Hypokalaemia and hypertension associated with use of liquorice flavoured chewing gum. BMJ 1997;314(7082):731–2.

36. Stormer FC, Reistad R, Alexander J. Glycyrrhizic acid in liquorice—evaluation of health hazard. Food Chem Toxicol 1993;31(4):303–12.

37. Lindley G. Was it something you ate? BMJ 2003;326:87.

38. Sigurjonsdottir HA, Franzson L, Manhem K, Ragnarsson J, Sigurdsson G, Wallerstedt S. Liquorice-induced rise in blood pressure: a linear dose-response relationship. J Hum Hypertens 2001;15(8):549–52.

39. Dobbins KR, Saul RF. Transient visual loss after licorice ingestion. J Neuroophthalmol 2000;20(1):38–41.

40. Olukoga A, Donaldson D. Liquorice and its health implications. J R Soc Health 2000;120(2):83–9.

41. Russo S, Mastropasqua M, Mosetti MA, Persegani C, Paggi A. Low doses of liquorice can induce hypertension encephalopathy. Am J Nephrol 2000;20(2):145–8.

42. Eriksson JW, Carlberg B, Hillorn V. Life-threatening ventricular tachycardia due to liquorice-induced hypokalaemia. J Intern Med 1999;245(3):307–10.

43. Feingold RM. Should we fear "health foods"? Arch Intern Med 1999;159(13):1502.

44. Lin SH, Chau T. A puzzling cause of hypokalaemia. Lancet 2002;360(9328):224.

45. Hussain RM. The sweet cake that reaches parts other cakes can't! Postgrad Med J 2003;79:115–6.

46. Armanini D, Scali M, Zennaro MC, Karbowiak I, Wallace C, Lewicka S, Vecsei P, Mantero F. The pathogenesis of pseudohyperaldosteronism from carbenoxolone. J Endocrinol Invest 1989;12(5):337–41.

47. Fukui M, Kitagawa Y, Nakamura N, Yoshikawa T. Glycyrrhizin and serum testosterone concentrations in male patients with type 2 diabetes. Diabetes Care 2003;26:2962.

48. Mosaddegh M, Naghibi F, Abbasi PR, Esmaeili S. The effect of liquorice extract on serum testosterone level in healthy male volunteers. J Pharm Pharmacol 2003;55(Suppl):S87–8.

49. Coxeter PD, McLachlan AJ, Duke CC, Roufogalis BD. Liquorice-drug interactions. Complement Med 2003;July/August:40–2.

50. Gattas Zaror V, Barrera Acevedo G, Yanez Soto E, Uauy-Dagach Imbarack R. Evaluacion de la tolerancia y aceptabilidad cronica de la harina de lupino (*Lupinus albus* var. multolupa) en la alimentacion de adultos jovenes. [Tolerance and chronic acceptability of lupine (*Lupinus albus* var. *multolupa*) flour for feeding of young adults.] Arch Latinoam Nutr 1990;40(4):490–502.

51. Gross R, Morales E, Gross U, von Baer E. Die Lupine, ein Beitrag zur Nahrungsversorgung in den Anden 3. Ernahrungsphysiologische Untersuchung mit dem Mehl der Susslupine (*Lupinus albus*). [Lupine, a contribution to the human food supply. 3. Nutritional physiological study with lupine (*Lupinus albus*) flour.] Z Ernahrungswiss 1976;15(4):391–5.

52. Antoun MD, Taha OM. Studies on Sudanese medicinal plants. II. Evaluation of an extract of *Lupinus termis* seeds in chronic eczema. J Nat Prod 1981;44(2):179–83.

53. Luque Marquez R, Gutierrez-Rave M, Infante Miranda F. Acute poisoning by lupine seed debittering water. Vet Hum Toxicol 1991;33(3):265–7.

54. Tsiodras S, Shin RK, Christian M, Shaw LM, Sass DA. Anticholinergic toxicity associated with lupine seeds as a home remedy for diabetes mellitus. Ann Emerg Med 1999;33(6):715–7.

55. Kunkle GA, Greiner EC. Dermatitis in horses and man caused by the straw itch mite. J Am Vet Med Assoc 1982;181(5):467–9.

56. Roberts JL, Hayashi JA. Exacerbation of SLE associated with alfalfa ingestion. N Engl J Med 1983;308(22):1361.

57. Alcocer-Varela J, Iglesias A, Llorente L, Alarcon-Segovia D. Effects of L-canavanine on T cells may explain the induction of systemic lupus erythematosus by alfalfa. Arthritis Rheum 1985;28(1):52–7.

58. Van Beneden CA, Keene WE, Strang RA, Werker DH, King AS, Mahon B, Hedberg K, Bell A, Kelly MT, Balan VK, Mac Kenzie WR, Fleming D. Multinational outbreak of Salmonella enterica serotype Newport infections due to contaminated alfalfa sprouts. JAMA 1999;281(2):158–62.

59. Backer HD, Mohle-Boetani JC, Werner SB, Abbott SL, Farrar J, Vugia DJ. High incidence of extra-intestinal infections in a Salmonella havana outbreak associated with alfalfa sprouts. Public Health Rep 2000;115(4):339–45.

60. Centers for Disease Control and Prevention (CDC). Outbreak of Salmonella serotype kottbus infections associated with eating alfalfa sprouts—Arizona, California, Colorado, and New Mexico, February–April 2001. MMWR Morb Mortal Wkly Rep 2002;51(1):7–9.

61. Winthrop KL, Palumbo MS, Farrar JA, Mohle-Boetani JC, Abbott S, Beatty ME, Inami G, Werner SB. Alfalfa sprouts and Salmonella kottbus infection: a multistate outbreak following inadequate seed disinfection with heat and chlorine. J Food Prot 2003;66(1):13–7.

62. Suslow TV, Wu J, Fett WF, Harris LJ. Detection and elimination of Salmonella mbandaka from naturally contaminated alfalfa seed by treatment with heat or calcium hypochlorite. J Food Prot 2002;65(3):452–8.

63. Proctor ME, Hamacher M, Tortorello ML, Archer JR, Davis JP. Multistate outbreak of Salmonella serovar muenchen infections associated with alfalfa sprouts grown from seeds pretreated with calcium hypochlorite. J Clin Microbiol 2001;39(10):3461–5.

64. Stratton J, Stefaniw L, Grimsrud K, Werker DH, Ellis A, Ashton E, Chui L, Blewett E, Ahmed R, Clark C, Rodgers F, Trottier L, Jensen B. Outbreak of Salmonella paratyphi B var java due to contaminated alfalfa sprouts in Alberta, British Columbia and Saskatchewan. Can Commun Dis Rep 2001;27(16):133–7.

65. Mahon BE, Ponka A, Hall WN, Komatsu K, Dietrich SE, Siitonen A, Cage G, Hayes PS, Lambert-Fair MA, Bean NH, Griffin PM, Slutsker L. An international outbreak of Salmonella infections caused by alfalfa sprouts grown from contaminated seeds. J Infect Dis 1997;175(4):876–82.

66. Inami GB, Moler SE. Detection and isolation of Salmonella from naturally contaminated alfalfa seeds following an outbreak investigation. J Food Prot 1999;62(6):662–4.

67. Stewart DS, Reineke KF, Ulaszek JM, Tortorello ML. Growth of Salmonella during sprouting of alfalfa seeds associated with salmonellosis outbreaks. J Food Prot 2001;64(5):618–22.

68. Czajka J, Batt CA. Verification of causal relationships between Listeria monocytogenes isolates implicated in food-borne outbreaks of listeriosis by randomly amplified polymorphic DNA patterns. J Clin Microbiol 1994;32(5):1280–7.

69. Breuer T, Benkel DH, Shapiro RL, Hall WN, Winnett MM, Linn MJ, Neimann J, Barrett TJ, Dietrich S, Downes FP, Toney DM, Pearson JL, Rolka H, Slutsker L, Griffin PMInvestigation Team. A multistate outbreak of Escherichia coli O157:H7 infections linked to alfalfa sprouts grown from contaminated seeds. Emerg Infect Dis 2001;7(6):977–82.

70. Anonymous. From the Centers for Disease Control and Prevention. Outbreaks of Escherichia coli O157:H7 infection associated with eating alfalfa sprouts—Michigan and Virginia, June–July 1997. JAMA 1997;278(10):809–10.

71. Hausen BM, Simatupang T, Bruhn G, Evers P, Koenig WA. Identification of new allergenic constituents and proof of evidence for coniferyl benzoate in balsam of Peru. Am J Contact Dermatitis 1995;6:199–208.

72. Temesvari E, Soos G, Podanyi B, Kovacs I, Nemeth I. Contact urticaria provoked by balsam of Peru. Contact Dermatitis 1978;4(2):65–8.

73. Cancian M, Fortina AB, Peserico A. Contact urticaria syndrome from constituents of balsam of Peru and fragrance mix in a patient with chronic urticaria. Contact Dermatitis 1999;41(5):300.

74. Pfutzner W, Thomas P, Niedermeier A, Pfeiffer C, Sander C, Przybilla B. Systemic contact dermatitis elicited by oral intake of Balsam of Peru. Acta Derm Venereol 2003;83(4):294–5.

75. Bruynzeel DP, van den Hoogenband HM, Koedijk F. Purpuric vasculitis-like eruption in a patient sensitive to balsam of Peru. Contact Dermatitis 1984;11(4):207–9.

76. Tanaka S, Matsumoto Y, Dlova N, Ostlere LS, Goldsmith PC, Rycroft RJ, Basketter DA, White IR, Banerjee P, McFadden JP. Immediate contact reactions to fragrance mix constituents and Myroxylon pereirae resin. Contact Dermatitis 2004;51(1):20–1.

77. Yong M, Cheong I. Jering-induced acute renal failure with blue urine. Trop Doct 1995;25(1):31.

78. H'ng PK, Nayar SK, Lau WM, Segasothy M. Acute renal failure following jering ingestion. Singapore Med J 1991;32(2):148–9.

79. Drew AK, Bensoussan A, Whyte IM, Dawson AH, Zhu X, Myers SP. Chinese herbal medicine toxicology database: monograph on Radix Sophorae Flavescentis, "ku shen". J Toxicol Clin Toxicol 2002;40(2):173–6.

80. Baber RJ, Templeman C, Morton T, Kelly GE, West L. Randomized placebo-controlled trial of an isoflavone supplement and menopausal symptoms in women. Climacteric 1999;2(2):85–92.

81. Heck AM, DeWitt BA, Lukes AL. Potential interactions between alternative therapies and warfarin. Am J Health Syst Pharm 2000;57(13):1221–7.

Gentianaceae

General Information

The genera in the family of Gentianaceae (Table 31) include centaury and various types of gentian.

Gentiana species

Gentiana (gentian) root is mutagenic in bacteria, which is due to the xanthone derivatives, gentisin and isogentisin. In experimental animals *Gentiana olivieri* is hypoglycemic, an effect that has been attributed to isoorientin, which is present in several species of *Gentiana* (1) and which also has antinociceptive and anti-inflammatory effects in animals (2). Some species of *Gentiana* inhibit monoamine oxidase (3) and acetylcholinesterase (4) in vitro.

Gentian violet is a purple dye that is obtained from coal tar; it is so called because its color resembles that of the flower and it has nothing to do with *Gentiana* species.

Table 31 The genera of Gentianaceae

Bartonia (screwstem)
Centaurium (centaury)
Cicendia (cicendia)
Enicostema (whitehead)
Eustoma (prairie gentian)
Frasera (green gentian)
Gentiana (gentian)
Gentianella (dwarf gentian)
Gentianopsis (fringed gentian)
Halenia (spurred gentian)
Lisianthius (lisianthius)
Lomatogonium (lomatogonium)
Obolaria (obolaria)
Sabatia (rose gentian)
Schultesia (wingcup)
Swertia (felwort)
Voyria (ghostplant)

Swertia species

Swertia species (felwort) are mutagenic in bacteria, because of the several xanthone derivatives that it contains. The clinical relevance of this remains to be established. They also have hypoglycemic effects in animals (5).

References

1. Sezik E, Aslan M, Yesilada E, Ito S. Hypoglycaemic activity of *Gentiana olivieri* and isolation of the active constituent through bioassay-directed fractionation techniques. Life Sci 2005;76(11):1223–38.
2. Kupeli E, Aslan M, Gurbuz I, Yesilada E. Evaluation of in vivo biological activity profile of isoorientin. Z Naturforsch [C] 2004;59(11–12):787–90.
3. Haraguchi H, Tanaka Y, Kabbash A, Fujioka T, Ishizu T, Yagi A. Monoamine oxidase inhibitors from *Gentiana lutea*. Phytochemistry 2004;65(15):2255–60.
4. Urbain A, Marston A, Queiroz EF, Ndjoko K, Hostettmann K. Xanthones from *Gentiana campestris* as new acetylcholinesterase inhibitors. Planta Med 2004;70(10):1011–4.
5. Grover JK, Yadav S, Vats V. Medicinal plants of India with anti-diabetic potential. J Ethnopharmacol 2002;81(1):81–100.

Ginkgoaceae

General Information

The family of Ginkgoaceae contains the single genus, *Ginkgo*, and a single species, *Ginkgo biloba*.

Ginkgo biloba

Ginkgo biloba (maidenhair tree, silver apricot) contains ginkgolides, which inhibit platelet-activating factor, reducing aggregability; this may contribute to bleeding disorders in patients taking *G. biloba*. *Ginkgo* has small beneficial effects in patients with intermittent claudication (1) and dementias (1,2).

Extracts from the leaves of *G. biloba* are marketed in some countries for the treatment of cerebral dysfunction and of intermittent claudication. In a review it was concluded that seven out of eight controlled trials of good quality showed positive effects of *G. biloba* compared with placebo on the following symptoms: memory difficulties, dizziness, tinnitus, headache, and emotional instability with anxiety (3). For intermittent claudication, the evidence for efficacy was judged unconvincing.

No serious adverse effects have been noted in any trial. However, *G. biloba* has been associated with gastrointestinal complaints, headache, and allergic skin reactions (SED-13, 538). Bleeding has been associated with chronic *G. biloba* ingestion because of its adverse effects on platelet aggregability, since the ginkgolide constituents of *G. biloba* inhibit platelet-activating factor.

Adverse effects
Ginkgo biloba has been associated with gastrointestinal complaints, headache, antiplatelet effects, and allergic skin reactions.

Nervous system
Ginkgo biloba can precipitate seizures in patients with well-controlled epilepsy (4).

- A 78-year-old man and an 84-year-old woman with previously well-controlled epilepsy presented with recurrent seizures (4). There were no obvious reasons for these events, and the investigator suspected self-medication with *G. biloba* extracts. Both patients had started taking *G. biloba* within 2 weeks of the start of the seizures. The herbal remedy was withdrawn and both patients remained seizure-free several months later. No other change of medication was made.

The author postulated that 4-0-methylpyridoxine, a constituent of *G. biloba* and a known neurotoxin, had caused the seizures.

Hematologic
Bleeding complications have been reported in patients taking extracts of *Ginkgo biloba* (5, 6), attributed to the antiplatelet properties of the extracts. In a systematic review of bleeding with *Ginkgo biloba* 15 reports were found in which there was a temporal association between administration and a bleeding event, including eight episodes of intracranial bleeding (7). In 13 cases additional risk factors for bleeding were identified. In only six cases did the bleeding did not recur after *Ginkgo* was withdrawn. Bleeding times were measured in three patients only and were prolonged.

In a placebo-controlled study in volunteers a special extract of *Ginkgo biloba*, EGb 761, was compared with placebo; there was no evidence of any effect on tests of primary hemostasis, coagulation, or platelet function (8). This particular extract could lack an effect on bleeding tendency, or routine laboratory tests do not reflect in vivo effects under all circumstances, or the previous anecdotal reports have been in patients who are in some way specifically susceptible.

Some studies (but not all) have shown that *Ginkgo biloba* inhibits platelet aggregation. It could therefore theoretically cause bleeding, particularly when used in combination with other antiplatelet drugs.

- A 71-year-old man had a fatal stroke due to intracerebral bleeding (9). His medical history was unremarkable, except for the fact that he had taken 40 mg bd of a *Ginkgo* extract for 2.5 years. Four weeks before his death he started taking ibuprofen (600 mg/day).

Whether this association was causal is speculative.

A rare case of retrobulbar hemorrhage was associated with chronic intake of *Ginkgo biloba* (10).

- A 65-year-old woman was admitted for routine lens implantation. Immediately after injection of 5 ml of local anesthetic, there was sudden proptosis, bruising of her lower lid, pain, and reduced vision. The problems turned out to be caused by retrobulbar bleeding. She had taken 3 × 40 mg of *Ginkgo biloba* extract daily for 2 years. Her blood count, prothrombin time, and partial thromboplastin time were normal.

A few new cases of bleeding have been reported in patients undergoing a variety of surgical interventions after taking extracts of *Ginkgo biloba* regularly (11, 12, 13). Contrary to previous belief, this augmented bleeding tendency may not be related to inhibition of platelet-activating factor by ginkgolides. The concentration required to inhibit PAF-mediated aggregation of human platelets exceeds by more than 100 times the peak values measured after the oral administration of *Ginkgo biloba* leaf extracts in recommended doses (14).

- A 59-year-old man developed bleeding complications after liver transplantation while taking *G. biloba* supplements (15). Postoperatively, he developed large hematomas in the subphrenic space and near the porta hepatis. They were drained and his hematocrit fell to 21%. Three weeks later he complained of blurring in the right eye, and a vitreous hemorrhage was diagnosed. He then admitted to taking an unknown amount of *G. biloba*, which was withdrawn. Subsequently, no further bleeding episodes occurred.

- A 78-year-old man developed progressivesided muscular weakness after a fall (16). A CT scan showed a subdural hematoma. He had taken 150 mg of *G. biloba* extract.
- A 56-year-old man had a stroke without apparent risk factors (17). A CT scan confirmed a right parietal hematoma. He had not taken any medications, except for a *G. biloba* extract (3 × 40 mg/day) which he had started 18 months before.
- A 34-year-old woman had a laparoscopic cholecystectomy and started bleeding into the surgical wound postoperatively (18). This led to a fall in hemoglobin from 16.5 to 12.4 g/dl. She was given blood transfusions and recovered uneventfully.
- Hyphema led to temporarily impaired vision in a man who had taken *G. biloba* for 2 weeks; no other cause was found; the lesion regressed after withdrawal and did not recur (19).
- A spontaneous cerebellar bleed occurred in a man who had taken *G. biloba* (20).

In each case the authors believed that self-medication with *G. biloba* had caused the bleeding through its effect on platelets. In some cases another factor that implicated the extract was the absence of recurrence after withdrawal, although this is a weak argument.

Spontaneous bilateral subdural hematomas and increased bleeding time have been associated with chronic ingestion of *G. biloba* (120 mg/day for 2 years) (21).

Skin

- A 40-year-old Afro-American woman developed an exfoliative rash and blistering and swelling of the tongue (22). A diagnosis of Stevens–Johnson syndrome was made. She had not taken any medications other than two doses of a formulation that contained *G. biloba*. Her condition responded to treatment with prednisolone, clotrimazole, and famotidine. *G. biloba* was withdrawn and no further events occurred. However, 5 months later she still had tenderness in the soles of the feet, peeling of the nails, and discoloration of the skin.

Formulations

Subarachnoid hemorrhage associated with *G. biloba* (23) stimulated a vigorous discussion on the differences between the usual extract sold over the counter and a ginkgolide mixture, and their respective (absence of) effects on platelet-aggregating factor and bleeding time (24,25). The dispute points to confusion that can arise when formulations of similar origin but with variable composition are available.

Drug overdose

Acute poisoning with *G. biloba* seeds is occasionally seen in East Asian countries.

- A 36-year-old woman, without any past or family history of epilepsy, developed frequent vomiting and generalized convulsions about 4 hours after taking about 70–80 ginkgo nuts, seeds of *G. biloba*, in an attempt to improve her health (26).
- A 2-year-old girl was admitted with vomiting, diarrhea, and irritability a few hours after eating a large quantity

of *G. biloba* seeds and then developed an afebrile generalized clonic seizure (27). Her serum concentration of 4-methoxypyridoxine, the putative toxic agent in the seed, was extremely high at 360 ng/ml.

Administration of pyridoxal phosphate, a competitive antagonist of 4-methoxypyridoxine, which inhibits the formation of gamma-aminobutyric acid (GABA), prevented further seizures in this case.

Drug-drug interactions

Anticoagulants

Gingko biloba extracts are increasingly being linked to a bleeding tendency, owing to an inhibitory action on platelet activating factor-mediated platelet aggregation. The danger is that over-the-counter herbal medicines are not taken into consideration when other drugs with antiplatelet effects, such as ibuprofen, are prescribed.

- A 78-year-old woman who had taken warfarin for 5 years, took *G. biloba* for 2 months, when she developed signs of a stroke, diagnosed as intracerebral hemorrhage (23).

Even though it is impossible to establish causality in this case, it is conceivable that the stroke was the result of overanticoagulation induced by the herbal medication (21).

- A 71-year-old man in good health had taken a concentrated extract of *Gingko biloba* for at least 2.5 years because of occasional dizziness (28). He then took ibuprofen for osteoarthritic hip pain. Ibuprofen inhibits thromboxane-dependent platelet aggregation. Four weeks later, he suffered a fatal massive intracranial bleeding.

In an open, crossover, randomized study, 12 healthy men took a single dose of warfarin 25 mg either alone or after pretreatment with *Ginkgo biloba* for 7 days; *Ginkgo* did not significantly affect clotting or the pharmacokinetics or pharmacodynamics of warfarin (29).

Anticonvulsants

- A 55-year-old man with epilepsy had a fatal breakthrough seizure. He was considered to be compliant with his anticonvulsant therapy but the autopsy report showed subtherapeutic serum concentrations of sodium valproate and phenytoin. He had also taken herbal supplements, including *Ginkgo biloba* (30).

Induction of CYP2C19 is one possible mechanism to explain the subtherapeutic concentrations of the anticonvulsant drugs in this case. *Ginkgo* nuts (but not the leaves) are also said to contain a potent neurotoxin, which can cause seizure activity.

References

1. Pittler MH, Ernst E. *Ginkgo biloba* extract for the treatment of intermittent claudication: a meta-analysis of randomized trials. Am J Med 2000;108(4):276–81.
2. Oken BS, Storzbach DM, Kaye JA. The efficacy of *Ginkgo biloba* on cognitive function in Alzheimer disease. Arch Neurol 1998;55(11):1409–15.
3. Kleijnen J, Knipschild P. *Ginkgo biloba*. Lancet 1992;340(8828):1136–9.

4. Granger AS. *Ginkgo biloba* precipitating epileptic seizures. Age Ageing 2001;30(6):523–5.

5. Benjamin J, Muir T, Briggs K, Pentland B. A case of cerebral haemorrhage—can *Ginkgo biloba* be implicated? Postgrad Med J 2201; 77:112–3.

6. Miller LG, Freeman B. Possible subdural hematoma associated with *Ginkgo biloba*. J Herbal Pharmacother 2002;2:57–63.

7. Bent S, Goldberg H, Padula A, Avins AL. Spontaneous bleeding associated with *Ginkgo biloba*: a case report and systematic review of the literature. J Gen Intern Med 2005;20:657–61.

8. Kohler S, Funk P, Kieser M. Influence of a 7-day treatment with *Ginkgo biloba* special extract EGb 761 on bleeding time and coagulation: a randomized, placebo-controlled double- blind study in healthy volunteers. Blood Coagul Fibrinolysis 2004;15:303–9.

9. Meisel C, Johne A, Roots I. Fatal intracerebral mass bleeding associated with *Ginkgo biloba* and ibuprofen. Artherosclerosis 2003;167:367.

10. Fong KCS, Kinnear PE. Retrobulbar haemorrhage associated with chronic *Ginkgo biloba* ingestion. Postgrad Med J 2003;79:531–2.

11. Destro MW, Speranzini MB, Cavalheiro Filho C, Destro T, Destro C. Bilateral haematoma after rhytidoplasty and blepharoplasty following chronic use of Ginkgo biloba. Br J Plast Surg 2005;58:100–1.

12. Bebbington A, Kulkarni R, Roberts P. Ginkgo biloba: persistent bleeding after total hip arthroplasty caused by herbal self-medication. J Arthroplasty 2005;20:125–6.

13. Yagmur E, Piatkowski A, Groger A, Pallua N, Gressner AM, Kiefer P. Bleeding complication under Gingko biloba medication. Am J Hematol 2005;79:343–4.

14. Koch E. Inhibition of platelet activating factor (PAF)-induced aggregation of human thrombocytes by ginkgolides: considerations on possible bleeding complications after oral intake of Ginkgo biloba extracts. Phytomedicine 2005;12:10–16.

15. Hauser D, Gayowski T, Singh N. Bleeding complications precipitated by unrecognized *Ginkgo biloba* use after liver transplantation. Transpl Int 2002;15(7):377–9.

16. Miller LG, Freeman B. Possible subdural hematoma associated with *Ginkgo biloba*. J Herb Pharmacother 2002;2(2):57–63.

17. Benjamin J, Muir T, Briggs K, Pentland B. A case of cerebral haemorrhage—can *Ginkgo biloba* be implicated? Postgrad Med J 2001;77(904):112–3.

18. Fessenden JM, Wittenborn W, Clarke L. *Ginkgo biloba*: a case report of herbal medicine and bleeding postoperatively from a laparoscopic cholecystectomy. Am Surg 2001;67(1):33–5.

19. Schneider C, Bord C, Misse P, Arnaud B, Schmitt-Bernard CF. Hyphéma spontané provoqué par l'extrait de *Ginkgo biloba*. [Spontaneous hyphema caused by *Ginkgo biloba* extract.] J Fr Ophtalmol 2002;25(7):731–2.

20. Purroy Garcia F, Molina C, Alvarez Sabin J. Hemorragia cerebelosa esponetanea asociada a la ingestion de *Ginkgo biloba*. [Spontaneous cerebellar haemorrhage associated with *Ginkgo biloba* ingestion.] Med Clin (Barc) 2002;119(15):596–7.

21. Rowin J, Lewis SL. Spontaneous bilateral subdural hematomas associated with chronic *Ginkgo biloba* ingestion. Neurology 1996;46(6):1775–6.

22. Davydov L, Stirling AL. Stevens–Johnson syndrome with *Ginkgo biloba*. J Herb Pharmacother 2001;1:65–9.

23. Vale S. Subarachnoid haemorrhage associated with *Ginkgo biloba*. Lancet 1998;352(9121):36.

24. Skogh M. Extracts of *Ginkgo biloba* and bleeding or haemorrhage. Lancet 1998;352(9134):1145–6.

25. Vale S. Reply. Lancet 1998;352(9134):1146.

26. Miwa H, Iijima M, Tanaka S, Mizuno Y. Generalized convulsions after consuming a large amount of gingko nuts. Epilepsia 2001;42(2):280–1.

27. Kajiyama Y, Fujii K, Takeuchi H, Manabe Y. Ginkgo seed poisoning. Pediatrics 2002;109(2):325–7.

28. Meisel C, Johne A, Roots I. Fatal intracerebral mass bleeding associated with *Ginkgo biloba* and ibuprofen. Atherosclerosis 2003;167:367.

29. Jiang X, Williams KM, Liauw WS, Ammit AJ, Roufogalis BD, Duke CC, Day RO, McLachlan AJ. Effect of ginkgo and ginger on the pharmacokinetics and pharmacodynamics of warfarin in healthy subjects. Br J Clin Pharmacol 2005;59:425–62.

30. Kupiec T, Raj V. Fatal seizures due to potential herb–drug interactions with Ginkgo biloba. J Anal Toxicol 2005;25:755–8.

Hippocastanaceae

General Information

The family of Hippocastanaceae contains the single genus *Aesculus* (buckeye, horse chestnut).

Aesculus species

Aescin is a complex mixture of triterpene saponins prepared from the seeds of the horse chestnut, *Aesculus hippocastanum*. It consists of a water-soluble fraction (alpha-aescin) and a water-insoluble fraction (beta-aescin).

Adverse effects

A systematic review of randomized trials of extracts of horse chestnut in chronic venous insufficiency showed that adverse effects are usually mild, for example pruritus, nausea, headache, dizziness, and gastrointestinal symptoms (1).

Urinary tract

Beta-aescin has been repeatedly associated with acute renal insufficiency when given intravenously in massive doses. Whether such effects can also occur after oral administration is unclear, as animal studies have shown poor absorption of beta-aescin from the gastrointestinal tract. (SEDA-3, 181) (SEDA-9, 190).

Skin

Contact dermatitis has been ascribed to aescin (2), as has contact urticaria (3).

Drug interactions

Beta-aescin can precipitate renal insufficiency when combined with aminoglycoside antibiotics (SEDA-3, 181) (SEDA-9, 190).

References

1. Pittler MH, Ernst E. Horse-chestnut seed extract for chronic venous insufficiency. A criteria-based systematic review. Arch Dermatol 1998;134(11):1356–60.
2. Comaish JS, Kersey PJ. Contact dermatitis to extract of horse chestnut (esculin). Contact Dermatitis 1980;6(2):150–1.
3. Escribano MM, Munoz-Bellido FJ, Velazquez E, Delgado J, Serrano P, Guardia J, Conde J. Contact urticaria due to aescin. Contact Dermatitis 1997;37(5):233.

Illiciaceae

General Information

The family of Illiciaceae contains a single genus, *Illicium*, the star anise species.

Illicium species

Illicium anisatum contains sesquiterpenoids, such as anisatin, anisotin, neoanisatin, and pseudoanisatin. *Illicium religiosum* (Japanese star anise) contains shikimic acid, anisatin and neoanisatin. *Illicium verum* (Chinese star anise) contains the monoterpenoid transanethole. Chinese star anise has been used to treat infant colic, but can be confused with Japanese star anise, which contains the neurotoxin anisatin.

Safrole is a mutagenic and animal carcinogenic monoterpenoid. It is the major component of oil of sassafras, and lesser quantities occur in essential oils from cinnamon, mace, nutmeg, and star anise. Some of its known or possible metabolites have mutagenic activity in bacteria and it has weak hepatocarcinogenic effects in rodents. Experiments in mice have suggested the possibility of transplacental and lactational carcinogenesis.

Adverse effects

When Japanese star anise was mixed into a commercially sold herbal tea, perhaps inadvertently consumption of the tea was associated with adverse events in 63 Dutch consumers (1). Their symptoms occurred 2–4 hours after they drank the tea and included general malaise, nausea, and vomiting. In 22 cases hospitalization was required, and 16 had generalized tonic-clonic seizures. All made a full recovery after withdrawal of the herbal tea. Anisatin is a non-competitive GABA receptor antagonist, which causes nervous system hyperactivity, and the authors believed that this mechanism explained the high rate of seizures in these patients.

Nervous system

Seven cases have been reported of infants aged 2 weeks to 3 months who developed neurological symptoms (clonus, myoclonus, increased deep tendon reflexes, nystagmus, vomiting, seizures) after ingesting the tea (2). All the symptoms resolved within 24 hours of dechallenge. Analyses of the herbal ingredients showed that most contained Japanese star anise (*Illicium anisatum*), which is known to be neurotoxic. The authors suggested that the problems could have been due to an overdose of *Illicium verum*, contamination with *Illicium anisatum*, or both.

- A 1-month-old girl developed status epilepticus after being given a large amount of star anis for colic (3).
- Two infants whose parents gave them star anise herbal tea developed tremors or spasms, hypertonia, hyperexcitability with crying, nystagmus, and vomiting (4). The Chinese star anise tea had been contaminated with Japanese star anise.

From February to September 2001, a matched case-control study was performed in infants aged under 3 months admitted to the pediatric emergency departments of two hospitals in Madrid (5). There were 23 cases, whose symptoms and signs were irritability, abnormal movements, vomiting, and nystagmus. The odds ratio for anise consumption was 18.0 (CI = 2.03, 631). Laboratory analyses showed contamination of *I. verum* by *I. anisatum*.

References

1. Johanns ES, van der Kolk LE, van Gemert HM, Sijben AE, Peters PW, de Vries I. Een epidemie van epileptische aanvallen na drinken van kruidenthee. [An epidemic of epileptic seizures after consumption of herbal tea.] Ned Tijdschr Geneeskd 2002;146(17):813–6.
2. Duchowny M, Garcia Peña BM. Chemical composition of Chinese star anise (*Illicium verum*) and neurotoxicity in infants. JAMA 2004;291:562–3.
3. Gil Campos M, Perez Navero JL, Ibarra De La Rosa I. Crisis convulsiva secundaria a intoxicacion por anis estrellado en un lactante. [Convulsive status secondary to star anise poisoning in a neonate.] An Esp Pediatr 2002;57(4):366–8.
4. Minodier P, Pommier P, Moulene E, Retornaz K, Prost N, Deharo L. Intoxication aiguë par la badiane chez le nourrisson. [Star anise poisoning in infants.] Arch Pediatr 2003;10(7):619–21.
5. Garzo Fernandez C, Gomez Pintado P, Barrasa Blanco A, Martinez Arrieta R, Ramirez Fernandez R, Ramon Rosa F. Grupo de Trabajo del Anis Estrellado. Casos de enfermedad de sintomatologia neurologica asociados al consumo de anis estrellado empleado como carminativo. [Cases of neurological symptoms associated with star anise consumption used as a carminative.] An Esp Pediatr 2002;57(4):290–4.

Iridaceae

General Information

The genera in the family of Iridaceae (Table 32) include *Crocus*, *Freesia*, and some types of lilies.

Table 32   The genera of Iridaceae

Acidanthera (acidanthera)
Alophia (alophia)
Aristea (aristea)
Belamcanda (belamcanda)
Calydorea (violet lily)
Chasmanthe (African cornflag)
Crocosmia (crocosmia)
Crocus (crocus)
Dietes (dietes)
Eleutherine (eleutherine)
Freesia (freesia)
Gladiolus (gladiolus)
Herbertia (herbertia)
Homeria (Cape tulip)
Iris (iris)
Ixia (African cornlily)
Libertia (libertia)
Nemastylis (pleatleaf)
Neomarica (neomarica)
Olsynium (grass widow)
Romulea (romulea)
Schizostylis (Kaffir lily)
Sisyrinchium (blue-eyed grass)
Sparaxis (wandflower)
Tigridia (peacock flower)
Trimezia (trimezia)
Watsonia (bugle-lily)

Crocus sativus

The stamens of *Crocus sativus* (Indian saffron) have been used primarily as a coloring and flavoring agent. Its potential use as an anticancer agent has been reviewed (1,2). No risks have been documented for daily doses up to 1.5 g, but 5 g is toxic, 10 g is abortive, and 20 g is lethal.

References

1. Abdullaev FI. Cancer chemopreventive and tumoricidal properties of saffron (Crocus sativus L). Exp Biol Med (Maywood) 2002;227(1):20–5.
2. Deng Y, Guo ZG, Zeng ZL, Wang Z. [Studies on the pharmacological effects of saffron (*Crocus sativus* L) — a review.] Zhongguo Zhong Yao Za Zhi 2002;27(8):565–8.

Juglandaceae

General Information

The family of Juglandaceae contains three genera:

1. *Carya* (hickory)
2. *Juglans* (walnut)
3. *Pterocarya* (pterocarya).

Juglans regia

The fresh fruit-shell of *Juglans regia* (English walnut) contains the naphthoquinone constituent juglone, which is mutagenic and possibly carcinogenic. The juglone content of dried shells has not yet been studied adequately.

In a placebo-controlled study in 60 Persian hyperlipidemic subjects, walnut oil encapsulated in 500 mg capsules reduced plasma triglyceride concentrations by 19–33% of baseline (1).

Adverse effects

Hyperpigmentation and contact dermatitis have been attributed to *J. regia* (2) as has dermatitis bullosa (3).

A brown orange pigmentation occurred in a patient who used the bark of *J. regia* for cleaning teeth (4).

References

1. Zibaeenezhad MJ, Rezaiezadeh M, Mowla A, Ayatollahi SM, Panjehshahin MR. Antihypertriglyceridemic effect of walnut oil. Angiology 2003;54(4):411–4.
2. Bonamonte D, Foti C, Angelini G. Hyperpigmentation and contact dermatitis due to *Juglans regia*. Contact Dermatitis 2001;44(2):101–12.
3. Barniske R. Dermatitis bullosa, ausgelost durch den Saft gruner Walnussfruchtschalen (*Juglans regia*). [Dermatitis bullosa, caused by the juice of green walnut shells (*Juglans regia*).] Dermatol Wochenschr 1957;135(8):189–92.
4. Ashri N, Gazi M. More unusual pigmentations of the gingiva. Oral Surg Oral Med Oral Pathol 1990;70(4):445–9.

Krameriaceae

General Information

The family of Krameriaceae contains the single genus *Krameria* (ratany). Contact dermatitis has been reported in a patient using an extract of *Krameria triandra* (1).

Reference

1. Goday Bujan JJ, Oleaga Morante JM, Yanguas Bayona I, Gonzalez Guemes M, Soloeta Arechavala R. Allergic contact dermatitis from Krameria triandra extract. Contact Dermatitis 1998;38(2):120–1.

Lamiaceae

General Information

The genera in the family of Lamiaceae (Table 33) include basil, catnip, germander, lavender, mints, origanum, rosemary, sage, skullcap, and thyme.

Table 33 The genera of Lamiaceae

Acanthomintha (thorn-mint)
Acinos (acinos)
Agastach (giant hyssop)
Ajuga (bugle)
Ballota (horehound)
Blephilia (pagoda-plant)
Brazoria (brazos-mint)
Calamintha (calamint)
Cedronella (cedronella)
Chaiturus (lion's tail)
Clinopodium (clinopodium)
Coleus (coleus)
Collinsonia (horsebalm)
Conradina (false rosemary)
Cunila (cunila)
Dicerandra (balm)
Dracocephalum (dragonhead)
Elsholtzia (elsholtzia)
Erythrochlamys (erythrochlamys)
Galeopsis (hemp nettle)
Glechoma (glechoma)
Haplostachys (haplostachys)
Hedeoma (false pennyroyal)
Hyptis (bushmint)
Hyssopus (hyssop)
Isanthus (fluxweed)
Lallemantia (lallemantia)
Lamiastrum (lamiastrum)
Lamium (dead nettle)
Lavandula (lavender)
Leonotis (lion's ear)
Leonurus (motherwort)
Lepechinia (pitchersage)
Leucas (leucas)
Lycopus (waterhorehound)
Macbridea (macridea)
Marrubium (horehound)
Marsypianthes (marsypianthes)
Meehania (meehania)
Melissa (balm)
Mentha (mint)
Moluccella (moluccella)
Monarda (beebalm)
Monardella (monardella)
Mosla (mosla)
Nepeta (catnip)
Ocimum (basil)
Origanum (origanum)
Orthosiphon (orthosiphon)
Perilla (perilla)
Perovskia (perovskia)

(Continued)

Phlomis (Jerusalem sage)
Phyllostegia (phyllostegia)
Physostegia (lion's heart)
Piloblephis (piloblephis)
Plectranthus (plectranthus)
Pogogyne (mesamint)
Pogostemon (pogostemon)
Poliomintha (rosemary mint)
Prunella (self heal)
Pycnanthemum (mountain mint)
Rhododon (sand mint)
Rosmarinus (rosemary)
Salazaria (bladder sage)
Salvia (sage)
Satureja (savory)
Scutellaria (skullcap)
Sideritis (ironwort)
Solenostemon (solenostemon)
Stachys (hedge nettle)
Stachydeoma (mock pennyroyal)
Stenogyne (stenogyne)
Synandra (synandra)
Teucrium (germander)
Thymus (thyme)
Trichostema (bluecurls)
Warnockia (brazos mint)

Hedeoma pulegoides and *Mentha pulegium*

The volatile oil of pennyroyal (prepared from *Hedeoma pulegoides* or *Mentha pulegium*) is a folk medicine used as an abortifacient. The doses that are required for this effect can cause serious symptoms, including vomiting, seizures, hallucinations, renal damage, hepatotoxicity and shock; deaths have also occurred [1,2].

Adverse effects
The hepatotoxicity of pulegone, the principal constituent of pennyroyal oil, has been demonstrated in animal studies and reported in humans [3].

Two cases of serious or fatal toxicity have been described in two infants who had been treated with herbal tea containing pennyroyal oil [4]. One infant developed fulminant liver failure with cerebral edema and necrosis; the other infant developed hepatic dysfunction and a severe epileptic encephalopathy.

Lavandula angustifolia

The volatile oil of *Lavandula angustifolia* (lavender) contains linralyl acetate and linalool, and lavender also contains coumarins. It has been used in aromatherapy to treat insomnia and headaches, and may have small beneficial effects [5].

Adverse effects
Patch tests with lavender oil from 1990 to 1998 in Japan were positive in 3.7% of cases [6]. In five of 11 positive cases in 1997 and eight of 15 positive cases in 1998, the patients had used dried lavender flowers in pillows, drawers, cabinets, or rooms.

- A 53-year-old patient with relapsing eczema had contact allergy to various essential oils used in aromatherapy (7). Sensitization was due to previous exposure to lavender, jasmine, and rosewood. Laurel, eucalyptus, and pomerance also produced positive tests, without previous exposure.

Contact dermatitis has been attributed to a lavender oil pillow (8), and to lavender in an analgesic gel (9).

Mentha piperita

Peppermint oil (*Menthae piperitae aetheroleum*), which contains cineol, liomonene, menthofuran, menthol, and menthone, is obtained from the fresh leaves of peppermint, *Mentha piperita* by steam distillation. As a calcium channel blocker, (−) menthol is responsible for the spasmolytic effect of peppermint oil, and it has been used as a carminative and antispasmodic for esophageal spasm and irritable bowel syndrome (10). The FDA has listed peppermint and peppermint oil as being "generally recognized as safe" (11). Limonene is covered in a separate monograph (p. 000).

Adverse effects
Gastrointestinal
Heartburn is one of the major adverse effects of oral peppermint oil, mostly because of inappropriate release of the oil in the upper gastrointestinal tract (12), resulting in relaxation of the lower esophageal sphincter, thus facilitating reflux. This effect is minimized by the use of modified-release formulations.

Skin
Contact dermatitis has been attributed to menthol in peppermint (13).

Drug administration route
The systemic availability of menthol is very low after transdermal administration; the average C_{max} after the application of eight patches in eight subjects was only 30 ng/ml; the mean terminal half-life was 4.7 hours (14).

The pharmacokinetics of peppermint oil and its effects on the gastrointestinal tract have been reviewed, with a focus on irritable bowel syndrome (15). In nine studies 269 subjects were exposed to peppermint oil either orally or by topical intraluminal (stomach or colon) administration in either single doses of 0.1–0.24 ml or daily for 2 weeks. With the exception of one study, in which peppermint oil potentiated neostigmine-stimulated colon activity (16), the data showed that it has a substantial spasmolytic effect. This effect begins as early as 0.5 minutes after topical (intestinal tract) administration and can last for up to 23 minutes. This is too short a duration of action for the treatment of, for example, irritable bowel syndrome. To expose the target organ, i.e. the large bowel, to a constant concentration of peppermint oil, in order to maintain the beneficial effect, a modified-release formulation is needed.

Drug-drug interactions
Bupropion
In 600 African–American smokers taking part in a study of smoking cessation with bupropion, menthol (n = 471)

and non-menthol (n = 129) smokers were compared (17). Menthol smokers were younger (41 versus 53 years), more likely to be female (74% versus 57%) and more likely to smoke their first cigarette within 30 minutes of waking up (82% versus 70%). Seven-day point-prevalence abstinence rates from smoking for menthol and non-menthol smokers respectively were 28% and 42% at 6 weeks and 21% and 27% at 6 months. Among those under 50 years old, non-menthol smokers were more likely to quit smoking (OR = 2.0; 95% CI = 1.03, 3.95) as those who took bupropion (OR = 2.12; 95% CI = 1.32, 3.39). The authors concluded that menthol attenuated the beneficial effects of bupropion.

Caffeine
The kinetics and effects of a single oral dose of caffeine 200 mg in coffee taken together with a single oral dose of menthol 100 mg or placebo capsules have been studied in a randomized, double-blind, two-way, crossover study in 11 healthy women (18). Co-administration of menthol increased the tmax of caffeine 44 to 76 minutes but did not significantly reduce the Cmax, AUC, terminal half-life, or oral clearance. Menthol reduced the reduction in heart rate due to caffeine. The authors concluded that a single oral dose of pure menthol 100 mg delayed caffeine absorption and blunted the heart rate response without altering caffeine metabolism.

Felodipine
The effects of menthol on a single oral dose of felodipine in a modified-release tablet (Plendil) 10 mg have been studied in a randomized, double-blind, two-way, crossover study in 10 healthy subjects (19). Felodipine was given at the start of the study with menthol 100 mg or placebo, and menthol 50, 25, and 25 mg or placebo were given at 2, 5, and 7 hours respectively. Menthol co-administration did not significantly change the pharmacokinetics of felodipine or its effects on blood pressure and heart rate.

Warfarin
A possible interaction of menthol cough drops (Halls) with warfarin has been reported (20).

- A 57-year-old white man awaiting cardioversion for atrial fibrillation was given warfarin. The dosage was adjusted to 7 mg/day and the target international normalized ratio (INR) was 2.28–2.68. About 1 week later his INR fell to 1.45. He reported that he had a flu-like illness during the previous week and had used menthol cough drops. No other potential causes for the reduced INR were found. The dosage of warfarin was increased to 53 mg/week. After withdrawal of the menthol cough drops, the dosage of warfarin returned to what it had been before and the INR remained stable.

An objective causality assessment suggested that the reduced INR was possibly related to the use of menthol cough drops during warfarin therapy. The mechanism was not elucidated, but menthol may have slowed the absorption of warfarin.

Salvia species

In China, the root of *Salvia miltiorrhiza* (danshen) has been used traditionally for the treatment of coronary diseases.

The leaf of *Salvia officinalis* (sage) contains 1.0–2.5% of essential oil, consisting of 35–60% of thujone.

Adverse effects

A patient taking warfarin and who had taken a decoction of *S. miltiorrhiza* presented with a prolonged bleeding time and melena (21) and other cases have been reported (22). Pharmacodynamic and pharmacokinetic studies in rats have shown that danshen increases the absorption rate constant, AUC, C_{max}, and half-lives of both *R*- and *S*-warfarin, and reduces their clearances and apparent volumes of distribution (23,24).

Thujone can cause toxicity when the herb is taken in overdose (more than 15 g per dose) or for a prolonged period. Pregnancy is listed as a contraindication to the use of the essential oil or alcoholic extracts of *S. officinalis*.

Scutellaria species

The *Scutellaria* species (skullcap) include over 40 members, which contain a variety of flavonoids and terpenoids, the latter including scutellariosides and smithiandienol. Some also contain pyrrolizidine alkaloids (see separate monograph), which have repeatedly been implicated in veno-occlusive disease of the liver.

Adverse effects

Although Western skullcap formulations are supposed to come from *Scutellaria lateriflora*, it is unclear whether it is responsible for the liver damage that has been associated with skullcap. In the UK, the American germander (*Teucrium canadense*) has been widely used to replace *S. lateriflora* in commercial skullcap materials and products. In one UK case of skullcap-associated hepatotoxicity, the material was found to come from *T. canadense*. This raises the possibility that other cases of skullcap toxicity may also have involved *Teucrium* rather than *Scutellaria* (25).

Respiratory

- A 53-year-old Japanese man, who had taken skullcap intermittently for hemorrhoids, developed recurrent interstitial pneumonitis (26). Re-challenge, after he had stopped taking the herbal remedy and had become symptom free, resulted in a high fever and signs and symptoms of interstitial pneumonitis. Transbronchial lung biopsy showed lymphocytic alveolitis with eosinophilic infiltration. The symptoms subsided again after withdrawal.

Liver

Skullcap has repeatedly been associated with hepatotoxicity, and veno-occlusive disease has been reported (27).

- A 28-year-old man presented with jaundice after taking six tablets of skullcap (together with zinc and pau d'arco) daily for the previous 6 months to help his

multiple sclerosis. His liver enzymes were raised and hepatitis A, B, and C serologies were negative. He developed progressive liver failure and received a transplant but died shortly after. His explanted liver showed fibrous stenosis and obliteration of most of the terminal venules with extensive perivenular fibrosis, indicative of veno-occlusive disease.

Teucrium species

Teucrium species (germander) contain a variety of iridoids, monoterpenoids, diterpenoids, and sesquiterpenoids.

Adverse effects
Liver

The hepatotoxicity of germander has been well documented. An animal study suggests that the hepatotoxicity resides in one or more reactive metabolites of its furano-diterpenoids (28).

In France, many cases of hepatitis have been associated with the normal use of *Teucrium chamaedrys* (wall germander). The frequency of this adverse effect has been estimated at one case in about 4000 months of treatment (29). Two cases of hepatitis in women who had been taking germander daily for 6 months have been reported from Canada (30). Although most cases are not serious, deaths have been reported (31), and progression to liver cirrhosis has also been described (32).

Seven patients developed hepatitis after taking *T. chamaedrys* and had no other cause of liver damage (33). The hepatitis was characterized by jaundice and a marked increase in serum transaminases 3–18 weeks after taking germander. Liver biopsies in three patients showed hepatic necrosis. After withdrawal of germander the jaundice disappeared within 8 weeks and recovery was complete in 1.5–6 months. In three cases, germander was followed by prompt recurrence of hepatitis.

- A 62-year-old man had taken a tea made from *T. capitatum* for 4 months when he developed anorexia, nausea, and malaise (34). He also noted dark urine and jaundice. He was admitted to hospital with acute icteric hepatitis. A liver biopsy showed bridging necrosis, inflammatory infiltration, and bile emboli. After withdrawal of the herbal tea he made a full recovery within 3 months.
- A 67-year-old man consumed large amounts of *Teucrium polium* tea daily for 6 months to normalize his blood lipids and developed acute cholestatic hepatitis (35). After withdrawal of the herbal remedy, his liver function returned to normal.

Several other cases have been reported (36–40).

References

1. Ciganda C, Laborde A. Herbal infusions used for induced abortion. J Toxicol Clin Toxicol 2003;41(3):235–9.
2. Anderson IB, Mullen WH, Meeker JE, Khojasteh-Bakht SC, Oishi S, Nelson SD, Blanc PD. Pennyroyal toxicity: measurement of toxic metabolite levels in two cases and review of the literature. Ann Intern Med 1996;124(8):726–34.

3. Sullivan JB Jr, Rumack BH, Thomas H Jr, Peterson RG, Bryson P. Pennyroyal oil poisoning and hepatotoxicity. JAMA 1979;242(26):2873–4.

4. Bakerink JA, Gospe SM Jr, Dimand RJ, Eldridge MW. Multiple organ failure after ingestion of pennyroyal oil from herbal tea in two infants. Pediatrics 1996;98(5):944–7.

5. Hardy M, Kirk-Smith MD, Stretch DD. Replacement of drug treatment for insomnia by ambient odour. Lancet 1995;346(8976):701.

6. Sugiura M, Hayakawa R, Kato Y, Sugiura K, Hashimoto R. Results of patch testing with lavender oil in Japan. Contact Dermatitis 2000;43(3):157–60.

7. Schaller M, Korting HC. Allergic airborne contact dermatitis from essential oils used in aromatherapy. Clin Exp Dermatol 1995;20(2):143–5.

8. Coulson IH, Khan AS. Facial "pillow" dermatitis due to lavender oil allergy. Contact Dermatitis 1999;41(2):111.

9. Rademaker M. Allergic contact dermatitis from lavender fragrance in Difflam gel. Contact Dermatitis 1994;31(1):58–9.

10. Pittler MH, Ernst E. Peppermint oil for irritable bowel syndrome: a critical review and metaanalysis. Am J Gastroenterol 1998;93(7):1131–5.

11. National Archives and Records Administration. Code of Federal Regulations. Title 21: Food and Drugs. Substances generally recognized as safe. 21.CFR.182.10 and 21.CFR.182.20 1998;April 1.

12. Sigmund CJ, McNally EF. The action of a carminative on the lower esophageal sphincter. Gastroenterology 1969;56:13–18.

13. Wilkinson SM, Beck MH. Allergic contact dermatitis from menthol in peppermint. Contact Dermatitis 1994;30(1):42–3.

14. Hansen T, Kunkel M, Weber A, James Kirkpatrick C. Osteonecrosis of the jaws in patients treated with bisphosphonates - histomorphologic analysis in comparison with infected osteoradionecrosis. J Oral Pathol Med 2006;35(3):155–60.

15. Grigoleit H-G, Grigoleit P. Gastrointestinal clinical pharmacology of peppermint oil. Phytomedicine 2005;12:607–11.

16. Rogers J, Tay HH, Misiewicz JJ. Peppermint oil. Lancet 1988;98–9.

17. Okuyemi KS, Ahluwalia JS, Ebersole-Robinson M, Catley D, Mayo MS, Resnicow K. Does menthol attenuate the effect of bupropion among African American smokers? Addiction 2003;98(10):1387–93.

18. Gelal A, Guven H, Balkan D, Artok L, Benowitz NL. Influence of menthol on caffeine disposition and pharmacodynamics in healthy female volunteers. Eur J Clin Pharmacol 2003;59(5–6):417–22.

19. Gelal A, Balkan D, Ozzeybek D, Kaplan YC, Gurler S, Guven H, Benowitz NL. Effect of menthol on the pharmacokinetics and pharmacodynamics of felodipine in healthy subjects. Eur J Clin Pharmacol 2005;60(11):785–90.

20. Arin MJ, Bate J, Krieg T, Hunzelmann N. Possible warfarin interaction with menthol cough drops. Ann Pharmacother 2005;39:S53–6.

21. Tam LS, Chan TY, Leung WK, Critchley JA. Warfarin interactions with Chinese traditional medicines: danshen and methyl salicylate medicated oil. Aust NZ J Med 1995;25(3):258.

22. Chan TY. Interaction between warfarin and danshen (Salvia miltiorrhiza). Ann Pharmacother 2001;35(4):501–4.

23. Chan K, Lo AC, Yeung JH, Woo KS. The effects of Danshen (Salvia miltiorrhiza) on warfarin pharmacodynamics and pharmacokinetics of warfarin enantiomers in rats. J Pharm Pharmacol 1995;47(5):402–6.

24. Lo AC, Chan K, Yeung JH, Woo KS. The effects of danshen (Salvia miltiorrhiza) on pharmacokinetics and

25. pharmacodynamics of warfarin in rats. Eur J Drug Metab Pharmacokinet 1992;17(4):257–62.

25. De Smet PA. Health risks of herbal remedies. Drug Saf 1995;13(2):81–93.

26. Takeshita K, Saisho Y, Kitamura K, Kaburagi N, Funabiki T, Inamura T, Oyamada Y, Asano K, Yamaguchi K. Pneumonitis induced by ou-gon (scullcap). Intern Med 2001;40(8):764–8.

27. Akatsu T, Santo RM, Nakayasu K, Kanai A. Oriental herbal medicine induced epithelial keratopathy. Br J Ophthalmol 2000;84(8):934.

28. Loeper J, Descatoire V, Letteron P, Moulis C, Degott C, Dansette P, Fau D, Pessayre D. Hepatotoxicity of germander in mice. Gastroenterology 1994;106(2):464–72.

29. Castot A, Larrey D. Hépatites observées au cours d'un traitement par un médicament ou une tisane contenant de la germandrée petit-chêne. Bilan des 26 cas rapportés aux Centres Régionaux de Pharmacovigilance. [Hepatitis observed during a treatment with a drug or tea containing wild germander. Evaluation of 26 cases reported to the Regional Centers of Pharmacovigilance.] Gastroenterol Clin Biol 1992;16(12):916–22.

30. Laliberte L, Villeneuve JP. Hepatitis after the use of germander, a herbal remedy. CMAJ 1996;154(11):1689–92.

31. Mostefa-Kara N, Pauwels A, Pines E, Biour M, Levy VG. Fatal hepatitis after herbal tea. Lancet 1992;340(8820):674.

32. Dao T, Peytier A, Galateau F, Valla A. Hépatite chronique cirrhogène à la germandrée petit-chêne. [Chronic cirrhogenic hepatitis induced by germander.] Gastroenterol Clin Biol 1993;17(8–9):609–10.

33. Larrey D, Vial T, Pauwels A, Castot A, Biour M, David M, Michel H. Hepatitis after germander (Teucrium chamaedrys) administration: another instance of herbal medicine hepatotoxicity. Ann Intern Med 1992;117(2):129–32.

34. Dourakis SP, Papanikolaou IS, Tzemanakis EN, Hadziyannis SJ. Acute hepatitis associated with herb (Teucrium capitatum L.) administration Eur J Gastroenterol Hepatol 2002;14(6):693–5.

35. Mazokopakis E, Lazaridou S, Tzardi M, Mixaki J, Diamantis I, Ganotakis E. Acute cholestatic hepatitis caused by Teucrium polium. Phytomedicine 2004;11:83–4.

36. Mazokopakis E, Lazaridou S, Tzardi M, Mixaki J, Diamantis I, Ganotakis E. Acute cholestatic hepatitis caused by Teucrium polium L. Phytomedicine 2004;11(1):83–4.

37. Polymeros D, Kamberoglou D, Tzias V. Acute cholestatic hepatitis caused by Teucrium polium (golden germander) with transient appearance of antimitochondrial antibody. J Clin Gastroenterol 2002;34(1):100–1.

38. Perez Alvarez J, Saez-Royuela F, Gento Pena E, Lopez Morante A, Velasco Oses A, Martin Lorente J. Hepatitis aguda por ingestion de infusiones con Teucrium chamaedrys. [Acute hepatitis due to ingestion of Teucrium chamaedrys infusions.] Gastroenterol Hepatol 2001;24(5):240–3.

39. Mattei A, Rucay P, Samuel D, Feray C, Reynes M, Bismuth H. Liver transplantation for severe acute liver failure after herbal medicine (Teucrium polium) administration. J Hepatol 1995;22(5):597.

40. Pauwels A, Thierman-Duffaud D, Azanowsky JM, Loiseau D, Biour M, Levy VG. Hépatite aiguë a la germandrée petit-chêne. Hepatotoxicité d'une plante medicinale. Deux observations. [Acute hepatitis caused by wild germander. Hepatotoxicity of herbal remedies. Two cases.] Gastroenterol Clin Biol 1992;16(1):92–5.

Lauraceae

General Information

The genera in the family of Lauraceae (Table 34) include cinnamon and sassafras.

Cinnamonum camphora

Camphor is a white crystalline substance, obtained from the tree *Cinnamonum camphora* (camphor tree), but the name has also been given to various volatile substances found in different aromatic plants.

Camphor has been used as a substance of abuse for many centuries, both by ingestion and by inhalation. Today it is found in many non-prescription vaporized or topical "cold cures," topical musculoskeletal anesthetic rubs, "cold sore" formulations, sunscreens, and mothballs. It has also been used to procure an abortion (1).

Adverse effects
The acute effects of ingesting camphor were described by Louis Lewin in "Phantastica" (1924):

> After ingestion of 1.2 grains [72 g] the following symptoms occur: an agreeable warmth of the skin, general nervous excitation, a desire to move, tickling of the skin, and a peculiar feeling of ecstasy similar to drunkenness. One addict said that "he saw his destiny full of great possibilities clearly and distinctly before his eyes." This state continued for one and a half hours. After ingestion of 2.4 grains there was an urgent desire to move. All movements were generally facilitated, and when walking the limbs were lifted far more than necessary. Intellectual thought was impossible. There was a flood of ideas, chasing each other with great rapidity, without any one being analysed. The individual lost perception of his identity. After vomiting, awareness returned, although distraction, forgetfulness and vacancy of mind persisted. On awaking the state of intoxication seemed

to have been extraordinarily long and full of events of which the subject did not remember any. After 3 hours he was able to pull himself together and return to full consciousness, but the effect on the brain was so potent that unconsciousness and convulsions again occurred within an hour and lasted for half an hour, after which the subject gradually regained his full mental faculties and normal muscle function.

Lewin also described the effects of chronic abuse:

> Loss of the sense of location and brief gaps in memory usually succeed gastric irritation and convulsions when camphor is taken habitually. The lost memories finally reappear, but in a very strange fashion, so that, according to the statement of an addict, all affairs, events, and things that he had forgotten seemed new, as if he had had no previous knowledge of them. Even after he recognized all the members of his family, the objects in his room seemed unfamiliar, as if they had just been given to him. In Slovakia convulsive states similar to epilepsy are so common in consumers of camphor that all cases of similar fits are directly attributed to it.

Nervous system
Convulsions can occur after transdermal absorption of camphor (2).

- A 20-month-old girl developed status epilepticus after ingesting camphor and required ventilation. She was treated with intravenous diazepam and phenobarbital and nasogastric-activated charcoal and made a complete neurological recovery (3).

Liver
Hepatotoxicity occurred in a 2-month-old baby after a camphor-containing cold remedy was applied to the skin; liver function tests returned to normal after the remedy was withdrawn (4).

Chronic ingestion of camphor can mimic Reye's syndrome (5).

Skin
Besides other components, cinnamon, which is derived from the bark of the *Cinnamonum camphora* tree, contains about 1–3% ethereal oils, cinnamic aldehyde being the main constituent, and about 35 other constituents, such as eugenol and cinnamic alcohol. The constituents of oil of cinnamon are common allergens or irritants, and occupational allergic contact dermatitis from oil of cinnamon has often been reported in bakers, confectioners, and cooks (6, 7).

- Contact with insoles containing cinnamon has been reported as having caused allergic contact dermatitis in a 47-year-old man, consistent with podopompholyx, 2 days after he had started to use new insoles containing cinnamon powder as an odor-neutralizing agent (8). There was no past history of plantar eczema. Treatment with topical glucocorticoids initially failed to clear the lesions. He developed erysipelas on the left foot 5 days after the onset of the eczema and was therefore given intravenous antibiotics. There was marked improvement

Table 34 The genera of Lauraceae

Aniba (aniba)
Beilschmiedia (beilschmiedia)
Cassytha (cassytha)
Cinnamomum (cinnamon)
Cryptocarya (cryptocarya)
Endiandra
Eusideroxylon (ironwood)
Laurus (laurel)
Licaria (licaria)
Lindera (spicebush)
Litsea (litsea)
Nectandra (sweetwood)
Ocotea (sweetwood)
Persea (bay)
Sassafras (sassafras)
Umbellularia (California laurel)

in the eczema and clearance of the erysipelas within 2 weeks. Patch testing with the European standard series showed positive reactions to fragrance mix (+++ at D2, further testing stopped), balsam of Peru (*Myroxylon pereirae* resin; + at D2 and D3), and thiomersal (+ at D2 and D3). Testing with components of the fragrance mix showed reactions to cinnamic aldehyde (cinnamal; ++ at D2 and +++ at D3) and cinnamic alcohol (+ at D2 and +++ at D3). There was also a strongly positive reaction with both the inside and the outside of the cinnamon insole (each +++ at D2, further testing stopped).

Contact dermatitis has been reported from the use of camphor in ear-drops (9). Photo-distributed skin eruptions have been reported after use of a sunscreen containing camphor (10).

Drug overdose

In children exposure to as little as 500 mg of camphor can be fatal (11). More commonly, a dose of 750–1000 mg is associated with seizures and death. Products that contain 10% camphor contain 500 mg in 5 ml.

After taking an unknown amount of a 10% camphor spirit (maximum dose 200 ml), a 54-year-old woman became comatose, having developed tonic-clonic seizures and respiratory failure (12). After gastric lavage, hemoperfusion was performed with amberlite XAD4 but did not alter either the pharmacokinetics of camphor or the course of the intoxication.

Laurus nobilis

Laurus nobilis (laurel) has well-known analgesic, diaphoretic, antipyretic, and diuretic effects (13), and is widely used in rheumatic, pyrexial, and infective disorders (14), as well as in the perfume and soap industries (13).

Adverse effects

Laurel oil obtained from the berries of *L. nobilis* is a potent skin sensitizer, owing to the presence of allergenic sesquiterpene lactones, and is usually seen in aromatherapists or their clients (15–17).

- A 55-year-old woman developed erythema and edema over her knees (15). She had applied laurel oil, obtained from a herbalist, to her knees to relieve joint pain 15 days earlier. After 3 days, the erythema and edema had begun to appear. She had erythema, edema, and papules over her patellae, and eczema around the eye. She was treated with an oral antihistamine and a topical glucocorticoid. Two days later, the lesions worsened and systemic glucocorticoid therapy was needed. The lesions started to heal, leaving slight postinflammatory hyperpigmentation. Patch-testing was performed with a European standard series and commercial laurel oil 1 month later. There was a +++ reaction to the oil only, and no reaction to either fragrance mix or *Myroxylon pereirae* resin in the standard series. The same formulation of laurel oil was negative on patch-testing in 15 control subjects.

Sassafras albidum

Sassafras albidum (sassafras) root contains 1–2% of volatile oil, which in turn consists largely of safrole, a weakly hepatocarcinogenic agent in laboratory animals. Some metabolites of safrole have mutagenic activity in bacteria and weak hepatocarcinogenic effects in rodents. The carcinogenic effect is primarily mediated by the formation of 1'-hydroxysafrole, followed by sulfonation to an unstable sulfate that reacts to form DNA adducts; these metabolites are formed by CYP2C9 and CYP2E1, the latter contributing three-fold more to the metabolic clearance than the former (18).

In one case sassafras tea caused sweating (19).

In Germany, the health authorities have proposed the withdrawal of sassafras-containing medicines, including homeopathic products up to D3, from the market (20). Of particular concern is the uncontrolled availability of sassafras oil because of its use in aromatherapy. Internal use of sassafras oil in recommended doses up to 12 drops/day can lead to a daily intake up to 0.2 g of safrole (21).

References

1. Rabl W, Katzgraber F, Steinlechner M. Camphor ingestion for abortion (case report). Forensic Sci Int 1997;89(1–2):137–40.
2. Piyaraly S, Boumahni B, Raudrant-Sigogne N, Edmar A, Renouil M, Mallet EC. Camphre percutane et convulsion chez un nouveau-né. [Percutaneous camphor and convulsions in a neonate.] Arch Pediatr 1998;5(2):205–6.
3. Emery DP, Corban JG. Camphor toxicity. J Paediatr Child Health 1999;35(1):105–6.
4. Uc A, Bishop WP, Sanders KD. Camphor hepatotoxicity. South Med J 2000;93(6):596–8.
5. Jimenez JF, Brown AL, Arnold WC, Byrne WJ. Chronic camphor ingestion mimicking Reye's syndrome. Gastroenterology 1983;84(2):394–8.
6. Kanerva L, Estlander T, Jolanki R. Occupational allergic contact dermatitis from spices. Contact Dermatitis 1996;35:157–62.
7. De Benito V, Alzaga R. Occupational allergic contact dermatitis from cassia (Chinese cinnamon) as a flavouring agent in coffee. Contact Dermatitis 1999;40:165.
8. Hartman K, Huyzelmann N. Allergic contact dermatitis from cinnamon as an odour-neutralizing agent in shoe insoles. Contact Dermatitis 2004;50:253–4.
9. Stevenson OE, Finch TM. Allergic contact dermatitis from rectified camphor oil in Earex ear drops. Contact Dermatitis 2003;49(1):51.
10. Marguery MC, Rakotondrazafy J, el Sayed F, Bayle-Lebey P, Journe F, Bazex J. Contact allergy to 3-(4' methylbenzylidene) camphor and contact and photocontact allergy to 4-isopropyl dibenzoylmethane. Photodermatol Photoimmunol Photomed 1996;11(5–6):209–12.
11. Love JN, Sammon M, Smereck J. Are one or two dangerous? Camphor exposure in toddlers. J Emerg Med 2004;27(1):49–54.
12. Koppel C, Martens F, Schirop T, Ibe K. Hemoperfusion in acute camphor poisoning. Intensive Care Med 1988; 14(4):431–3.
13. Ilisulu K. In: Ilaç ve Baharat Bitkileri. 1st ed. 1992:63–75 Ankara.
14. Yesilada E, Ustun O, Sezik E, Takaishi Y, Ono Y, Honda G. Inhibitory effects of Turkish folk remedies on

inflammatory cytokines: interleukin-1alpha, interleukin-1beta and tumor necrosis factor alpha. J Ethnopharmacol 1997;58(1):59–73.

15. Ozden MG, Oztas P, Oztas MO, Onder M. Allergic contact dermatitis from Laurus nobilis (laurel) oil. Contact Dermatitis 2001;45(3):178.

16. Schaller M, Korting HC. Allergic airborne contact dermatitis from essential oils used in aromatherapy. Clin Exp Dermatol 1995;20(2):143–5.

17. Keane FM, Smith HR, White IR, Rycroft RJ. Occupational allergic contact dermatitis in two aromatherapists. Contact Dermatitis 2000;43(1):49–51.

18. Ueng YF, Hsieh CH, Don MJ, Chi CW, Ho LK. Identification of the main human cytochrome P450 enzymes involved in safrole 1′-hydroxylation. Chem Res Toxicol 2004;17(8):1151–6.

19. Haines JD Jr. Sassafras tea and diaphoresis. Postgrad Med 1991;90(4):75–6.

20. Arzneimittelkommission der Deutschen Apotheker. Vorinformation Sassafras-haltige Arzneimittel. Dtsch Apoth Ztg 1995;135:366–8.

21. De Smet PAGM. Een alternatieve olie met een luchtje. Pharm Weekbl 1994;129:258.

Liliaceae

General Information

The genera in the family of Liliaceae (Table 35) include various types of lily, amarylis, asphodel, crocus, daffodil, fritillary, hyacinth, onions (including garlic), snowdrop, and tulip.

Table 35   The genera of Liliaceae

Aletris (colicroot)
Allium (onion)
Alstroemeria (lily of the Incas)
Amaryllis (amaryllis)
Amianthium (amianthium)
Androstephium (funnel lily)
Asparagus (asparagus)
Asphodelus (asphodel)
Astelia (pineapple grass)
Bloomeria (golden star)
Brodiaea (brodiaea)
Calochortus (mariposa lily)
Camassia (camas)
Chamaelirium (chamaelirium)
Chionodoxa (chionodoxa)
Chlorogalum (soap plant)
Chlorophytum (chlorophytum)
Clintonia (blue bead)
Colchicum (crocus)
Convallaria (lily of the valley)
Cooperia (rain lily)
Cordyline (cordyline)
Crinum (swamp lily)
Curculigo (curculigo)
Dasylirion (sotol)
Dianella (dianella)
Dichelostemma (snake lily)
Disporum (fairy bells)
Echeandia (echeandia)
Eremocrinum (eremocrinum)
Erythronium (fawn lily)
Eucharis (Amazon lily)
Fritillaria (fritillary)
Gagea (gagea)
Galanthus (snowdrop)
Gloriosa (flame lily)
Habranthus (copper lily)
Harperocallis (harperocallis)
Hastingsia (rush lily)
Helonias (helonias)
Hemerocallis (day lily)
Hesperocallis (desert lily)
Hippeastrum (hippeastrum)
Hosta (plantain lily)
Hyacinthoides (hyacinthoides)
Hyacinthus (hyacinth)
Hymenocallis (spider lily)
Hypoxis (star grass)
Kniphofia (red hot poker)
Leucocrinum (star lily)

(Continued)

Leucojum (snowflake)
Lilium (lily)
Liriope (lily turf)
Lloydia (alp lily)
Lophiola (lophiola)
Lycoris (lycoris)
Maianthemum (may flower)
Medeola (Indian cucumber)
Melanthium (bunch flower)
Merendera
Milla (milla)
Muilla (muilla)
Muscari (grape hyacinth)
Narcissus (daffodil)
Narthecium (asphodel)
Nolina (bear grass)
Nothoscordum (false garlic)
Odontostomum (Hartweg's doll's-lily)
Ophiopogon (ophiopogon)
Ornithogalum (star of Bethlehem)
Ornithoglossum
Pleea (pleea)
Pleomele (hala pepe)
Polygonatum (Solomon's seal)
Ruscus (broom)
Schoenocaulon (feathershank)
Schoenolirion (sunnybell)
Scilla (scilla)
Scoliopus (fetid adder's tongue)
Stenanthium (featherbells)
Sternbergia (winter daffodil)
Streptopus (twisted stalk)
Tofieldia (tofieldia)
Tricyrtis (tricyrtis)
Trillium (trillium)
Tristagma (springstar)
Triteleia (triteleia)
Triteleiopsis (Baja lily)
Tulipa (tulip)
Uvularia (bellwort)
Veratrum (false hellebore)
Xerophyllum (bear grass)
Zephyranthes (zephyr lily)
Zigadenus (death camas)

Food allergy to spices accounts for 2% of all cases of food allergies but 6.4% of cases in adults. Prick tests to native spices in 589 patients with food allergies showed frequent sensitization to the Liliaceae garlic, onion, and chive (4.6% of prick tests in children, 7.7% of prick tests in adults) (1).

Allium sativum

Allium sativum (garlic, camphor of the poor, da suan, poor man's treacle, rustic treacle, stinking rose) contains a variety of amino acids and steroids, including ajoene, alliin, allicin, glutamyl-*S*-allylcysteine, and glutamyl-*S*-(2-carboxy-1-propyl)-cysteinglycine. Traditionally it has been used as an antiseptic, diaphoretic, diuretic, expectorant, and stimulant and in the treatment of asthma, hoarseness, cough, difficulty in breathing, chronic bronchitis, leprosy, tubercular consumption, whooping-cough,

153

worms, epilepsy, rheumatism, dropsy, and hysteria. In modern times it has been used as a hypolipidemic (2), although its effect is small (3).

Adverse effects
Respiratory
Occupational inhalation of garlic powder can lead to asthma (4).

Nervous system
A spontaneous spinal epidural hematoma resulting in paraplegia in an 87-year-old patient was attributed to chronic excessive use of garlic cloves (5).

Hematologic
Bleeding due to impaired platelet function has been attributed to garlic (6).

- A 54-year-old woman underwent strabismus surgery and had bilateral retrobulbar hemorrhages intraoperatively. In the absence of other possible causes, the authors thought that the bleeding had been due to garlic pills prescribed by a naturopath. On the day of surgery, she had taken five pills, equivalent to about 5 g of fresh garlic bulb. Platelet function, measured 2 weeks later, was normal.

Garlic has well-documented effects on platelet aggregation (7).

Skin
Topical administration of garlic can lead to allergic contact dermatitis or burn-like skin lesions (8).

- A 50-year-old Romanian man was advised by his herbalist to treat his asthma with a compress of freshly crushed garlic (9). He wore the compress on his forehead overnight and subsequently developed second-degree burns in this area. Specific IgE RAST tests for garlic were negative. He was treated conservatively and made an uneventful recovery.
- Two Korean patients used topical garlic for pruritus and subsequently developed irritant contact dermatitis of the treated skin areas (10). Withdrawal resulted in full recovery.
- A 60-year-old Eastern European man developed partial thickness burns to both feet after having applied crushed raw garlic to his feet for 12 hours (11). He was initially treated with Silvadene cream and sent home, but 3 days later developed a low-grade fever and wound erythema. He was treated with intravenous nafcillin and Accuzyme debridement. He recovered uneventfully within 3 days.

Garlic contact dermatitis is a type IV allergic reaction limited to the epidermis, thought to be due to diallyldisulfide, allicin, and allylpropyldisulfide (12). However, partial or full thickness skin injury due to garlic is less common and is probably not a true allergic reaction.

Of about 1000 patients with occupational skin diseases, five had occupational allergic contact dermatitis from spices (13). They were chefs or workers in kitchens, coffee rooms, and restaurants. In all cases the dermatitis affected the hands. The causative spices were garlic, cinnamon, ginger, allspice, and clove. The same patients had positive patch-test reactions to carrot, lettuce, and tomato.

Drug-drug interactions
The potential interactions between garlic and prescribed drugs have been reviewed (14). The medicines that might thus be affected include ritonavir, saquinavir, paracetamol, warfarin, and chlorpropamide.

The effect of garlic on cytochrome P450 enzymes has been studied in 14 subjects, using the probe substances dextromethorphan and alprazolam (15). The results suggested that garlic extracts are unlikely to alter the disposition of co-administered medications whose metabolism primarily depends on CYP2D6 or CYP3A4.

Anticoagulants
Garlic has been proposed to reduce the effects of the anticoagulant fluindione (16).

- An 82-year-old man developed a reduced international normalized ratio after he started to take garlic tablets while also taking fluindione. He had started taking fluindione 1 year earlier because of chronic atrial fibrillation; his co-medications included enalapril, furosemide, and pravastatin. His international normalized ratio remained between 2 and 3 during 1 year of fluindione treatment and then suddenly decreased. Despite an increase in fluindione dosage (mean dose 6.6-10 mg) it remained below 2 for 12 consecutive days. Garlic was withdrawn and the international normalized ratio returned to its previous level within 4 days.

The antiplatelet activity of garlic might lead to over-anticoagulation when garlic and warfarin are administered concomitantly (17). However, evidence of such an effect is scanty (18).

Colchicum autumnale

Colchicum autumnale (autumn crocus) and other *Colchicum* species belong to the family known as the Colchicaceae, a proposed subdivision of the Liliaceae. They contain colchicine and related alkaloids. Other members of the Colchicaceae include *Gloriosa* species and *Merenda* species. *C. autumnale* (autumn crocus) is the traditional source of colchicine, which is covered in a separate monograph.

Another plant subdivision, the Wurmbaeoideae, includes *Androcymbium* species, *Iphigenia* species, and *Wurmbea* species, which also contain colchicine. *Iphigenia indica* (shan cigu) is a traditional Chinese herbal medicine (19).

Ruscus aculeatus

Ruscus aculeatus (butcher's broom, knee holy, knee holly, knee holm, Jew's myrtle, sweet broom, pettigree) has been used topically for vasoconstrictor treatment of varicose veins and hemorrhoids (20), and for chronic venous insufficiency, both alone (21,22) and in the combination known as Cyclo 3 fort, marketed in France, which contains an extract of *R. aculeatus* 150 mg, hesperidin methyl chalcone 150 mg, ascorbic acid 100 mg, and metesculetol.

In a meta-analysis of the efficacy of Cyclo 3 fort in patients with chronic venous insufficiency 20 double-blind, randomized, placebo-controlled studies and 5 randomized comparison studies in 10 246 subjects were included (23). Cyclo 3 fort significantly reduced the severity of pain, cramps, heaviness, and paresthesia compared with placebo. There were also significant reductions in venous capacity and severity of edema.

Adverse effects
Gastrointestinal
Chronic diarrhea has been described with Cyclo 3 fort and attributed to a disturbance of gastrointestinal motility (SEDA-16, 205) (SEDA-17, 244). However, the mechanism may be immunological, since lymphocytic colitis has occasionally been reported (24).

Skin
R. aculeatus can cause allergic contact dermatitis (25).

Veratrum species
The rhizome and root of *Veratrum album* (white hellebore) and the rhizome of *Veratrum viride* (green hellebore) contain many alkaloids, including hypotensive ester alkaloids and jervine. *Veratrum californicum* contains the alkaloids cyclopamine, cycloposine, and jervine.

Adverse effects
Among the major toxic symptoms of the veratrum alkaloids are vomiting, hypotension, and bradycardia.

V. californicum is teratogenic activity in livestock (26).

Drug overdose
Veratrum poisoning can cause heartburn and vomiting, bradydysrhythmias, atrioventricular dissociation, and vasodilatation with hypotension (27,28). It can be fatal (29).

Of 12 patients, aged 20–80 years, with acute hellebore (*V. album*) intoxication, 10 had a sinus bradycardia at about 40/minute, shortening of the PR interval down to 0.08–0.12 seconds and of the QT_c interval down to 0.32–0.36 seconds, transient right and incomplete left bundle-branch block, atrial and ventricular extra beats, nodal rhythm ($n = 1$), and altered ventricular repolarization (30). The authors suggested that the bradycardia due to *V. album* is caused by a reflex increase in vagal tone and responds to atropine.

Five patients with acute accidental poisoning with *V. album* rapidly developed nausea, vomiting, abdominal pain, hypotension, and bradycardia (31). In four cases the electrocardiogram showed sinus bradycardia and in one there was complete atrioventricular block with an ectopic atrial bradycardia and an intermittent idioventricular rhythm. Symptomatic treatment and/or atropine led to recovery within a few hours.

References

1. Moneret-Vautrin DA, Morisset M, Lemerdy P, Croizier A, Kanny G. Food allergy and IgE sensitization caused by spices: CICBAA data (based on 589 cases of food allergy). Allerg Immunol (Paris) 2002;34(4):135–40.
2. Neil A, Silagy C. Garlic: its cardio-protective properties. Curr Opin Lipidol 1994;5(1):6–10.
3. Stevinson C, Pittler MH, Ernst E. Garlic for treating hypercholesterolemia. A meta-analysis of randomized clinical trials. Ann Intern Med 2000;133(6):420–9.
4. Canduela V, Mongil I, Carrascosa M, Docio S, Cagigas P. Garlic: always good for the health? Br J Dermatol 1995;132(1):161–2.
5. Rose KD, Croissant PD, Parliament CF, Levin MB. Spontaneous spinal epidural hematoma with associated platelet dysfunction from excessive garlic ingestion: a case report. Neurosurgery 1990;26(5):880–2.
6. Carden SM, Good WV, Carden PA, Good RM. Garlic and the strabismus surgeon. Clin Experiment Ophthalmol 2002;30(4):303–4.
7. Briggs WH, Xiao H, Parkin KL, Shen C, Goldman IL. Differential inhibition of human platelet aggregation by selected *Allium* thiosulfinates. J Agric Food Chem 2000;48(11):5731–5.
8. Lee TY, Lam TH. Contact dermatitis due to topical treatment with garlic in Hong Kong. Contact Dermatitis 1991;24(3):193–6.
9. Baruchin AM, Sagi A, Yoffe B, Ronen M. Garlic burns. Burns 2001;27(7):781–2.
10. Yim YS, Park CW, Lee CH. Two cases of irritant contact dermatitis due to garlic. Korean J Dermatol 2001; 39:86–9.
11. Dietz DM, Varcelotti JR, Stahlfeld KR. Garlic burns: a not-so-rare complication of a naturopathic remedy? Burns 2004;30:612–3.
12. Papageorgiou C, Corbet JP, Menezes-Brandao F, Pecegueiro M, Benezra C. Allergic contact dermatitis to garlic (*Allium sativum*). Identification of the allergens: the role of mono-, di-, and trisulfides present in garlic. Arch Dermatol Res 1983;275:229–34.
13. Kanerva L, Estlander T, Jolanki R. Occupational allergic contact dermatitis from spices. Contact Dermatitis 1996;35(3):157–62.
14. Coxeter PD, McLachlan AJ, Duke CC, Roufogalis BD. Garlic-drug interactions. Complement Med 2003;Nov/Dec:57–9.
15. Markowitz JS, DeVane CL, Chavin KD, Taylor RM, Ruan Y, Donovan JL. Effects of garlic (*Allium sativum* L.) supplementation on cytochrome P450 2D6 and 3A4 activity in healthy volunteers Clin Pharmacol Ther 2003;74:170–7.
16. Pathak A, Léger P, Bagheri H, Senard J-M, Boccalon H, Montastruc J-L. Garlic interaction with fluindione: a case report. Therapie 2003;58:380–1.
17. Sunter WH. Warfarin and garlic. Pharm J 1991;246:722.
18. Vaes LP, Chyka PA. Interactions of warfarin with garlic, ginger, ginkgo, or ginseng: nature of the evidence. Ann Pharmacother 2000;34(12):1478–82.
19. De Smet PA. Health risks of herbal remedies. Drug Saf 1995;13(2):81–93.
20. MacKay D. Hemorrhoids and varicose veins: a review of treatment options. Altern Med Rev 2001;6(2):126–40.
21. Vanscheidt W, Jost V, Wolna P, Lucker PW, Muller A, Theurer C, Patz B, Grutzner KI. Efficacy and safety of a Butcher's broom preparation (*Ruscus aculeatus* L. extract) compared to placebo in patients suffering from chronic venous insufficiency. Arzneimittelforschung 2002; 52(4):243–50.
22. Beltramino R, Penenory A, Buceta AM. An open-label, randomized multicenter study comparing the efficacy and safety of Cyclo 3 Fort versus hydroxyethyl rutoside in chronic venous lymphatic insufficiency. Angiology 2000;51(7):535–44.
23. Boyle P, Diehm C, Robertson C. Meta-analysis of clinical trials of Cyclo 3 Fort in the treatment of chronic venous insufficiency. Int Angiol 2003;22(3):250–62.

24. Tysk C. Lakemedelsutlost enterokolit. Viktig differential-diagnos vid utredning av diarre och tarmblodning. [Drug-induced enterocolitis. Important differential diagnosis in the investigation of diarrhea and intestinal hemorrhage.] Lakartidningen 2000;97(21):2606–10.
25. Landa N, Aguirre A, Goday J, Raton JA, Diaz-Perez JL. Allergic contact dermatitis from a vasoconstrictor cream. Contact Dermatitis 1990;22(5):290–1.
26. James LF. Teratological research at the USDA-ARS poisonous plant research laboratory. J Nat Toxins 1999; 8(1):63–80.
27. Festa M, Andreetto B, Ballaris MA, Panio A, Piervittori R. Un caso di avvelenamento da veratro. [A case of veratrum poisoning.] Minerva Anestesiol 1996;62(5):195–6.
28. Quatrehomme G, Bertrand F, Chauvet C, Ollier A. Intoxication from *Veratrum album*. Hum Exp Toxicol 1993;12(2):111–5.
29. Gaillard Y, Pepin G. LC-EI-MS determination of veratridine and cevadine in two fatal cases of *Veratrum album* poisoning. J Anal Toxicol 2001;25(6):481–5.
30. Marinov A, Koev P, Mirchev N. [Electrocardiographic studies of patients with acute hellebore (*Veratrum album*) poisoning.] Vutr Boles 1987;26(6):36–9.
31. Garnier R, Carlier P, Hoffelt J, Savidan A. Intoxication aiguë alimentaire par l'ellebore blanc (*Veratum album* L.). Données cliniques et analytiques. A propos de 5 cas. [Acute dietary poisoning by white hellebore (*Veratrum album* L.). Clinical and analytical data. A propos of 5 cases.] Ann Med Interne (Paris) 1985;136(2):125–8.

Loganiaceae

General Information

The genera in the family of Loganiaceae (Table 36) include strychnos and trumpet flower.

Strychnos nux-vomica

The dried ripe seeds of *Strychnos nux-vomica* (nux vomica) contain the alkaloids strychnine and brucine, together with traces of other alkaloids.

Table 36 The genera of Loganiaceae

Fagraea
Gelsemium (trumpet flower)
Labordia (labordia)
Mitreola (hornpod)
Spigelia (pinkroot)
Strychnos (strychnos)

Adverse effects

Strychnine is a powerful convulsant, which can cause serious and even lethal poisoning [1].

Maqianzi, the dried ripe seed of *S. nux-vomica* contains 1.0–1.4% each of strychnine and brucine. It has been used as a herbal remedy for rheumatism, musculoskeletal injuries, and limb paralysis.

- A 42-year-old woman with neck pain took 15 g of maqianzi in two doses 7 hours apart (recommended dose 0.3–0.6 g) [2]. One hour after she took the second dose she suddenly developed tonic contractions of all her limbs and carpopedal spasm lasting 5 minutes, difficulty in breathing, chest discomfort, and perioral numbness. She complained of muscle pain and tiredness and had hyperventilation and weakness of all four limbs. All her symptoms gradually subsided over the next few hours.

References

1. Wang Z, Zhao J, Xing J, He Y, Guo D. Analysis of strychnine and brucine in postmortem specimens by RP-HPLC: a case report of fatal intoxication. J Anal Toxicol 2004;28(2):141–4.
2. Chan TY. Herbal medicine causing likely strychnine poisoning. Hum Exp Toxicol 2002;21(8):467–8.

Lycopodiaceae

General Information

The family of Lycopodiaceae contains three genera of clubmoss, *Huperzia*, *Lycopodium*, and *Lycopodiella*.

Lycopodium serratum

Lycopodium serratum (clubmoss, Jin bu huan) has been used in Chinese medicine for more than 1000 years. It contains the alkaloid serratidine and triterpenoids, such as oxolycoclavinol, oxoserratenetriol, tohogeninol, and tohogenol.

Adverse effects

Jin bu huan can cause liver damage (1).

- A 49-year-old man developed signs of hepatitis after taking three tablets of Jin bu huan per day for 2 months for insomnia. No other potential causes for the liver damage could be identified. A liver biopsy showed chronic hepatitis with moderate portal and parenchymal lymphocytic inflammation and focal necrosis. The patient stopped taking the herbal remedy and his liver function normalized.

There have been other reports of hepatotoxic effects of Jin bu huan (2).

Overdosage of Jin bu huan can cause central nervous system and respiratory depression with rapid life-threatening bradycardia (3,4).

References

1. Picciotto A, Campo N, Brizzolara R, Giusto R, Guido G, Sinelli N, Lapertosa G, Celle G. Chronic hepatitis induced by Jin Bu Huan. J Hepatol 1998;28(1):165–7.
2. McRae CA, Agarwal K, Mutimer D, Bassendine MF. Hepatitis associated with Chinese herbs. Eur J Gastroenterol Hepatol 2002;14(5):559–62.
3. Centers for Disease Control and Prevention (CDC). Jin bu huan toxicity in adults—Los Angeles, 1993. MMWR Morb Mortal Wkly Rep 1993;42(47):920–2.
4. Centers for Disease Control and Prevention (CDC). Jin bu huan toxicity in children—Colorado, 1993. MMWR Morb Mortal Wkly Rep 1993;42(33):633–6.

Malvaceae

General Information

The genera in the family of Malvaceae (Table 37) include cotton, okra, and various types of mallow.

Gossypium species

Gossypol occurs in certain species of *Gossypium* (cotton), mostly in the seeds and root bark. Clinical studies have confirmed its efficacy as a male contraceptive agent (1).

Gossypol is a racemic mixture of (+)-gossypol and (−)-gossypol. When tested separately in male rats in a dosage of 30 mg/kg orally for 14 days, only the latter enantiomer had a clear anti-fertility effect. Furthermore, (+)-gossypol has less acute toxicity than (−)-gossypol after intraperitoneal administration to mice. In a human study, the enantiomers showed pharmacokinetic differences in half-life; less than 5 hours for (−)-gossypol compared with 133 hours for the (+)-enantiomer.

Adverse effects
The reported adverse effects of gossypol include fatigue, changes in appetite, transient rises in alanine transaminase, and hypokalemia.

Because of the adverse effects of gossypol, The Special Program of Research, Development and Research Training in Human Reproduction (HRP) at the World Health Organization (WHO) decided that gossypol is not acceptable as an antifertility drug (2).

Electrolyte balance
Gossypol causes potassium loss, perhaps by a direct toxic effect on the renal tubules (3), which can cause renal tubular acidosis (4). Hypokalemic paralysis (5) has been reported in about 1% of 8806 volunteers and in China was more common in areas of low dietary potassium intake, such as Nanjing (6); muscular weakness and severe fatigue are prodromal signs (7). Neither potassium supplementation nor triamterene prevented potassium loss in patients taking gossypol (8).

Fertility
Of 151 men from Brazil, Nigeria, Kenya, and China who received gossypol 15 mg/day for 12 or 16 weeks and were then randomized to either 7.5 or 10 mg/day for 40 weeks, 81 had suppression of spermatogenesis (9). Testicular volume decreased, but normalized after withdrawal. Of 19 subjects who took 7.5 mg/day, 12 recovered sperm counts over 20 million/ml within 12 months of discontinuing gossypol; of 24 who took 10 mg/day, sperm counts recovered in only 10. Eight remained azoospermic 1 year after stopping gossypol.

Psoralea corylifolia

Infusions prepared from the seeds of *Psoralea corylifolia* (bakuchi) can cause photosensitivity due to the presence of psoralens (10).

Table 37 The genera of Malvaceae

Abelmoschus (okra)
Abutilon (Indian mallow)
Alcea (hollyhock)
Allosidastrum (mock fanpetals)
Allowissadula (false Indian mallow)
Althaea (marshmallow)
Anoda (anoda)
Bastardia (bastardia)
Batesimalva (gaymallow)
Billieturnera (false fanpetals)
Callirhoe (poppymallow)
Cienfuegosia (flymallow)
Eremalche (mallow)
Fioria (fanleaf)
Freyxellia (fryxellwort)
Gossypium (cotton)
Herissantia (herissantia)
Hibiscadelphus (hibiscadelphus)
Hibiscus (rose mallow)
Horsfordia (velvet mallow)
Iliamna (wild hollyhock)
Kokia (tree cotton)
Kosteletzkya (kosteletzkya)
Krapovickasia (krapovickasia)
Lavatera (tree mallow)
Malachra (leaf bract)
Malacothamnus (bush mallow)
Malva (mallow)
Malvastrum (false mallow)
Malvaviscus (wax mallow)
Malvella (mallow)
Meximalva (meximalva)
Modiola (bristle mallow)
Napaea (napaea)
Pavonia (swamp mallow)
Pseudabutilon (pseudabutilon)
Rhynchosida (rhynchosida)
Sida (fanpetals)
Sidalcea (checkerbloom)
Sidastrum (sand mallow)
Sphaeralcea (globe mallow)
Thespesia (thespesia)
Urena (urena)
Wissadula (wissadula)

References

1. Gu ZP, Mao BY, Wang YX, Zhang RA, Tan YZ, Chen ZX, Cao L, You GD, Segal SJ. Low dose gossypol for male contraception. Asian J Androl 2000;2(4):283–7.
2. Waites GM, Wang C, Griffin PD. Gossypol: reasons for its failure to be accepted as a safe, reversible male antifertility drug. Int J Androl 1998;21(1):8–12.
3. Wang C, Yeung RT. Gossypol and hypokalaemia. Contraception 1985;32(3):237–52.
4. Gao H, Yang ZS, Jin SX. [Primary observations on distal renal tubule acidosis in 177 cases caused by gossypol intoxication.]Zhonghua Nei Ke Za Zhi 1985;24(7):419–21447.
5. Fan LP, Ren CQ, Hou TX. [Emergency treatment and care of hypokalemia paralysis due to gossypol poisoning.] Zhonghua Hu Li Za Zhi 1995;30(9):522–5.
6. Qian SZ. Gossypol–hypokalaemia interrelationships. Int J Androl 1985;8(4):313–24.

7. Qian SZ, Jing GW, Wu XY, Xu Y, Li YQ, Zhou ZH. Gossypol related hypokalemia. Clinicopharmacologic studies. Chin Med J (Engl) 1980;93(7):477–82.

8. Liu GZ, Ch'iu-Hinton K, Cao JA, Zhu CX, Li BY. Effects of K salt or a potassium blocker on gossypol-related hypokalemia. Contraception 1988;37(2):111–7.

9. Coutinho EM, Athayde C, Atta G, Gu ZP, Chen ZW, Sang GW, Emuveyan E, Adekunle AO, Mati J, Otubu J, Reidenberg MM, Segal SJ. Gossypol blood levels and inhibition of spermatogenesis in men taking gossypol as a contraceptive. A multicenter, international, dose-finding study. Contraception 2000;61(1):61–7.

10. Maurice PD, Cream JJ. The dangers of herbalism. BMJ 1989;299(6709):1204.

Meliaceae

General Information

The genera in the family of Meliaceae (Table 38) include mangrove and different types of mahogany.

Table 38 The genera of Meliaceae

Aglaia (Chinese rice flower)
Azadirachta (azadirachta)
Carapa (carapa)
Cedrela (cedrela)
Dysoxylum (mahogany)
Guarea (guarea)
Khaya (African mahogany)
Lansium (lansium)
Melia (melia)
Swietenia (mahogany)
Toona (redcedar)
Trichilia (trichilia)
Xylocarpus (mangrove)

Azadirachta indica

Azadirachta indica (bead tree, pride of China, margosa, neem or nim tree, holy tree, indiar, lilac tree) contains a variety of steroids, sesquiterpenoids, ester terpenoids, tetranortriterpenoids, and triterpenoids.

Adverse effects

Margosa oil, a long-chain fatty acid that is extracted from the seeds of the neem tree, has been implicated as a potential cause of Reye's syndrome in 13 infants who developed vomiting, drowsiness, metabolic acidosis, a polymorphonuclear leukocytosis, and encephalopathy within hours of taking margosa oil; liver biopsy in one infant showed pronounced fatty infiltration of the liver (1). In experimental animals given margosa oil there was also fatty infiltration of the proximal renal tubules and cerebral edema. Electron microscopy showed mitochondrial damage.

Reference

1. Sinniah D, Baskaran G. Margosa oil poisoning as a cause of Reye's syndrome. Lancet 1981;1(8218):487–9.

Menispermaceae

General Information

The genera in the family of Menispermaceae (Table 39) include chondrodendron and moonseed.

Stephania species

Various species of *Corydalis* (fumewort, Fumariaceae) and *Stephania* (Jin bu huan) contain tetrahydropalmatine (1), which is the active constituent in Chinese "Jin Bu Huan Anodyne" tablets, sold on the Western market. The package insert suggested that *Polygala chinensis* is the source plant, but in reality this alkaloid comes from a species of *Stephania*. Both l-tetrahydropalmatine and its racemic dl-form are used in Chinese medicine as analgesic and hypnotic agents.

Stephania has been adulterated with *Aristolochia* (2,3), and confusion is possible because of the similarity in their Chinese names: han fang ji (*Stephania*) and guang fang ji (*Aristolochia*).

Table 39 The genera of Menispermaceae

Calycocarpum (calycocarpum)
Chondrodendron (chondrodendron)
Cissampelos (cissampelos)
Cocculus (coral bead)
Dioscoreophyllum (dioscoreophyllum)
Fibraurea (fibraurea)
Hyperbaena (hyperbaena)
Jateorhiza (jateorhiza)
Menispermum (moonseed)
Stephania

Adverse effects

The reported adverse effects of Jin bu huan include vertigo, fatigue, nausea and drowsiness, which could make users unfit for driving. Life-threatening bradycardia and respiratory depression occur in small children after unintentional overdosage (4) and acute hepatitis in adults (5).

Nine adults acutely poisoned with tetrahydropalmatine recovered quickly after mild neurological disturbances; tetrahydropalmatine was metabolized rapidly and excreted as polar metabolites in the urine (6).

References

1. Zhu XZ. Development of natural products as drugs acting on central nervous system. Mem Inst Oswaldo Cruz 1991;86(Suppl 2):173–5.
2. Nortier JL, Vanherweghem JL. Renal interstitial fibrosis and urothelial carcinoma associated with the use of a Chinese herb (*Aristolochia fangchi*). Toxicology 2002;181–182:577–80.
3. Vanhaelen M, Vanhaelen-Fastre R, But P, Vanherweghem JL. Identification of aristolochic acid in Chinese herbs. Lancet 1994;343(8890):174.
4. Horowitz RS, Gomez H, Moore LL, Fulton B, Feldhaus K, Brent J, Stermitz FR, Beck JJ, Alessi JR, De Smet PAGMCenters for Disease Control and Prevention (CDC). Jin bu huan toxicity in children—Colorado, 1993. MMWR Morb Mortal Wkly Rep 1993;42(33):633–6.
5. Woolf GM, Petrovic LM, Rojter SE, Wainwright S, Villamil FG, Katkov WN, Michieletti P, Wanless IR, Stermitz FR, Beck JJ, Vierling JM. Acute hepatitis associated with the Chinese herbal product jin bu huan. Ann Intern Med 1994;121(10):729–35.
6. Lai CK, Chan AY. Tetrahydropalmatine poisoning: diagnoses of nine adult overdoses based on toxicology screens by HPLC with diode-array detection and gas chromatography-mass spectrometry. Clin Chem 1999;45(2):229–36.

Myristicaceae

General Information

There are three genera of the family of Myristicaceae:

- *Horsfieldia*
- *Myristica* (nutmeg)
- *Virola* (virola).

Myristica fragrans

Myristica fragrans (nutmeg) has long been a highly desired spice. In the seventeenth century physicians claimed that a nutmeg pomander was a cure for the bloody flux and the sweating sickness, later called the black plague, and nutmeg was also supposed to have aphrodisiac properties (1). Mace, the dried shell of the nutmeg, is used as a spice; it should not be confused with chemical Mace, which is a form of tear gas containing 1% chloracetophenone (CN) gas in a solvent of sec-butanol, propylene glycol, cyclohexene, and dipropylene glycol methyl ether; some formulations also include oleoresin *Capsicum* (from peppers).

Myristicin is methoxysafrole, the principal aromatic constituent of the volatile oil of nutmeg. Myristicin is also found in several members of the carrot family (Apiaceae, formerly Umbelliferae), such as *Petroselinum crispum* (parsley) (2). Safrole is a mutagenic and animal carcinogenic monoterpenoid. It is the major component of oil of sassafras, and lesser quantities occur in essential oils from cinnamon, mace, nutmeg, and star anise. Some of its known or possible metabolites have mutagenic activity in bacteria and it has weak hepatocarcinogenic effects in rodents. Experiments in mice have suggested the possibility of transplacental and lactational carcinogenesis. The carcinogenic effect is primarily mediated by the formation of 1′-hydroxysafrole, followed by sulfonation to an unstable sulfate that reacts to form DNA adducts; these metabolites are formed by CYP2C9 and CYP2E1, the latter contributing three-fold more to the metabolic clearance than the former (3).

Adverse effects
Respiratory
Asthma has been attributed to nutmeg (4).

- A 27-year-old man developed rhinitis and asthma 1 year after starting to prepare a certain kind of sausage, having a previous allergy to coconut, banana, and kiwi and allergic rhinitis to horses, cats, dogs, and cows. There was a positive immediate skin prick test with paprika (dry powder of *Capsicum annuum*), coriander (*Coriandrum sativum*), and mace (shell of *M. fragrans*) at concentrations of 10%. There were specific IgE antibodies to paprika, coriander, and mace.

Nervous system
Large doses of nutmeg seed can cause nausea, vomiting, flushing, dry mouth, tachycardia, nervous system stimulation possibly with epileptiform convulsions, miosis, mydriasis, euphoria, and hallucinations (5).

- A 16-year-old youth who had taken nutmeg for recreational purposes developed a number of neurological symptoms and signs along with non-specific electrocardiographic changes and anticholinergic-type symptoms (6).

Psychiatric
Nutmeg is sometimes abused for its hallucinogenic potential; less than one tablespoon can be enough to produce severe symptoms similar to those seen in anticholinergic poisoning (7,8). It can also cause psychosis, both acute (9,10) and chronic (11).

- A 13-year-old girl took 15–24 g of nutmeg in gelatin capsules over 3 hours and smoked two joints of marijuana (12). She developed bizarre behavior and visual, auditory, and tactile hallucinations, nausea, gagging, hot/cold sensations, and blurred vision, followed by numbness, double, and "triple" vision, headache, and drowsiness. There was nystagmus, muscle weakness, and ataxia. She received activated charcoal 50 g and recovered within 2 days.

Deaths from acute poisoning have been reported (13).

Psychological
Nutmeg is the dried kernel of the evergreen tree *Myristica fragrans*. At sufficiently high doses it has psychoactive effects.

- A 23-year-old man developed delusions and hallucinations (14). His medical history was unremarkable. For several months he had been regularly taking nutmeg 5 g/day. He improved after withdrawal of nutmeg and was treated with olanzapine. After 6 months he was still taking antipsychotic medication but could resume work.

Skin
Among 55 patients with suspected contact dermatitis, skin patch tests that were positive at concentrations of both 10% and 25% were most common with ginger ($n = 7$), nutmeg ($n = 5$), and oregano ($n = 4$); other spices produced no responses or one positive response (15). Positive reactions at only one concentration were more likely at 25%: nutmeg ($n = 5$), ginger and cayenne ($n = 4$), curry, cumin, and cinnamon ($n = 3$), turmeric, coriander, and sage ($n = 2$), oregano ($n = 1$), and basil and clove ($n = 0$).

Contact allergy has been reported with mace (16).

References

1. Milton G. Nathaniel's Nutmeg. London: Hodder and Stoughton, 1999.
2. Hallstrom H, Thuvander A. Toxicological evaluation of myristicin. Nat Toxins 1997;5(5):186–92.
3. Ueng YF, Hsieh CH, Don MJ, Chi CW, Ho LK. Identification of the main human cytochrome P450 enzymes

involved in safrole 1′-hydroxylation. Chem Res Toxicol 2004;17(8):1151–6.

4. Sastre J, Olmo M, Novalvos A, Ibanez D, Lahoz C. Occupational asthma due to different spices. Allergy 1996;51(2):117–20.

5. Servan J, Chochon F, Duclos H. Hallucinations après ingestion volontaire de noix de muscade: une toxicomanie méconnue. [Hallucinations after voluntary ingestion of nutmeg: an unrecognized drug abuse.] Rev Neurol (Paris) 1998;154(10):708.

6. McKenna A, Nordt SP, Ryan J. Acute nutmeg poisoning. Eur J Emerg Med 2004;11(4):240–1.

7. Lavy G. Nutmeg intoxication in pregnancy. A case report. J Reprod Med 1987;32(1):63–4.

8. Abernethy MK, Becker LB. Acute nutmeg intoxication. Am J Emerg Med 1992;10(5):429–30.

9. Kelly BD, Gavin BE, Clarke M, Lane A, Larkin C. Nutmeg and psychosis. Schizophr Res 2003;60(1):95–6.

10. Dinakar HS. Acute psychosis associated with nutmeg toxicity. Med Times 1977;105(12):63–4.

11. Brenner N, Frank OS, Knight E. Chronic nutmeg psychosis. J R Soc Med 1993;86(3):179–80.

12. Sangalli BC, Chiang W. Toxicology of nutmeg abuse. J Toxicol Clin Toxicol 2000;38(6):671–8.

13. Stein U, Greyer H, Hentschel H. Nutmeg (myristicin) poisoning—report on a fatal case and a series of cases recorded by a poison information centre. Forensic Sci Int 2001; 118(1):87–90.

14. Kelly BD, Gavin BE, Clarke M, Lane A, Larkin C. Nutmeg and psychosis. Schizophr Res 2003;60:95–6.

15. Futrell JM, Rietschel RL. Spice allergy evaluated by results of patch tests. Cutis 1993;52(5):288–90.

16. Frazier CA. Contact allergy to mace. JAMA 1976;236(22): 2526.

Myrtaceae

General Information

The family of Myrtaceae (Table 40) includes guava and various types of gums.

Eugenia caryophyllus (oil of cloves)

Oil of cloves contains about 80% eugenol, its active component. Initially used to treat gastrointestinal disturbances, it is now used to treat toothache.

Adverse effects

Liver

It has been suggested that eugenol-induced hepatotoxicity is similar to that seen with paracetamol poisoning.

- A 15-month-old boy accidentally took 10–20 ml of oil of cloves (1). On arrival at the emergency department 1 hour later he was agitated and tachypneic, with biphasic stridor. Serial measurements of alanine transaminase activities showed evolving hepatic impairment 15 hours after ingestion. After 24 hours, the transaminase activity was in excess of 13 000 U/l, with blood urea and creatinine concentrations of 11.8 mmol/l and 134 μmol/l respectively, indicating evolving acute renal insufficiency. He improved slowly over the next 8 days, during which his urine output and blood urea returned to normal, although the transaminase activity remained moderately raised.

Table 40 The genera of Myrtaceae

Baeckea (baeckea)
Callistemon (bottle brush)
Calothamnus (net bush)
Calyptranthes (mountain bay)
Chamelaucium (chamelaucium)
Corymbia (corymbia)
Eucalyptus (gum)
Eugenia (stopper)
Feijoa (feijoa)
Gomidesia (gomidesia)
Leptospermum (tea tree)
Lophostemon (lophostemon)
Marlierea (marlierea)
Melaleuca (melaleuca)
Metrosideros (lehua)
Myrcia (rodwood)
Myrcianthes (myrcianthes)
Myrciaria (guava berry)
Myrtus (myrtus)
Pimenta (pimenta)
Pseudanamomis (pseudanamomis)
Psidium (guava)
Rhodomyrtus (rhodomyrtus)
Siphoneugena (siphoneugena)
Syncarpia (turpentine tree)
Syzygium (syzygium)
Tristaniopsis (tristaniopsis)

The authors pointed out that a substantial number of cases involving aromatherapy oils are on record. However, most are cases of poisoning, rather than true adverse effects following correct usage. Nevertheless, these cases do stress that such oils are not free of risks.

Eucalyptus species

Eucalyptus oil has been used in different medications to relieve symptoms of asthma and rhinitis and is a constituent of cold remedies and ointments designed to be rubbed on to the chest or applied around and even into the nostrils (Vicks Vaporub, Obat Madjan, etc.). Applied to the skin it acts as a rubefacient and counterirritant, but substantial concentrations of volatile oil can be inhaled from the skin. Eucalyptus oil has also been used systemically to treat asthma.

Adverse effects

Fever and headache occurred in a heavy consumer of eucalyptus extract (2).

Respiratory

Asthma can be exacerbated by eucalyptus (3).

- In a 30-year-old woman the symptoms of asthma and rhinoconjunctivitis were exacerbated by eucalyptus pollens and by ingestion of an infusion containing eucalyptus. There were specific IgE antibodies to eucalyptus pollens but not to common aeroallergens.

Eucalyptus oil has been reported to cause vocal cord dysfunction (4).

- A 46-year-old woman with vocal cord dysfunction associated with exposure to eucalyptus underwent inhalation challenges consisting of water, ammonia, pine oil, and a combination of eucalyptus (dried leaves) and ammonia. Vocal cord dysfunction occurred within minutes of exposure to eucalyptus.

Skin

Contact urticaria has been ascribed to eucalyptus pollen (5).

Drug overdose

Eucalyptus oil has caused death in doses of up to 560 ml in adults and 15 ml in children.

- A 73-year-old woman who deliberately took 200–250 ml of eucalyptus oil was found unconscious in her home after she had vomited and had been incontinent of urine and feces. Despite intensive treatment she developed pneumonitis and aspiration pneumonia to which she ultimately succumbed (6).

In a telephone survey of 109 parents or guardians of children under 5 years who had been involved in actual or suspected ingestion of eucalyptus oil, 90 incidents had involved vaporizer solutions, 15 eucalyptus oil formulations, and the remainder other medicinal products containing eucalyptus oil (7).

In a retrospective analysis of case histories of 109 children admitted to the Royal Children's Hospital, Melbourne, between 1 January 1981 and 31 December 1992 with a diagnosis of eucalyptus oil poisoning, there were clinical effects in 59%; 31 had depression of consciousness, 27 were drowsy, three were unconscious after taking known or estimated volumes of 5–10 ml, and one was unconscious with hypoventilation after taking an estimated 75 ml (8). Vomiting occurred in 37%, ataxia in 15%, and pulmonary disease in 11%. No treatment was given to 12%; ipecac or oral activated charcoal was given to 21%, nasogastric charcoal to 57%, and gastric lavage without anesthesia to 4% and under anesthesia to 6%. All recovered.

Of 42 children with oral eucalyptus oil poisoning, 33 were entirely asymptomatic (9). This group included all of the four children who were reported to have taken more than 30 ml of eucalyptus oil. Only two of the others had symptoms or clinical signs on presentation to hospital. No child required advanced life support. There was no correlation between the amount of eucalyptus oil taken and the presence of symptoms.

Systemic eucalyptus oil toxicity can result from topical application.

- A 6-year-old girl presented with slurred speech, ataxia and muscle weakness progressing to unconsciousness following the widespread application of a home remedy for urticaria containing eucalyptus oil. Six hours after removal of the topical preparation her symptoms had resolved and there were no long-term sequelae (10).

Drug administration route

Systemic effects of eucalyptus oil applied locally have been reported (10).

- A 6-year-old girl was treated with a home-made concoction mainly containing 50 ml of eucalyptus oil (consisting of 80–85% cineole oil) for pruritus. As the mixture seemed to relieve her symptoms, the parents applied it more and more generously to the girl's skin. She subsequently developed symptoms of intoxication: slurred speech, unsteady gait, and drowsiness. Eventually she lost consciousness and was unrousable. She was admitted to hospital, where several tests were negative, but her urine sample contained components of eucalyptus oil. She made a full recovery within 24 hours after withdrawal of the external herbal treatment.

Melaleuca species

The undiluted essential oil from the leaves of *Melaleuca alternifolia* (tea tree), which contains the sesquiterpene viridiflorene and the monoterpenoids eucalyptol, limonene, and terpinen-4-ol, has been used as a topical natural cure for bacterial and fungal skin infections.

Adverse effects
Nervous system

- A 23-month-old boy became confused and was unable to walk 30 minutes after ingesting less than 10 ml of T36-C7, a commercial product containing 100% tea tree oil (11). His condition improved and he was asymptomatic within 5 hours of ingestion.

Skin

Tea tree oil can have adverse effects on the skin (12). Several patients developed an allergic contact dermatitis (13), which was most commonly caused by the constituent d-limonene (14). Internal use of half a teaspoonful of the oil can result in a dramatic rash (15), whereas half a teacup can induce coma followed by a semiconscious state with hallucinations (16). Less than 10 ml is sufficient to produce serious signs of toxicity in small children.

- In one case, allergic contact dermatitis to tea tree oil presented with an extensive erythema multiforme-like reaction (17). However, a skin biopsy from a target-like lesion showed a spongiotic dermatitis without the features of erythema multiforme. Five months after treatment with systemic and topical glucocorticoids, patch testing elicited a 3+ reaction to old, oxidized tea tree oil, a 2+ reaction to fresh tea tree oil, a 2+ reaction to colophony, a 1+ reaction to abitol, and a 1+ reaction to balsam of Peru.

- An 18-year-old woman developed linear IgA bullous dermatosis after applying tea tree oil to her umbilical region after piercing (18). Patch testing was positive to tea tree oil. She was treated with topical clobetasol and recovered fully after 5 days.

Of 1216 patients who were patch tested, 14 with eczema had used products (creams, hair products, and essential oils) containing tea tree oil (19). They were patch tested for a standard panel of allergens, topical emulgents, perfumes, plants, topical medications, metal, gloves, topical disinfectants and preservatives, dental products, and rubber derivatives. Seven patients had allergic contact dermatitis due to tea tree oil. Two of them also had delayed type IV hypersensitivity toward fragrance-mix or colophony, suggesting the possibility of cross-reactions or allergic group reactions caused by contamination of the colophony with the volatile fraction of turpentines.

Immunologic

Of 1017 subjects, 16 had positive skin tests to *Melaleuca quinquenervia* pollen extract (20). Six of them were subjected to double-blind nasal challenge with the pollen extract and four to single-blind bronchial challenge; 11 received 34 different *Melaleuca* odor challenges (blossoms, bark, and leaves) through a closed system for up to 30 minutes. Four inhaled an odor from cajeput oil (derived from *Melaleuca* leaves) for 1 hour. One of six nasal challenges and one of four bronchial challenges were positive. All the odor challenges with blossoms, bark, leaves, and cajeput oil were negative. A radioallergosorbent test for *Melaleuca* pollen extract correlated with the skin test results. The authors concluded that the *Melaleuca* tree is not a significant source of aeroallergens and explained the few positive results on the basis of cross-reactivity with pollen extracts from a proven aeroallergen, *Bahia* grass pollen.

Anaphylaxis has been attributed to topical tea tree oil.

- A 38-year-old man applied tea tree oil to psoriatic lesions on his legs and immediately developed erythema at the sites of application (21). Within minutes he developed throat constriction and generalized pruritus and subsequently all the signs of anaphylaxis. He was treated accordingly and made a swift, full recovery.

Syzygium aromaticum

Syzygium aromaticum (clove) is used as an aromatic spice.

Adverse effects

Of about 1000 patients with occupational skin diseases, five had occupational allergic contact dermatitis from spices (22). They were chefs, or workers in kitchens, coffee rooms, and restaurants. In all cases the dermatitis affected the hands. The causative spices were garlic, cinnamon, ginger, allspice, and clove. The same patients had positive patch test reactions to carrot, lettuce, and tomato.

References

1. Janes SE, Price CS, Thomas D. Essential oil poisoning: N-acetylcysteine for eugenol-induced hepatic failure and analysis of a national database. Eur J Pediatr 2005; 164(8): 520–2.
2. Tascini C, Ferranti S, Gemignani G, Messina F, Menichetti F. Clinical microbiological case: fever and headache in a heavy consumer of eucalyptus extract. Clin Microbiol Infect 2002;8(7):437445–6.
3. Galdi E, Perfetti L, Calcagno G, Marcotulli MC, Moscato G. Exacerbation of asthma related to eucalyptus pollens and to herb infusion containing eucalyptus. Monaldi Arch Chest Dis 2003;59(3):220–1.
4. Huggins JT, Kaplan A, Martin-Harris B, Sahn SA. Eucalyptus as a specific irritant causing vocal cord dysfunction. Ann Allergy Asthma Immunol 2004;93(3):299–303.
5. Vidal C, Cabeza N. Contact urticaria due to eucalyptus pollen. Contact Dermatitis 1992;26(4):265.
6. Anpalahan M, Le Couteur DG. Deliberate self-poisoning with eucalyptus oil in an elderly woman. Aust NZ J Med 1998;28(1):58.
7. Day LM, Ozanne-Smith J, Parsons BJ, Dobbin M, Tibballs J. Eucalyptus oil poisoning among young children: mechanisms of access and the potential for prevention. Aust NZ J Public Health 1997;21(3):297–302.
8. Tibballs J. Clinical effects and management of eucalyptus oil ingestion in infants and young children. Med J Aust 1995;163(4):177–80.
9. Webb NJ, Pitt WR. Eucalyptus oil poisoning in childhood: 41 cases in south-east Queensland. J Paediatr Child Health 1993;29(5):368–71.
10. Darben T, Cominos B, Lee CT. Topical eucalyptus oil poisoning. Australas J Dermatol 1998;39(4):265–7.
11. Jacobs MR, Hornfeldt CS. Melaleuca oil poisoning. J Toxicol Clin Toxicol 1994;32(4):461–4.
12. Crawford GH, Sciacca JR, James WD. Tea tree oil: cutaneous effects of the extracted oil of Melaleuca alternifolia. Dermatitis 2004;15(2):59–66.
13. van der Valk PG, de Groot AC, Bruynzeel DP, Coenraads PJ, Weijland JW. Allergisch contacteczeem voor "tea tree"-olie. [Allergic contact eczema due to "tea tree" oil.] Ned Tijdschr Geneeskd 1994;138(16):823–5.
14. Knight TE, Hausen BM. Melaleuca oil (tea tree oil) dermatitis. J Am Acad Dermatol 1994;30(3):423–7.
15. Elliott C. Tea tree oil poisoning. Med J Aust 1993;159(11–12):830–1.
16. Seawright A. Tea tree oil poisoning. Med J Aust 1993;159:831.
17. Khanna M, Qasem K, Sasseville D. Allergic contact dermatitis to tea tree oil with erythema multiforme-like id reaction. Am J Contact Dermat 2000;11(4):238–42.
18. Perrett CM, Evans AV, Russell-Jones R. Tea tree oil dermatitis associated with linear IgA disease. Clin Exp Dermatol 2003; 28: 167–70.
19. Fritz TM, Burg G, Krasovec M. Dermatite de contact allergique aux cosmetiques a base de Melaleuca alternifolia (tea tree oil). [Allergic contact dermatitis to cosmetics containing Melaleuca alternifolia (tea tree oil).] Ann Dermatol Venereol 2001;128(2):123–6.
20. Stablein JJ, Bucholtz GA, Lockey RF. Melaleuca tree and respiratory disease. Ann Allergy Asthma Immunol 2002;89(5):523–30.
21. Mozelsio NB, Harris KE, McGrath KG, Grammer LC. Immediate systemic hypersensitivity reaction associated with topical application of Australian tea tree oil. Allergy Asthma Proc 2003; 24: 73–5.
22. Kanerva L, Estlander T, Jolanki R. Occupational allergic contact dermatitis from spices. Contact Dermatitis 1996;35(3):157–62.

Onagraceae

General Information

The genera in the family of Onagraceae (Table 41) include evening primrose and fuchsia.

Oenothera biennis

The seeds of *Oenothera biennis* (evening primrose, fever plant, king's cure-all, night willow herb, scabish, sundrop, tree primrose) yield evening primrose oil, which contains gamma-linolenic acid and has been used in various disorders, such as atopic eczema, premenstrual syndrome, and benign breast pain, but is probably not efficacious (1,2).

Table 41 The genera of Onagraceae

Calylophus (sundrops)
Camissonia (suncup)
Chamerion (fireweed)
Circaea (enchanter's nightshade)
Clarkia (clarkia)
Epilobium (willowherb)
Fuchsia (fuchsia)
Gaura (beeblossom)
Gayophytum (groundsmoke)
Ludwigia (primrose willow)
Oenothera (evening primrose)
Stenosiphon (stenosiphon)

Adverse effects

Evening primrose oil has generally only minor adverse effects, such as nausea, diarrhea, and headache.

In three patients evening primrose oil triggered temporal lobe epilepsy (3).

References

1. Huntley AL, Ernst E. A systematic review of herbal medicinal products for the treatment of menopausal symptoms. Menopause 2003;10(5):465–76.
2. Budeiri D, Li Wan Po A, Dornan JC. Is evening primrose oil of value in the treatment of premenstrual syndrome? Control Clin Trials 1996;17(1):60–8.
3. Vaddadi KS. The use of gamma-linolenic acid and linoleic acid to differentiate between temporal lobe epilepsy and schizophrenia. Prostaglandins Med 1981;6(4):375–9.

Papaveraceae

General Information

The genera in the family of Papaveraceae (Table 42) include a variety of poppies and the greater celandine.

Chelidonium majus

Chelidonium majus (celandine, common celandine, greater celandine) contains a number of alkaloids, including chelidonine, chelerythrine, chelidocystatin, coptisine, sanguinarine, berberine, and sparteine.

Greater celandine was traditionally used to improve eyesight and in modern times has been used as a mild sedative, and antispasmodic in the treatment of bronchitis, whooping cough, asthma, jaundice, gallstones, and gallbladder pain. The latex is used topically to treat warts, ringworm, and corns. A semisynthetic thiophosphate derivative of alkaloids from *C. majus*, called Ukrain, has cytotoxic and cytostatic effects on tumor cells (1).

Adverse effects

The Australian Complementary Medicines Evaluation Committee (CMEC) has advised that all oral products containing *Chelidonium majus* (greater celandine) should contain a label with the warning that they should be used under the supervision of health-care professionals; consumers with a history of liver disease should seek advice from a health-care professional before starting to use such a product and to stop using it if particular symptoms occur (2). This recommendation follows the CMEC's careful examination of all available evidence linking ingestion of *Chelidonium majus* with moderate to severe reversible acute hepatitis in a relatively small number of individuals worldwide. The mechanism underlying the hepatotoxic effect needs to be elucidated. Pending further information, the Therapeutic Goods Administration (TGA) has advised

Table 42 The genera of Papaveraceae

Arctomecon (bear poppy)
Argemone (prickly poppy)
Bocconia (bocconia)
Canbya (pygmy poppy)
Chelidonium (celandine)
Dendromecon (tree poppy)
Eschscholzia (California poppy)
Glaucium (horn poppy)
Hunnemannia (hunnemannia)
Macleaya (macleaya)
Meconella (fairy poppy)
Papaver (poppy)
Platystemon (cream cups)
Platystigma (queen poppy)
Roemeria (roemeria)
Romneya (Matilija poppy)
Sanguinaria (bloodroot)
Stylophorum (stylophorum)
Stylomecon (wind poppy)

---

health-care professionals to watch for signs of liver toxicity associated with the use of *Chelidonium majus*, which has traditionally been used to treat a range of conditions, including liver disorders, and is available internationally.

Hematologic

Hemolytic anemia has been reported after the oral use of a celandine extract; there was intravascular hemolysis, renal insufficiency, liver cytolysis, and thrombocytopenia; a direct antiglobulin test was positive (3).

Liver

Hepatitis has been attributed to celandine (4).

- A 42-year-old woman had been admitted twice to hospital with hepatitis (5). No toxic agent could be identified during the first episode, but detailed questioning during the second showed that in both cases she had self-prescribed a commercially available medication of common celandine. Withdrawal of this herbal remedy was followed by an unremarkable recovery.

- A 42-year-old woman developed acute hepatitis several weeks after taking a herbal formulation containing greater celandine and curcuma root for a skin complaint (6). After withdrawal recovery was rapid and hepatic function returned to normal within 2 months.

Ten cases of acute hepatitis induced by formulations of greater celandine were observed over 2 years in a German University hospital (7). Perhaps ironically, this product is popular in Germany for gastric and gall-bladder problems. In five cases there was marked cholestasis but no liver failure. After withdrawal of the product, the symptoms subsided and the liver enzymes normalized within 2–6 months. Unintentional rechallenge led to a further episode of acute hepatitis in one patient.

In addition to about 15 published cases, some 40 cases of liver damage from *C. majus* have been reported to the German regulatory authorities (8). The course of the hepatitis can be severe and can include cholestasis and fibrosis, but acute liver failure has not been observed. Based on these data, celandine has been banned for oral use in several countries.

Skin

Contact sensitivity has been attributed to *C. majus* (9).

Susceptibility factors

Some authors caution that the use of celandine in children should be discouraged because of an early fatal colitis in a 3-year-old boy (10); however, this report does not provide convincing evidence that the victim had indeed taken *C. majus*.

Papaver somniferum

Papaver somniferum (opium poppy) contains a variety of opioid and related alkaloids, including codeine, morphine, noscapine, papaverine, and thebaine. Crude opium is the air-dried latex obtained by incising the unripe capsules of *P. somniferum*. Paregoric is ammoniated tincture of opium (Scotch paregoric) or camphorated tincture of opium (English paregoric). The use of these formulations has largely been replaced by use of the purified compounds.

Adverse effects

The adverse effects of herbal opium are generally the same as those of the pure compounds (see SED-15).

Respiratory

Of 28 workers in a pharmaceutical factory that produced morphine and other alkaloids extracted from the shells of *P. somniferum*, six had symptoms of sensitization and positive skin tests (11). A bronchial provocation test was positive in four of them, and in all six there was a specific IgE, detected by ELISA and RAST tests using an aqueous extract of *P. somniferum*.

Nervous system

- A young man dependent on "Kompot" or "Polish heroin", a domestic product produced from poppy straw or the juice of poppy heads (*P. somniferum*) and given intravenously, developed Guillain–Barré syndrome after severe intoxication induced by home-made heroin, barbiturates, and benzodiazepines (12).

"Kompot" contains variable amounts of morphine, heroin (diacetylmorphine), 3-monoacetylmorphine, 6-monoacetylmorphine, acetylcodeine, and codeine, in addition to papaverine, thebaine, and narcotine.

References

1. Uglyanitsa KN, Nefyodov LI, Doroshenko YM, Nowicky JW, Volchek IV, Brzosko WJ, Hodysh YJ. Ukrain: a novel antitumor drug. Drugs Exp Clin Res 2000;26(5–6):341–56.

2. Anonymous. *Chelidonium majus. Statement to advice use under supervision.* WHO Pharm Newslett 2003; 4:4.

3. Pinto Garcia V, Vicente PR, Barez A, Soto I, Candas MA, Coma A. Anemia hemolitica inducida por *Chelidonium majus*. [Hemolytic anemia induced by *Chelidonium majus*. Clinical case.] Sangre (Barc) 1990;35(5):401–3.

4. Stickel F, Poschl G, Seitz HK, Waldherr R, Hahn EG, Schuppan D. Acute hepatitis induced by greater celandine (*Chelidonium majus*). Scand J Gastroenterol 2003;38(5):565–8.

5. Strahl S, Ehret V, Dahm HH, Maier KP. Nedrotisierende Hepatitis nach Einnahme pflanzlicher Heilmittel. [Necrotizing hepatitis after taking herbal remedies.] Dtsch Med Wochenschr 1998;123(47):1410–4.

6. Crijns AP, de Smet PA, van den Heuvel M, Schot BW, Haagsma EB. Acute hepatitis na gebruik van een plantaardig preparaat met stinkende gouwe (*Chelidonium majus*). [Acute hepatitis after use of a herbal preparation with greater celandine (*Chelidonium majus*).] Ned Tijdschr Geneeskd 2002;146(3):124–8.

7. Benninger J, Schneider HT, Schuppan D, Kirchner T, Hahn EG. Acute hepatitis induced by greater celandine (*Chelidonium majus*). Gastroenterology 1999;117(5):1234–7.

8. De Smet PA. Safety concerns about kava not unique. Lancet 2002;360(9342):1336.

9. Etxenagusia MA, Anda M, Gonzalez-Mahave I, Fernandez E, Fernandez de Corres L. Contact dermatitis from *Chelidonium majus* (greater celandine). Contact Dermatitis 2000;43(1):47.

10. Koopman H. Tödliche Schöllkraut-Vergiftung (*Chelidonium majus*). Vergiftungsf lle 1937;8:93–8.

11. Moneo I, Alday E, Ramos C, Curiel G. Occupational asthma caused by Papaver somniferum. Allergol Immunopathol (Madr) 1993;21(4):145–8.

12. Gawlikowski T, Winnik L. Zespol Guillain–Barré jako wynik zatrucia mieszanego "kompotem" i lekami. [Guillain–Barré syndrome as a result of poisoning with a mixture of "kompot" (Polish heroin) and drugs.] Przegl Lek 2001;58(4):357–8.

Passifloraceae

General Information

The family of Passifloraceae contains the single genus *Passiflora*.

Passiflora incarnata

Passiflora incarnata (apricot vine, grenadille, passion flower, passion vine) is widely touted as a herbal sedative and anxiolytic (1). It contains harman alkaloids.

Adverse effects

Passion flower has reportedly caused prolongation of the QT interval (2).

- A 34-year-old woman developed nausea, vomiting, drowsiness, a prolonged QT interval, and episodes of non-sustained ventricular tachycardia after self-medication with passion flower for 1 day. She made a full recovery after withdrawal of the passion flower.

The authors suggested that the adverse event had been caused by harman alkaloids from *P. incarnata*.

Five patients developed altered consciousness after taking the herbal product Relaxir for insomnia and restlessness, produced mainly from the fruit of the passion flower (3).

References

1. Krenn L. Die Passionsblume (*Passiflora incarnata* L.)—ein bewahrtes pflanzliches Sedativum. [Passion Flower (*Passiflora incarnata* L.)—a reliable herbal sedative.] Wien Med Wochenschr 2002;152(15-16):404–6.
2. Fisher AA, Purcell P, Le Couteur DG. Toxicity of *Passiflora incarnata* L. J Toxicol Clin Toxicol 2000;38(1):63–6.
3. Solbakken AM, Rorbakken G, Gundersen T. Natur medisin som rusmiddel. [Nature medicine as intoxicant.] Tidsskr Nor Laegeforen 1997;117(8):1140–1.

Pedaliaceae

General Information

The genera in the family of Pedaliaceae (Table 43) include sesame.

Harpagophytum species

Harpagophytum procumbens (devil's claw, grapple plant, wood spider) contains the iridoids harpagoside and procumbide. It is used to treat pain in the joints and lower back, although the quality of trials demonstrating efficacy is poor and there is variability from formulation to formulation (1).

Table 43 The genera of Pedaliaceae

Ceratotheca (ceratotheca)
Craniolaria (craniolaria)
Harpagophytum (devil's claw)
Ibicella (yellow unicorn plant)
Martynia (martynia)
Proboscidea (unicorn plant)
Sesamum (sesame)

Adverse effects

It is sometimes stated that *H. procumbens* should be avoided during pregnancy because of its supposed abortifacient effect, but the oxytoxic properties of the plant remain to be verified. There may be an interaction with warfarin (2).

References

1. Chrubasik S, Conradt C, Black A. The quality of clinical trials with *Harpagophytum procumbens*. Phytomedicine 2003;10(6–7):613–23.
2. Izzo AA, Di Carlo G, Borrelli F, Ernst E. Cardiovascular pharmacotherapy and herbal medicines: the risk of drug interaction. Int J Cardiol 2005;98(1):1–14.

Phytolaccaceae

General Information

The genera in the family of Phytolaccaceae (Table 44) include pokeweed.

Phytolacca americana

Phytolacca americana (pokeweed) contains powerful mitogens (pokeweed mitogen), including phytolacain, used to study cell function.

Table 44 The genera of Phytolaccaceae

Agdestis (agdestis)
Gisekia (gisekia)
Petiveria (petiveria)
Phytolacca (pokeweed)
Rivina (rivina)
Stegnosperma (stegnosperma)
Trichostigma (alpine clubrush)

Adverse effects

Severe emesis and diarrhea, accompanied by tachycardia, have been observed after ingestion of raw leaves and after drinking tea prepared from the powdered root of *P. americana* (pokeweed).

In one case Mobitz type I heart block was associated with vomiting due to pokeweed, which resolved after intravenous promethazine (1). The authors suggested that the heart block had been due to increased vagal tone associated with severe gastrointestinal colic.

Reference

1. Hamilton RJ, Shih RD, Hoffman RS. Mobitz type I heart block after pokeweed ingestion. Vet Hum Toxicol 1995; 37(1):66–7.

Piperaceae

General Information

The family of Piperaceae contains three genera:

1. *Lepianthes* (lepianthes)
2. *Peperomia* (peperomia)
3. *Piper* (pepper).

Piper methysticum

Kava or awa (*Piper methysticum*) is a plant that grows in the South Pacific Islands. Eight species have been identified by the local cultivators on the basis of its habitat, such as mountainous versus lowland, in shade versus in full sun. In the Pacific Islands kava has been widely consumed for hundreds of years as a traditional ceremonial beverage and for its mood-altering and stress relieving properties (1). Traditional preparations use aqueous emulsions of the crushed or dried roots or lower stems of the kava shrub, and its pharmacological properties have been attributed to a group of components collectively know as kava pyrones or kava lactones.

However, in the last 10 years there has been an expanding global market for herbal formulations containing lipid kava extracts in the west, largely evolving in Germany and spreading to the American and European markets. Thus, Western formulations may contain compounds that are not present in the drink that is commonly prepared in the South Pacific and may therefore have different adverse effects (2). This may explain the risk of liver damage that has been seen in the West.

In the UK, three licensed medicines and a large number of unlicensed herbal remedies containing kava were available until they were banned in 2004. These products were marketed for the treatment of mild anxiety, for which kava seems to be effective (3), insomnia, premenstrual syndrome, and stress, and were sold over the counter as complementary medicines or as dietary supplements and nutraceuticals.

Preparations of *Piper methysticum* (kava) have been withdrawn from many European and North American markets because of safety concerns, and in particular hepatotoxicity. Depending on the preparation and dose used, adverse effects have been reported in 1.5–2.3% of subjects (4).

Kava contains the pyridine alkaloid pipermethystine, the chalcone flavokavain, the norsesquiterpenoid dihydrokavain, and the pyran tetrahydroyangonin.

Adverse effects

The acute and chronic effects of kava were first described by the 19th-century German toxicologist Louis Lewin (5) in his book *Phantastica* (1924).

A carefully prepared kava beverage taken in small quantity occasioned only slight and agreeable modifications of sensibility. In this form it is a stimulating beverage after the imbibition whereof hardships can be endured more easily. It refreshes the fatigued body and brightens and sharpens the intellectual faculties. Appetite is augmented, especially if it is taken half an hour before meals···After doses that are not too strong a state of happy carelessness, content, and well-being appears without any physical or mental excitation. It is a real euphoric state which is accompanied by an increased muscular efficiency. At the beginning speech is fluent and lively and the hearing becomes more sensible to subtle impressions. Kava has a soothing effect. Those who drink it are never choleric, angry, aggressive and noisy, as in the case of alcohol. Both natives and whites look upon it as a sedative in case of accidents. Reason and consciousness remain unaffected. After the consumption of greater quantities, however, the limbs become weary, the muscles seem out of control of the will, the gait is slow and unsteady, and the subject appears half drunk. An urgent desire to lie down manifests itself. The eye sees objects before it but is unable to identify them with exactness. In the same way the ears hear everything, but the individual is unable to account for what he hears. Everything becomes more and more diffuse. The drinker succumbs to fatigue, and experiences a desire to sleep which is stronger than all other impressions. He becomes somnolent and finally falls asleep. Many Europeans have themselves experienced this action of kava which paralyses the senses like magic and finally leads to deep sleep. Frequently a state of somnolent torpor accompanied by incoherent dreams and occasionally erotic visions remains without sleep supervening.

The sleep is similar to that produced by alcohol, out of which the individual can be awakened only with difficulty. If moderate quantities have been consumed it occurs twenty to thirty minutes later, and lasts from two to eight hours according to the degree of habituation of the subject. If the beverage is concentrated, that is contains a large amount of the resinous components of kava, intoxication comes on much more rapidly. The drinkers are found lying in the very places where they have been drinking. Occasionally a short state of nervous trembling occurs before they fall asleep. No excitation precedes these symptoms.

The kava drinker is incessantly tormented with the craving for his favourite beverage, which he cannot prepare for himself. It is a repugnant spectacle to see old and white-haired people, degenerate through prolonged abuse of the drug, going from house to house in order to beg for freshly prepared kava and often meeting with a refusal. Mental weakness has also been stated to follow from kavaism. It is said that old kava habitues have red, inflamed, bloodshot eyes, dull, bleary, and diminished in their functions. They become extremely emaciated, their hands tremble, and finally they cannot bring the drinking vessel to their mouths. Numerous cutaneous diseases of the natives of the South Seas, especially a kind of scaly eruption which results in a parchment-like state of the skin, have been attributed to the abuse of kava.

Cardiovascular

Tachycardia and electrocardiographic abnormalities have been reported in heavy users of kava (Ulbricht). Whether these observations represent true adverse effects or are due to other factors is unclear.

Respiratory

Shortness of breath has been reported in heavy aboriginal kava users; pulmonary hypertension has been proposed as a possible cause (Ulbricht).

Nervous system

Neurological adverse effects are uncommon after kava ingestion. However, torticollis, oculogyric crises, and oral dyskinesias in young to middle-aged people, exacerbation of the symptoms of Parkinson's disease, and three episodes of abnormal body movements have been reported (Ulbricht). Three cases of meningism have been reported after kava consumption, two of them with focal manifestations (Ulbricht).

In a pilot study, 24 patients with stress-induced insomnia were treated for 6 weeks with kava 120 mg/day, followed by a 2-week washout period and then treatment with valerian 600 mg/day for another 6 weeks (6). Stress was measured in three areas, including social, personal, and life events, and insomnia was assessed by evaluating the time taken to fall asleep, the number of hours slept, and waking mood. Total stress severity and insomnia were significantly improved by both compounds, with no significant differences. The most commonly reported adverse events were vivid dreams with valerian (16%) and dizziness with kava (12%).

In 24 patients treated with kava for generalized anxiety disorder for 4 weeks in an open, crossover, randomized trial, two dosage schedules were compared: 120 mg od and 45 mg tds (7). There were significant reductions in mean Hamilton Anxiety Rating Scale scores, irrespective of dose schedule, treatment order, or sex. The impact of adverse effects was relatively low, and only one patient had to withdraw from the study (tds schedule) because of nausea. There was daytime drowsiness in 33% of patients taking the thrice-daily regimen compared with 9% in those taking a once-daily dose.

Parkinsonism has been attributed to kava (8).

- A 45-year-old woman with a family history of essential tremor developed severe and persistent parkinsonism after taking kava extract for anxiety for 10 days. Her symptoms improved with anticholinergic drugs.

The authors concluded that kava derivatives can produce severe parkinsonism in individuals with a genetic susceptibility.

Sensory systems

There has been a case of impaired accommodation and convergence following one-time use of kava (Ulbricht).

Liver

Following its more widespread introduction in the west in 1999, several cases of severe hepatotoxicity in people using kava-containing herbal products were reported in Europe and the USA, including hepatitis, cirrhosis, and liver failure (3). By late 2002 there were 10 reports of patients requiring liver transplants (eight in Europe and two in the USA; one died after transplantation) after hepatic failure associated with the use of kava-containing products. Although these reports are of varied quality and the association is not completely convincing, this has led to various worldwide regulatory measures, ranging from a total ban of kava-containing products to consumer advisory warnings about the adverse effects of kava.

The WHO Global Database for Adverse Drug Reactions has received altogether 23 case reports of liver injury in suspected connection with the use of kava from the UK, Switzerland, Germany, the USA, and Canada; most have come from Germany. In addition, the database contains 26 reports of a variety of hypersensitivity reactions.

- A 39-year-old woman had been treated for toxic hepatitis of unknown cause (9). When she represented with hepatitis-like symptoms and high liver enzymes, toxic hepatitis was diagnosed. Other causes for the liver pathology were excluded and it was noted that before both exacerbations, she had self-prescribed kava. When the kava was withdrawn, the liver pathology normalized and she made an uneventful recovery.
- A 60-year-old patient, who had taken no medications other than kava extract, developed liver and kidney failure and progressive encephalopathy (10). Viral, metabolic, and autoimmune causes were excluded. Liver biopsy was consistent with toxic liver damage. The patient eventually received an orthotopic liver transplant and made a good recovery.
- A 50-year-old man developed jaundice. He had noticed fatigue for a month, a "tanned" skin, and dark urine (11). The medical history was unremarkable, apart from slight anxiety, for which he had been taking three or four capsules of kava extract daily for 2 months. He took no other drugs and did not consume alcohol. Liver function tests showed very large increases in transaminases. He subsequently developed stage IV encephalopathy but made a good recovery after liver transplantation.
- A 34-year-old woman developed toxic hepatitis after taking kava-kava (12). Ultrasound showed an enlarged echogenic liver, and histology showed centrilobular necrosis and periportal inflammation. After withdrawal of the kava the changes resolved completely. This case illustrates the high hepatotoxic potential of kava-kava.
- A 33-year-old woman took a kava extract equivalent to 210 mg of kavalactones daily for 3 weeks (13). She developed malaise, loss of appetite, and jaundice. Her liver enzymes were raised 3-fold to 60-fold. Viral hepatitis was excluded and liver biopsy confirmed toxic hepatitis. Kava was withdrawn, and within 8 weeks the liver enzymes returned to normal. A lymphocyte transformation test showed strong concentration-dependent T cell reactivity to kava. Phenotyping of CYP2D6 activity showed that she was a poor metabolizer.

The authors of the last report concluded that the liver damage in this case was due to an immune-mediated reaction, possibly mediated by a reactive metabolite of kava, even though she was a poor metabolizer.

In an analysis of 19 German reports it was concluded that a probably causal relation could be established in only one patient (14). However, in a review of 29 cases reported between 1990 and 2002, nine patients developed fulminant liver failure and three died; the rest made full recoveries (15). The authors considered that immunoallergic and idiosyncratic factors had been responsible. Further cases of severe liver damage have been associated with the use of kava (16).

Kava has been associated with toxic liver damage in six cases reported from Switzerland (17). In one patient, the liver damage was so extensive that liver transplantation became necessary. Histological data from four patients were consistent with an allergic mechanism. In several cases, other medications with hepatotoxic potential had been taken concurrently. Symptoms generally occurred at between 3 weeks and 4 months and involved daily doses that contained kavapyrones 60–210 mg. Most instances involved acetone extracts. The leading kava extract, Laitan, was subsequently withdrawn from the Swiss market.

Australia's Therapeutic Goods Administration initiated a voluntary recall of all complementary medicines containing kava after the death of a woman who used a medicine containing kava (18). Sponsors and retailers were asked to remove all products containing kava from the market immediately. Consumers were advised to discard kava-containing products in their possession.

The Federal Institute of Germany has withdrawn all products that contain kava and kavaine from the German market because of the risk of hepatotoxicity and insufficiently proven efficacy. The regulation included homeopathic products with dilutions up to D4. The German regulation applies to all kava-containing pharmaceutical formulations. Moreover, following a provisional opinion from the UK Committee on Safety of Medicines (CSM), the Medicines and Healthcare products Regulatory Agency (MHRA) has consulted on a proposal to prohibit the sale, supply, or importation of unlicensed medicinal products containing kava in the UK. The CSM reviewed the issue of kava-associated liver toxicity following the emergence of safety concerns in Europe. At that time, stocks of kava were voluntarily withdrawn by the herbal sector while the safety concerns were under investigation. Currently the MHRA is aware of 68 cases worldwide of suspected kava-associated liver problems, including 6 cases of liver failure that resulted in transplant, and 3 deaths. In the UK there have been three reports of kava-associated liver toxicity. The CSM has advised consumers to stop taking medicinal products containing kava, and to seek medical advice if they feel unwell or have concerns about possible liver problems.

Finally, the FDA has advised consumers of the potential for liver injury by kava-containing dietary supplements. People who have liver disease or liver problems or who are taking medicines that can affect the liver have been advised to consult a physician before using kava-containing supplements (19). Consumers who use kava-containing dietary supplements and who have signs of illness associated with liver disease should also consult a physician. The FDA has issued a letter to healthcare professionals informing them of the consumer advice and has urged consumers and healthcare professionals to report injuries that may be related to the use of kava.

The incidence of liver damage with kava seems to be less than one case per million daily doses (20). The mechanism of the effect is currently unclear. It has been suggested that supervised, monitored, short-term medication with kava would still do more good than harm (21). However, the FDA will continue to investigate the relation, if any, between the use of dietary supplements containing kava and liver damage.

The mechanism of kava toxicity remains to be elucidated, but it may be caused by a compound that is extracted in lipid extracts and not in aqueous extracts or a metabolite of such a compound (22). Pipermethystine, found in leaves and stem peelings, and kavalactones may contribute (23). The dose and duration of administration that increases the risk of liver damage is not clear. Histological studies have shown portal inflammation with lymphocytes and eosinophils. An idiosyncratic immune response to a reactive metabolite has been suggested as a possible cause. Genetic differences in liver metabolism of kava lactones also need to be examined.

Gastrointestinal

Mild and infrequent epigastric pain and nausea have been reported with use of kava (Ulbricht).

Hematologic

Increased erythrocyte volume, reduced platelet volume, reduced lymphocyte count, and reduced serum albumin have been reported in chronic heavy kava users (Ulbricht). Kavain has antiplatelet effects, due to inhibition of cyclo-oxygenase and thromboxane synthesis.

Skin

Allergic skin reactions (24) and other adverse effects, including dermatological problems and neurological symptoms, have been described in individuals using such formulations. "Kava dermopathy", a dry, scaly skin and/or yellow skin discoloration, which seems to be reversible after withdrawal, has been reported in chronic heavy users. It resembles pellagra but is unrelated to niacin deficiency (25). It can cover underlying liver damage and therefore requires a thorough clinical assessment. Regular kava use can also cause urticaria (26).

- A 36-year-old woman developed a generalized rash, severe itching with erythema, and papules 4 days after discontinuing a kava formulation (Antares), which she had taken in a dosage of 120 mg/day for 3 weeks (27). The condition improved with glucocorticoids and antihistamines, but the itching lasted several weeks. Patch tests with the kava extract were positive.

Musculoskeletal

Rhabdomyolysis has been attributed to kava (28).

- A 29-year-old man developed severe diffuse muscle pain and passed dark urine a few hours after taking a herbal combination product containing guaraná 500 mg, Ginkgo biloba 200 mg, and kava 100 mg. His blood creatine kinase activity and myoglobin concentration were raised and there were no signs of an underlying metabolic myopathy. His condition improved within 6 weeks.

The authors suggested that the methylxanthine-like effects of guaraná and the antidopaminergic and neuro-muscular blocking properties of kava had caused the rhabdomyolysis in this patient.

Rhabdomyolysis has also been reported after ingestion of the herbal combination preparation Guaranaginko Plus (28). Each flacon contains guaraná 500 mg, *Ginko biloba* extract 200 mg, and kava 100 mg.

- A 29-year-old man took a combination preparation containing *Ginkgo biloba*, guaraná, and kava and developed rhabdomyolysis He had diffuse severe muscle pain and passed dark urine after an overnight fast. He had been a regular body builder from age 18 to age 27 years, had just resumed weight training at low intensities, and was taking Guaranaginko Plus. His creatine kinase activity was 100 500 IU/l and myoglobin 10 000 ng/ml. There were no renal complications, and the muscle pain and creatine kinase gradually subsided over 6 weeks. An underlying metabolic myopathy as the cause of myoglobinuria was excluded.

Drug overdose

- A 34-year-old Tongan man complained of sore eyes, headache, generalized muscle weakness, and abdominal pain (29). He was disoriented and hallucinating. His family reported that he had been drinking large quantities of kava daily for about 14 years. Chronic kava intoxication was treated with intravenous Plasmalyte (a crystalloid solution) and he recovered within a day.

References

1. Anonymous. Safety issues involving herbal medicines: kava as a case study. WHO Pharm Newslett 2003; 5: 8–9.
2. Jamieson DD, Duffield PH, Cheng D, Duffield AM. Comparison of the central nervous system activity of the aqueous and lipid extract of kava (*Piper methysticum*). Arch Int Pharmacodyn Ther 1989;301:66–80.
3. Stevinson C, Huntley A, Ernst E. A systematic review of the safety of kava extract in the treatment of anxiety. Drug Saf 2002;25(4):251–61.
4. Ulbricht C, Basch E, Boon H, Ernst E, Hammerness P, Sollars D, Tsourounis C, Woods J, Bent S. Safety review of kava (*Piper methysticum*) by the Natural Standard Research Collaboration. Expert Opin Drug Saf 2005; 4(4): 779–94.
5. Aronson JK. Louis Lewin—Meyler's predecessor. In: Aronson JK, editor. Side Effects of Drugs Annual 27. Amsterdam: Elsevier, 2004:xxv–xxix.
6. Wheatley D. Kava and valerian in the treatment of stress-induced insomnia. Phytother Res 2001;15(6):549–51.
7. Wheatley D. Kava-kava (LI 150) in the treatment of generalized anxiety disorder. Prim Care Psychiatry 2001;7:97–100.
8. Meseguer E, Taboada R, Sanchez V, Mena MA, Campos V, Garcia De Yebenes J. Life-threatening parkinsonism induced by kava-kava. Mov Disord 2002;17(1):195–6.
9. Strahl S, Ehret V, Dahm HH, Maier KP. Nedrotisierende Hepatitis nach Einnahme pflanzlicher Heilmittel. [Necrotizing hepatitis after taking herbal remedies.] Dtsch Med Wochenschr 1998;123(47):1410–4.
10. Kraft M, Spahn TW, Menzel J, Senninger N, Dietl KH, Herbst H, Domschke W, Lerch MM. Fulminantes Leberversagen nach Einnahme des planzlichen Antidepressivums Kava-Kava. [Fulminant liver failure after administration of the herbal antidepressant kava-kava.] Dtsch Med Wochenschr 2001;126(36):970–2.
11. Escher M, Desmeules J, Giostra E, Mentha G. Hepatitis associated with kava, a herbal remedy for anxiety. BMJ 2001;322(7279):139.
12. Weise B, Wiese M, Plotner A, Ruf BR. Toxic hepatitis after intake of kava-kava. Vergauungskrankheiten 2002;20:166–9.
13. Russmann S, Lauterburg BH, Helbling A. Kava hepatotoxicity. Ann Intern Med 2001;135(1):68–9.
14. Teschke R, Gaus W, Loew D. Kava extracts: safety and risks including rare hepatotoxicity. Phytomedicine 2003;10:440–6.
15. Stickel F, Baumüller H-M, Seitz K, Vasilakis D, Seitz G, Seitz H, Schuppan D. Hepatitis induced by Kava (*Piper methysticum rhizoma*). J Hepatol 2003; 39: 62–7.
16. Gow PJ, Connelly NJ, Hill RL, Crowley P, Angus PW. Fatal fulminant hepatic failure induced by a natural therapy containing kava. Med J Aust 2003;178:442–3.
17. Stoller A. Leberschädigungen unter Kava-Extrakten. Schweiz Arztez 2000;24:1335–6.
18. Anonymous. Kava-kava. More withdrawals due to hepatotoxic risks. WHO Pharmaceuticals Newslett 2002;3:4–5.
19. Anonymous. Kava-kava. Further investigations into *Piper methysticum* and liver injury. WHO Pharmaceuticals Newslett 2002;2:2–3.
20. Ernst E. Safety concerns about kava. Lancet 2002; 359(9320):1865.
21. Teschke R. Hepatotoxizität durch Kava–Kava. Deutsch Arzteblatt 2002;99:A3411–8.
22. Singh YN, Devkota AK. Aqueous kava extracts do not affect liver function tests in rats. Planta Med 2003; 69: 496–9.
23. Nerurkar PV, Dragull K, Tang CS. In vitro toxicity of kava alkaloid, pipermethystine, in HepG2 cells compared to kavalactones. Toxicol Sci 2004;79(1):106–11.
24. Suss R, Lehmann P. Haematogenes Kontaktekzem durch pflanzliche Medikamente am Beispiel des Kavawurzel-extraktes. [Hematogenous contact eczema cause by phytogenic drugs exemplified by kava root extract.] Hautarzt 1996;47(6):459–61.
25. Ruze P. Kava-induced dermopathy: a niacin deficiency? Lancet 1990;335(8703):1442–5.
26. Grace R. Kava-induced urticaria. J Am Acad Dermatol 2005; 53(5): 906.
27. Schmidt P, Boehncke WH. Delayed-type hypersensitivity reaction to kava-kava extract. Contact Dermatitis 2000;42(6):363–4.
28. Donadio V, Bonsi P, Zele I, Monari L, Liguori R, Vetrugno R, Albani F, Montagna P. Myoglobinuria after ingestion of extracts of guaraná, *Ginkgo biloba* and kava. Neurol Sci 2000; 21(2): 124.
29. Chanwai LG. Kava toxicity. Emerg Med 2000; 12: 142–5.

Plantaginaceae

General Information

The family of Plantaginaceae contains two genera:

1. *Littorella* (littorella)
2. *Plantago* (plantain).

Plantago species

Plantago species have traditionally been used for many purposes in folk medicine (1,2). Plantain seeds are widely used as bulk laxatives under the names of "psyllium" (from *Plantago psyllium* or *Plantago indica*) and "ispaghula" (from *Plantago ovata*).

Psyllium husk combined with microencapsulated paraffin has been compared with standard psyllium for the treatment of constipation in a randomized, double-blind study (3). There was a significant increase in the weekly number of defecations with the combined formulation, which was well tolerated; no adverse effects were reported.

The efficacy, speed of action, and acceptability of ispaghula husk, lactulose, and other laxatives in the treatment of simple constipation in 394 patients have been studied by 65 general practitioners (4). Ispaghula was used by 224 patients and other laxatives by 170. After 4 weeks of treatment ispaghula husk was assessed by the GPs to be superior to the other laxatives. In patients' assessment, ispaghula users had a higher proportion of normal stools and less soiling than patients using other laxatives. Diarrhea and abdominal pain and gripes and were less common with ispaghula. Distension, flatulence, indigestion, and nausea were equally frequent in the two groups.

Esophageal obstruction by a bezoar after ingestion of psyllium has been reported (5).

- A 69-year-old man with Parkinson's disease developed severe dysphagia after taking granules of the bulk laxative Perdiem (82% psyllium and 18% senna formulated as granules). Disimpaction of the bezoar was performed via a rigid endoscope under general anaesthesia.

Occupational exposure to *Plantago* species has resulted in sensitization, with symptoms ranging from rhinitis and lacrimation to more severe respiratory compromise. This problem arises in a more serious form among the personnel of pharmaceutical factories processing psyllium (SEDA-16, 426), and eosinophilia has also been recorded. The allergen appears to reside in the endosperm or embryonic seed components and not in the husk, which is the laxative component; in principle, therefore, it should be feasible to supply a non-antigenic form of purified psyllium husk (SEDA-17, 423).

Ingestion of psyllium has been associated with rare cases of generalized urticarial rash and anaphylactic shock (6,7). The possibility that the intestinal absorption of lithium and other drugs may be inhibited by psyllium should also be considered (8).

References

1. Grigorescu E, Stanescu U, Basceanu V, Aur MM. Controlul fitochimic si microbiologic al unor specii utilizate in medicina populara. II. *Plantago lanceolata* L., *Plantago media* L., *Plantago major*. [Phytochemical and microbiological control of some plant species used in folk medicine. II. *Plantago lanceolata* L., *Plantago media* L., *Plantago major* L.] Rev Med Chir Soc Med Nat Ias 1973;77(4):835–41.
2. Samuelsen AB. The traditional uses, chemical constituents and biological activities of *Plantago major* L. A review. J Ethnopharmacol 2000;71(1–2):1–21.
3. Chicouri MJ. Estudo clinico do psyllium husk associado a parafina microencapsulada no tratamento da constipacao intestinal essencial. Rev Bras Med 2001;58:672–6.
4. Dettmar PW, Sykes J. A multi-centre, general practice comparison of ispaghula husk with lactulose and other laxatives in the treatment of simple constipation. Curr Med Res Opin 1998;14(4):227–33.
5. Shulman LM, Minagar A, Weiner WJ. Perdiem causing esophageal obstruction in Parkinson's disease. Neurology 1999;52(3):670–1.
6. Lantner RR, Espiritu BR, Zumerchik P, Tobin MC. Anaphylaxis following ingestion of a psyllium-containing cereal. JAMA 1990;264(19):2534–6.
7. Spence JD, Huff MW, Heidenheim P, Viswanatha A, Munoz C, Lindsay R, Wolfe B, Mills D. Combination therapy with colestipol and psyllium mucilloid in patients with hyperlipidemia. Ann Intern Med 1995;123(7):493–9.
8. Perlman BB. Interaction between lithium salts and ispaghula husk. Lancet 1990;335(8686):416.

Poaceae

General Information

The genera in the family of Poaceae (Table 45) include bamboo, barley, cane, fescue, lawn grass, oats, pappus, and rice.

Table 45   The genera of Poaceae

Achnatherum (needle grass)
Achnella (rice grass)
Acrachne (goose grass)
Aegilops (goat grass)
Aegopogon (relax grass)
Agropyron (wheat grass)
Agropogon (agropogon)
Agrostis (bent grass)
Aira (hair grass)
Allolepis (Texas salt)
Alloteropsis (summer grass)
Alopecurus (foxtail)
Ammophila (beach grass)
Ampelodesmos (Mauritanian grass)
Amphicarpum (maidencane)
Amphibromus (wallaby grass)
Andropogon (blue stem)
Anthaenantia (silky scale)
Anthephora (oldfield grass)
Anthoxanthum (vernal grass)
Apera (silkybent)
Apluda (Mauritian grass)
Arctagrostis (polar grass)
Arctophila (pendant grass)
Aristida (three awn)
Arrhenatherum (oat grass)
Arthraxon (carp grass)
Arthrostylidium (climbing bamboo)
Arundinaria (cane)
Arundinella (rabo de gato)
Arundo (giant reed)
Avena (oat)
Axonopus (carpet grass)
Bambusa (bamboo)
Beckmannia (slough grass)
Blepharidachne (desert grass)
Blepharoneuron (drop seed)
Bothriochloa (beard grass)
Bouteloua (grama)
Brachiaria (signal grass)
Brachyelytrum (short husk)
Brachypodium (false brome)
Briza (quaking grass)
Bromus (brome)
Buchloe (buffalo grass)
Calamagrostis (reed grass)
Calamovilfa (sand reed)
Calammophila (calammophila)
Catabrosa (whorl grass)
Cathestecum (false grama)
Cenchrus (sandbur)

(Continued)

Chasmanthium (wood oats)
Chloris (windmill grass)
Chrysopogon (false beard grass)
Chusquea (chusquea bamboo)
Cinna (wood reed)
Cladoraphis (bristly love grass)
Coelorachis (joint tail grass)
Coix (Job's tears)
Coleanthus (moss grass)
Cortaderia (pampas grass)
Corynephorus (club awn grass)
Cottea (cotta grass)
Crypsis (prickle grass)
Ctenium (toothache grass)
Cutandia (Memphis grass)
Cymbopogon (lemon grass)
Cynodon (Bermuda grass)
Cynosurus (dogstail grass)
Dactylis (orchard grass)
Dactyloctenium (crowfoot grass)
Danthonia (oat grass)
Dasyochloa (woolly grass)
Dasypyrum (mosquito grass)
Deschampsia (hair grass)
Desmazeria (fern grass)
Diarrhena (beak grain)
Dichanthelium (rosette grass)
Dichanthium (blue stem)
Dichelachne (plume grass)
Diectomis (folded leaf grass)
Digitaria (crab grass)
Dinebra (viper grass)
Dissanthelium (Catalina grass)
Dissochondrus (false brittle grass)
Distichlis (salt grass)
Dupontia (tundra grass)
Echinochloa (cockspur grass)
Ehrharta (veldt grass)
Eleusine (goose grass)
Elionurus (balsam scale grass)
Elyhordeum (barley)
Elyleymus (wild rye)
Elymus (wild rye)
Enneapogon (feather pappus grass)
Enteropogon (umbrella grass)
Entolasia (entolasia)
Eragrostis (love grass)
Eremochloa (centipede grass)
Eremopyrum (false wheat grass)
Eriochloa (cup grass)
Eriochrysis (moco de pavo)
Erioneuron (woolly grass)
Euclasta (mock bluestem)
Eustachys (finger grass)
Festuca (fescue)
Fingerhuthia (Zulu fescue)
Garnotia (lawn grass)
Gastridium (nit grass)
Gaudinia (fragile oat)
Glyceria (manna grass)
Gymnopogon (skeleton grass)
Gynerium (wild cane)
Hackelochloa (pitscale grass)
Hainardia (barb grass)
Helictotrichon (alpine oat grass)

(Continued)

Table 45 (Continued)

Hemarthria (joint grass)
Hesperostipa (needle and thread)
Heteropogon (tanglehead)
Hierochloe (sweet grass)
Hilaria (curly mesquite)
Holcus (velvet grass)
Hordeum (barley)
Hymenachne (marsh grass)
Hyparrhenia (thatching grass)
Hypogynium (West Indian bluestem)
Ichnanthus (bed grass)
Imperata (satin tail)
Isachne (blood grass)
Ischaemum (muraina grass)
Ixophorus (Central America grass)
Jarava (rice grass)
Karroochloa (South African oat grass)
Koeleria (June grass)
Lagurus (hare's tail grass)
Lamarckia (goldentop grass)
Lasiacis (small cane)
Leersia (cut grass)
Leptochloa (sprangle top)
Leptochloopsis (limestone grass)
Leptocoryphium (lanilla)
Lepturus (thin tail)
Leucopoa (spikeit> (toothache grass)
Cutandia (Memphis grass)
Cymbopogon (lemon grass)
Cynodon (Bermuda grass)
Cynosurus (dogstail grass)
Dactylis (orchard grass)
Dactyloctenium (crowfoot grass)
Danthonia (oat grass)
Dasyochloa (woolly grass)
Dasypyrum (mosquito grass)
Deschampsia (hair grass)
Desmazeria (fern grass)
Diarrhena (beak grain)
Dichanthelium (rosette grass)
Dichanthium (blue stem)
Dichelachne (plume grass)
Diectomis (folded leaf grass)
Digitaria (crab grass)
Dinebra (viper grass)
Dissanthelium (Catalina grass)
Dissochondrus (false brittle grass)
Distichlis (salt grass)
Dupontia (tundra grass)
Echinochloa (cockspur grass)
Ehrharta (veldt grass)
Eleusine (goose grass)
Elionurus (balsam scale grass)
Elyhordeum (barley)
Elyleymus (wild rye)
Elymus (wild rye)
Enneapogon (feather pappus grass)
Enteropogon (umbrella grass)
Entolasia (entolasia)
Eragrostis (love grass)
Eremochloa (centipede grass)
Eremopyrum (false wheat grass)
Eriochloa (cup grass)
Eriochrysis (moco de pavo)

Erioneuron (woolly grass)
Euclasta (mock bluestem)
Eustachys (finger grass)
Festuca (fescue)
Fingerhuthia (Zulu fescue)
Garnotia (lawn grass)
Gastridium (nit grass)
Gaudinia (fragile oat)
Glyceria (manna grass)
Gymnopogon (skeleton grass)
Gynerium (wild cane)
Hackelochloa (pitscale grass)
Hainardia (barb grass)
Helictotrichon (alpine oat grass)
Hemarthria (joint grass)
Hesperostipa (needle and thread)
Heteropogon (tanglehead)
Hierochloe (sweet grass)
Hilaria (curly mesquite)
Holcus (velvet grass)
Hordeum (barley)
Hymenachne (marsh grass)
Hyparrhenia (thatching grass)
Hypogynium (West Indian bluestem)
Ichnanthus (bed grass)
Imperata (satin tail)
Isachne (blood grass)
Ischaemum (muraina grass)
Ixophorus (Central America grass)
Jarava (rice grass)
Karroochloa (South African oat grass)
Koeleria (June grass)
Lagurus (hare's tail grass)
Lamarckia (goldentop grass)
Lasiacis (small cane)
Leersia (cut grass)
Leptochloa (sprangle top)
Leptochloopsis (limestone grass)
Leptocoryphium (lanilla)
Lepturus (thin tail)
Leucopoa (spike fescue)
Leymus (wild rye)
Limnodea (Ozark grass)
Lithachne (diente de perro)
Lolium (rye grass)
Luziola (water grass)
Lycurus (wolf's tail)
Melica (melic grass)
Melinis (stink grass)
Mibora (sand grass)
Microchloa (small grass)
Microstegium (brown top)
Milium (millet grass)
Miscanthus (silver grass)
Molinia (moor grass)
Monanthochloe (shore grass)
Monroa (false buffalo grass)
Muhlenbergia (muhly)
Nardus (mat grass)
Nassella (tussock grass)
Neeragrostis (creeping love grass)
Neostapfia (Colusa grass)
Neyraudia (neyraudia)
Olyra (carrycillo)
Opizia (opizia)

(Continued)

(Continued)

Table 45 (Continued)

Oplismenus (basket grass)
Orcuttia (Orcutt grass)
Oryza (rice)
Oryzopsis (rice grass)
Panicum (panic grass)
Pappophorum (pappus grass)
Parapholis (sickle grass)
Pascopyrum (wheat grass)
Paspalidium (watercrown grass)
Paspalum (crown grass)
Pennisetum (fountain grass)
Phalaris (canary grass)
Phanopyrum (savannah panic grass)
Pharus (stalk grass)
Phippsia (ice grass)
Phleum (timothy)
Phragmites (reed)
Phyllostachys (bamboo)
Piptatherum (rice grass)
Piptochaetium (spear grass)
Pleuraphis (galleta grass)
Pleuropogon (semaphore grass)
Poa (blue grass)
Polypogon (rabbit's foot grass)
Polytrias (Java grass)
Psathyrostachys (wild rye)
Pseudoroegneria (wheat grass)
Pseudelymus (foxtail wheat grass)
Pseudosasa (arrow bamboo)
Ptilagrostis (false needle grass)
Puccinellia (alkali grass)
Redfieldia (blowout grass)
Reimarochloa (reimar grass)
Rostraria (hair grass)
Rottboellia (itch grass)
Rytidosperma (wallaby grass)
Saccharum (sugar cane)
Sacciolepis (cupscale grass)
Sasa (broad leaf bamboo)
Schedonnardus (tumble grass)
Schismus (Mediterranean grass)
Schizachne (false melic)
Schizachyrium (little bluestem)
Schizostachyum (Polynesian 'ohe)
Sclerochloa (hard grass)
Scleropogon (burro grass)
Scolochloa (river grass)
Scribneria (Scribner's grass)
Secale (rye)
Setaria (bristle grass)
Sinocalamus (wide leaf bamboo)
Sorghastrum (Indian grass)
Sorghum (sorghum)
Spartina (cord grass)
Sphenopholis (wedge scale)
Sporobolus (drop seed)

(*Continued*)

Steinchisma (gaping grass)
Stenotaphrum (St. Augustine grass)
Swallenia (dune grass)
Taeniatherum (Medusa head)
Themeda (kangaroo grass)
Thinopyrum (wheat grass)
Thuarea (Kuroiwa grass)
Torreyochloa (false manna grass)
Trachypogon (crinkle awn grass)
Tragus (burr grass)
Trichoneura (Silveus' grass)
Tridens (tridens)
Triplasis (sand grass)
Tripogon (five minute grass)
Tripsacum (gama grass)
Trisetum (oat grass)
Triticum (wheat)
Tuctoria (spiral grass)
Uniola (sea oats)
Urochloa (signal grass)
Vahlodea (hair grass)
Vaseyochloa (Texas grass)
Ventenata (North Africa grass)
Vetiveria (vetiver grass)
Vulpia (fescue)
Willkommia (willkommia)
Zea (corn)
Zizania (wild rice)
Zizaniopsis (cut grass)
Zoysia (lawn grass)

Anthoxanthum odoratum

Anthoxanthum odoratum (sweet vernal grass) contains anticoagulant coumarins, which cause bleeding in cattle that consume the grass.

Adverse effects

In 125 Turkish patients with rhinitis and/or symptoms of asthma reactivity to *Dermatophagoides* as an indoor allergen was 50% ($n = 63$); in pollen allergic patients ($n = 100$) sensitivity to Poaceae was the most common (69%), and among them positivity to *A. odoratum* was 45% (1). Sensitivity to grass pollen was the same in patients from urban and rural areas (72 versus 71%).

Reference

1. Harmanci E, Metintas E. The type of sensitization to pollens in allergic patients in Eskisehir (Anatolia), Turkey. Allergol Immunopathol (Madr) 2000;28(2):63–6.

Polygonaceae

General Information

The genera in the family of Polygonaceae (Table 46) include buckwheat and rhubarb.

Polygonum species

Some species of *Polygonum* (knotweed) contain stilbene phytoestrogens, including resveratrol. *Polygonum tinctorium* contains indirubin, an isomer of indigo, which inhibits interferon-gamma production by human myelomonocytic HBL-38 cells and interferon-gamma and interleukin-6 production by murine splenocytes with no effect on the proliferation of either cells (1). *Polygonum multiflorum* contains compounds that inhibit calmodulin-depleted erythrocyte calcium-dependent ATPase (2). *Polygonum pennsylvanicum* contains vanicosides, glycosides that inhibit protein kinase C (3).

Adverse effects

Polygonum multiflorum is also known by the Chinese name He shou wu, and may be an ingredient of other traditional health products including, Shou wu pian, Shou wu wan, and Shen min. It is marketed for the relief of a variety of conditions, including early graying of the hair and baldness.

Table 46 The genera of Polygonaceae

Antenoron (antenoron)
Antigonon (antigonon)
Aristocapsa (spiny cape)
Brunnichia (buckwheat vine)
Centrostegia (centrostegia)
Chorizanthe (spineflower)
Coccoloba (coccoloba)
Dedeckera (July gold)
Dodecahema (spineflower)
Emex (three-corner jack)
Eriogonum (buckwheat)
Fagopyrum (buckwheat)
Gilmania (golden carpet)
Goodmania (spine cape)
Hollisteria (hollisteria)
Homalocladium (homalocladium)
Koenigia (koenigia)
Lastarriaea (lastarriaea)
Mucronea (spine flower)
Muehlenbeckia (maidenhair vine)
Nemacaulis (cottonheads)
Oxyria (mountain sorrel)
Oxytheca (oxytheca)
Polygonella (jointweed)
Polygonum (knotweed)
Pterostegia (pterostegia)
Rheum (rhubarb)
Rumex (dock)
Stenogonum (buckwheat)
Systenotheca (spineflower)

Liver

Polygonum multiflorum (witch hazel) is often recommended for topical use but in traditional Chinese medicine it is also used orally. Although *Polygonum multiflorum* has been recommended for "enrichment of the liver" it has itself been implicated in liver damage (4).

- A 5-year-old previously healthy Caucasian girl developed jaundice, dark urine, and pale stools (5). Liver biochemistry showed a raised serum bilirubin concentration and raised liver enzymes. After 1 month, the jaundice disappeared and the liver function tests normalized. However, 1 month later she relapsed with jaundice. Serum bilirubin and liver enzymes and were again raised. She had used Shou wu pian three tablets daily for 4 months before the first episode of jaundice and had stopped using it after the first presentation. However, she had again taken it in a lower dose for a further month. The remedy was withdrawn and after 5 months, her liver function normalized and she made an uneventful recovery.
- An 18-year-old man developed liver failure during long-term use of the herbal mixture "RespirActin" which contained numerous herbs, including witch hazel (6). He required a liver transplant and made a good recovery.
- *Polygonum multiflorum* (Shou wu Pian) was prescribed by a Chinese herbalist for a 46-year-old woman with graying hair (7). After taking it for 2 weeks she developed signs and symptoms of hepatitis. The history revealed no plausible cause for hepatitis and viral infection was ruled out. After withdrawal of the *P. multiflorum* her liver enzymes normalized and she recovered fully.
- A 31-year-old pregnant Chinese woman developed hepatitis after consuming Shou wu Pian; tests for viral hepatitis were negative and there was no evidence of other systemic disease (8).

After receiving several reports of liver disorders, including jaundice and hepatitis, the UK Medicines and Healthcare products Regulatory Agency (MHRA) issued a warning about the risks of using *Polygonum multiflorum*, advising members of the public who experience symptoms of liver disorders while taking products that contain *Polygonum multiflorum* to see their doctor and to stop taking the product immediately if a liver disorder is diagnosed (9).

Rheum palmatum

Rheum palmatum (rhubarb) contains the anthranoids sennosides and rhein. In a study of patients taking regular doses of rhubarb-containing Kampo medicines (extracts or decoctions) and patients taking excess doses there was tolerance to initial stimulant pain in the abdomen during excess use (10). The authors proposed that the absence of tenderness on pressure over the umbilical region could predict increasing or excess use of rhubarb.

In 14 616 patients who used various Kampo medicines, some of which contained rhubarb, there was no association between the use of rhubarb and the development of gastric carcinoma (11).

References

1. Kunikata T, Tatefuji T, Aga H, Iwaki K, Ikeda M, Kurimoto M. Indirubin inhibits inflammatory reactions in delayed-type hypersensitivity. Eur J Pharmacol 2000;410(1):93–100.

2. Grech JN, Li Q, Roufogalis BD, Duck CC. Novel Ca(2+)-ATPase inhibitors from the dried root tubers of *Polygonum multiflorum*. J Nat Prod 1994;57(12):1682–7.

3. Zimmermann ML, Sneden AT. Vanicosides A and B, protein kinase C inhibitors from *Polygonum pensylvanicum*. J Nat Prod 1994;57(2):236–42.

4. Mazzanti G, Battinelli L, Daniele C, Mastroianni CM, Lichtner M, Coletta S, Costantini S. New case of acute hepatitis following the consumption of Shou Wu Pian, a Chinese herbal product derived from *Polygonum multiflorum*. Ann Intern Med 2004;140(7):W30.

5. Panis B, Wong DR, Hooymans PM, De Smet PA, Rosias PP. Recurrent toxic hepatitis in a Caucasian girl related to the use of Shou-Wu-Pian, a Chinese herbal preparation. J Pediatr Gastroenterol Nutr 2005; 41(2): 256–8.

6. Health Canada. RespirActin. Canadian Adverse Drug Reaction Newsletter 2003;13:4Jan.

7. Park GJ, Mann SP, Ngu MC. Acute hepatitis induced by Shou-Wu-Pian, a herbal product derived from Polygonum multiflorum. J Gastroenterol Hepatol 2001;16(1):115–7.

8. But PP, Tomlinson B, Lee KL. Hepatitis related to the Chinese medicine Shou-wu-pian manufactured from *Polygonum multiflorum*. Vet Hum Toxicol 1996;38(4):280–2.

9. Anonymous. *Polygonum multiflorum*. Risk of liver effects. WHO Newslett 2006; 3:4.

10. Mantani N, Kogure T, Sakai S, Kainuma M, Kasahara Y, Niizawa A, Shimada Y, Terasawa K. A comparative study between excess-dose users and regular-dose users of rhubarb contained in Kampo medicines. Phytomedicine 2002;9(5):373–6.

11. Mantani N, Sekiya N, Sakai S, Kogure T, Shimada Y, Terasawa K. Rhubarb use in patients treated with Kampo medicines—a risk for gastric cancer? Yakugaku Zasshi 2002;122(6):403–5.

Ranunculaceae

General Information

The genera in the family of Ranunculaceae (Table 47) include anemone, buttercup, columbine, hellebore, larkspur, marsh marigold, and pasque flower.

Aconitum species

Species of aconite contain a variety of diterpene alkaloids; *Aconitum napellus* (monkshood) contains isonapelline, luciculine, and napelline.

Adverse effects

Aconite roots can produce serious heart failure. Among the other symptoms of aconite poisoning are numbing of mouth and tongue, gastrointestinal disturbances, muscular weakness, incoordination, and vertigo (1). A review from Hong Kong reported 17 cases of aconite poisoning after the administration of Chinese herbal mixtures. The toxicity of raw aconite can be reduced substantially by decoction, as this process leads to a change in alkaloid composition (2).

Cimicifuga racemosa

Cimicifuga racemosa (black bugbane, black cohosh, black snakeroot, rattleroot, rattletop, rattleweed) contains a variety of cycloartane triterpene glycosides, some of which have cytotoxic effects (3).

Table 47 The genera of Ranunculaceae

Aconitum (monkshood)
Actaea (baneberry)
Adonis (pheasant's eye)
Anemone (anemone)
Aquilegia (columbine)
Caltha (marsh marigold)
Ceratocephala (curveseed butterwort)
Cimicifuga (bugbane)
Clematis (leather flower)
Consolida (knight's-spur)
Coptis (gold thread)
Delphinium (larkspur)
Enemion (false rue anemone)
Eranthis (eranthis)
Helleborus (hellebore)
Hepatica (hepatica)
Hydrastis (hydrastis)
Kumlienia (false buttercup)
Myosurus (mousetail)
Nigella (nigella)
Pulsatilla (pasque flower)
Ranunculus (buttercup)
Thalictrum (meadow rue)
Trautvetteria (bugbane)
Trollius (globe flower)
Xanthorhiza (yellow root)

Cimicifuga racemosa has been used to relieve symptoms of menopause, although with little evidence of efficacy (4).

Adverse effects

A systematic review of all clinical data on the safety of *Cimicifuga racemosa* found only a slight risk of mild, transient adverse effects (5). The authors concluded that, taken for a limited length of time, this remedy is reasonably safe.

Liver

Cimicifuga racemosa contains diterpenoids that cause liver damage in animals, either via reactive metabolites or by an autoimmune mechanism, although causality was not established beyond doubt. *Cimicifuga racemosa* has been reported to cause hepatitis (6, 7). However, evaluation of reports of acute liver disease is difficult, owing to the use of combination preparations and failure to analyse suspected products for purity and identity. Causality has not therefore been established.

- A 57-year-old woman developed autoimmune hepatitis 3 weeks after taking black cohosh tablets (8).

The authors believed that a casual relation was likely but no brand or dose or any other characterization of the herbal remedy was obtained. Causality in this case may therefore have been related to something in the tablets but not necessarily to black cohosh itself.

- A 52-year-old woman was hospitalized with acute liver failure(9). She had taken a herbal mixture containing *Cimicifuga racemosa*, ground ivy, and three other medicinal herbs for 3 months. She had no risk factors for hepatitis and had taken no other medicines. Her condition deteriorated and she developed hepatic encephalopathy as well as hepatorenal failure. She underwent liver transplantation with an uneventful postoperative course. Analysis of the herbal mixture revealed no undeclared constituents.

The authors thought that either *Cimicifuga racemosa* or ground ivy could have caused this adverse event.

- A 47-year-old woman took an extract of *C. racemosa* for 1 week to treat menopausal symptoms; she developed jaundice and raised liver enzymes (10). No other causes of liver damage were found. She required liver transplantation.

The Committee on Herbal Medicinal Products of the European Medicines Agency (EMEA) has reviewed case reports of hepatotoxicity in patients taking *Cimicifuga racemosa* root, and considers that there is a potential association between hepatotoxicity and herbal medicines containing *Cimicifuga* (11). The Committee reviewed 16 of the 42 case reports of hepatotoxicity to assess if *Cimicifuga* is linked to liver damage. Five cases were excluded, seven were thought to be unlikely to be related, and in four cases there was a temporal association between the start of *Cimicifuga* treatment and the occurrence of the hepatic reaction. EMEA has advised patients to stop using *Cimicifuga*, to consult their doctor immediately if symptoms of liver injury develop, and to inform their doctor if they are using herbal medicine products. EMEA has advised health-care professionals to ask their patients about the use of *Cimicifuga*-containing products and to report suspected hepatic reactions to their national adverse reactions reporting schemes.

Delphinium species

Delphinium species contain complex diterpenoid alkaloids that cause acute intoxication and death in cattle (12). The alkaloids and their concentrations vary with the species and plant part involved, which causes variability in toxicity. In *Delphinium consolida* (larkspur) there are toxic alkaloids in the non-medicinal plant parts (root, seed, herb), but they are purportedly absent in the medicinal part (the flower).

Hydrastis canadensis

The root of *Hydrastis canadensis* (golden seal) contains isoquinoline alkaloids, including the quaternary base berberine and the tertiary base hydrastine. As the latter can stimulate uterine contractions, it is prudent to avoid golden seal root during pregnancy, even though the actual risk of premature labor still has to be verified.

Adverse effects

In man berberine has positive inotropic, negative chronotropic, antidysrhythmic, and vasodilator properties (13).

Cardiovascular

There is experimental evidence that berberine can cause arterial hypotension (14,15).

Nervous system

Berberine displaces bilirubin from albumin and there is therefore a risk of kernicterus in jaundiced neonates (16).

Gastrointestinal

In a study of the effect of berberine in acute watery diarrhea, oral doses of 400 mg were well tolerated, except for complaints about its bitter taste and a few instances of transient nausea and abdominal discomfort. However, patients with cholera given tetracycline plus berberine were more ill, suffered longer from diarrhea, and required larger volumes of intravenous fluid than those given tetracycline alone (17).

Skin

- A 32-year-old woman had a phototoxic reaction after taking a dietary supplement containing ginseng, golden seal, bee pollen, and other ingredients (18). She had a pruritic, erythematous rash, localized to the sun-exposed surfaces of her neck and limbs. She had no significant past medical history and was not taking any other medications. The skin rash slowly resolved after withdrawal of the supplement and treatment with subcutaneous and topical glucocorticoids.

Although the individual ingredients in this dietary supplement have not been associated with cases of photosensitivity, it is possible that the combination of ingredients may have interacted to cause this effect.

Drug interactions

Indinavir

Golden seal root inhibits various isoforms of cytochrome P450, including CYP3A4 (19). However, in a crossover study, goldenseal root (1140 mg bd for 14 days) had no effect on the pharmacokinetics of a single oral dose of indinavir 800 mg (20).

Pulsatilla species

Pulsatilla (pasque flower) species are widely used in homeopathic medicine. Some of them contain podophyllotoxins. High doses of *Pulsatilla vulgaris* (meadow windflower) can irritate the kidneys and urinary tract. Pregnancy is a contraindication.

Ranunculus damascenus

Ranunculus damascenus (buttercup) is used topically for abscess drainage, hemorrhoids, and burns.

Adverse effects

Ranunculus damascenus has been reported to cause skin damage (21).

- A 45-year-old Turkish woman developed open wounds on the abdomen, right knee, and neck. She had used buttercup topically and orally for pain relief. She was treated with antibiotics and adequate wound care. Complete healing was achieved within 10 days.

The authors argued that protoanemonin, a constituent of *R. damascenus*, which inhibits mitosis in plants, had caused this severe reaction.

References

1. Kelly SP. Aconite poisoning. Med J Aust 1990;153(8):499.
2. Hikino H, Yamada C, Nakamura K, Sato H, Ohizumi Y. [Change of alkaloid composition and acute toxicity of Aconitum roots during processing.]Yakugaku Zasshi 1977;97(4):359–66.
3. Watanabe K, Mimaki Y, Sakagami H, Sashida Y. Cycloartane glycosides from the rhizomes of Cimicifuga racemosa and their cytotoxic activities. Chem Pharm Bull (Tokyo) 2002;50(1):121–5.
4. Borrelli F, Ernst E. Cimicifuga racemosa: a systematic review of its clinical efficacy. Eur J Clin Pharmacol 2002;58(4):235–41.
5. Huntley A, Ernst E. A systematic review of the safety of black cohosh. Menopause 2003; 10:58–64.
6. Low Dog T. Menopause: a review of botanical dietary supplements. Am J Med 2005;118(Suppl 12B): 98–108.
7. Mahady GB. Black cohosh (*Actaea/Cimicifuga racemosa*): review of the clinical data for safety and efficacy in menopausal symptoms. Treat Endocrinol 2005;4(3): 177–84.

8. Cohen SM, O'Connor AM, Hart J, Merel NH, Te HS. Autoimmune hepatitis associated with the use of black cohosh: a case study. Menopause 2004; 11:575–7.

9. Lontos S, Jones RM, Angus PW, Gow PJ. Acute liver failure associated with the use of herbal preparations containing black cohosh. Med J Aust 2003; 179:390–1.

10. Whiting PW, Clouston A, Kerlin P. Black cohosh and other herbal remedies associated with acute hepatitis. Med J Aust 2002;177(8):440–3.

11. Anonymous. *Cimicifuga racemosa* (black cohosh). Concerns of liver injury. WHO Newslett 2006;4:1.

12. Puschner B, Booth MC, Tor ER, Odermatt A. Diterpenoid alkaloid toxicosis in cattle in the Swiss Alps. Vet Hum Toxicol 2002;44(1):8–10.

13. Lau CW, Yao XQ, Chen ZY, Ko WH, Huang Y. Cardiovascular actions of berberine. Cardiovasc Drug Rev 2001;19(3):234–44.

14. Sabir M, Bhide NK. Study of some pharmacological actions of berberine. Indian J Physiol Pharmacol 1971;15(3):111–32.

15. Chun YT, Yip TT, Lau KL, Kong YC, Sankawa U. A biochemical study on the hypotensive effect of berberine in rats. Gen Pharmacol 1979;10(3):177–82.

16. Chan E. Displacement of bilirubin from albumin by berberine. Biol Neonate 1993;63(4):201–8.

17. Khin-Maung-U, Myo-Khin, Nyunt-Nyunt-Wai, Aye-Kyaw, Tin-U. Clinical trial of berberine in acute watery diarrhoea. BMJ (Clin Res Ed) 1985;291(6509):1601–5.

18. Palanisamy A, Haller C, Olson KR. Photosensitivity reaction in a woman using an herbal supplement containing ginseng, goldenseal, and bee pollen. J Toxicol Clin Toxicol 2003;41(6):865–7.

19. Foster BC, Vandenhoek S, Hana J, Krantis A, Akhtar MH, Bryan M, Budzinski JW, Ramputh A, Arnason JT. In vitro inhibition of human cytochrome P450-mediated metabolism of marker substrates by natural products. Phytomedicine 2003;10(4):334–42.

20. Sandhu RS, Prescilla RP, Simonelli TM, Edwards DJ. Influence of goldenseal root on the pharmacokinetics of indinavir. J Clin Pharmacol 2003;43(11):1283–8.

21. Metin A, Calka O, Behcet L, Yildirim E. Phytodermatitis from Ranunculus damascenus. Contact Dermatitis 2001;44(3):183.

Rhamnaceae

General Information

The genera in the family of Rhamnaceae (Table 48) include buckthorn and jujube.

Rhamnus purshianus

The bark of *Rhamnus purshiana* (cascara sagrada) contains laxative anthranoid derivatives, which occur primarily in various laxative herbs (such as aloe, cascara sagrada, medicinal rhubarb, and senna) in the form of free anthraquinones, anthrones, dianthrones, and/or *O*- and *C*-glycosides derived from these substances.

Adverse effects

The anthranoids produce harmless discoloration of the urine. Depending on intrinsic activity and dose, they can also produce abdominal discomfort and cramps, nausea, violent purgation, and dehydration. They can be distributed into breast milk, but not always in sufficient amounts to affect the suckling infant. Long-term use can result in electrolyte disturbances and in atony and dilatation of the colon.

Respiratory

Cascara sagrada can cause IgE-mediated occupational asthma and rhinitis [1].

Liver

Cascara sagrada has been reported to cause liver damage [2].

- A 48-year-old man developed cholestatic hepatitis and hypertension shortly after he started to use the herbal laxative cascara sagrada. He took one capsule (425 mg of aged cascara sagrada bark) tds for 3 days and subsequently developed right upper quadrant pain, nausea, abdominal bloating, anorexia, and jaundice. The cascara was withdrawn, but his symptoms persisted and his liver function tests were abnormal. One week later, he developed ascites and jaundice and underwent liver biopsy, which showed moderately severe portal inflammation, intracanalicular bile stasis, portal bridging fibrosis, and mild steatosis. He gradually improved without specific treatment and 3 months later his ascites and jaundice had resolved.

Tumorigenicity

Several anthranoid derivatives (notably the aglycones aloe-emodin, chrysophanol, emodin, and physcion) are genotoxic in bacterial and/or mammalian test systems (SEDA-12, 409), and two anthranoid compounds (the synthetic laxative dantron and the naturally occurring l-hydroxyanthraquinone) have carcinogenic activity in rodents. In an epidemiological study, chronic abusers of anthranoid laxatives (identified by the presence of pseudomelanosis coli) had an increased relative risk of 3.04 (95% CI = 1.18, 4.90) for colorectal cancer [3]. The German health authorities therefore restricted the indication of herbal anthranoid laxatives to constipation which has not responded to bulk-forming therapy (which rules out their inclusion in slimming aids). In addition, they imposed restrictions on the laxative use of anthranoid-containing herbs (for example not to be used for more than 1–2 weeks without medical advice, not to be used in children under 12 years of age, and not to be used during pregnancy and lactation) [3,4].

Zizyphus jujuba

The fruit of *Ziziphus jujuba* (dazao) is often consumed in Eastern Asia as food or as a tonic and sedative.

Adverse effects

Angioedema has been described after the oral ingestion of dazao [5].

Table 48 The genera of Rhamnaceae

Adolphia (prickbush)
Alphitonia (alphitonia)
Auerodendron (auerodendron)
Berchemia (supplejack)
Ceanothus (ceanothus)
Colubrina (nakedwood)
Condalia (snakewood)
Frangula (buckthorn)
Gouania (chewstick)
Hovenia (hovenia)
Karwinskia (karwinskia)
Krugiodendron (krugiodendron)
Maesopsis (umbrella-tree)
Paliurus (Jeruselem thorn)
Reynosia (darlingplum)
Rhamnus (buckthorn)
Sageretia (mock buckthorn)
Smythea
Ziziphus (jujube)

References

1. Giavina-Bianchi PF Jr, Castro FF, Machado ML, Duarte AJ. Occupational respiratory allergic disease induced by Passiflora alata and Rhamnus purshiana. Ann Allergy Asthma Immunol 1997;79(5):449–54.
2. Nadir A, Reddy D, Van Thiel DH. Cascara sagrada-induced intrahepatic cholestasis causing portal hypertension: case report and review of herbal hepatotoxicity. Am J Gastroenterol 2000;95(12):3634–7.
3. Siegers CP, von Hertzberg-Lottin E, Otte M, Schneider B. Anthranoid laxative abuse—a risk for colorectal cancer? Gut 1993;34(8):1099–101.
4. Kommission E. Aufbereitungsmonographien. Dtsch Apoth Ztg 1993;133:2791–4.
5. Chan TY, Chan AY, Critchley JA. Hospital admissions due to adverse reactions to Chinese herbal medicines. J Trop Med Hyg 1992;95(4):296–8.

Rosaceae

General Information

The genera in the family of Rosaceae (Table 49) include avens, cotoneasters, hawthorn, and roses, and a variety of fruits such as apples, loquats, pears, plums, quince, and strawberries.

Table 49   The genera of Rosaceae

Acaena (acaena)
Adenostoma (chamise)
Agrimonia (agrimony)
Alchemilla (lady's mantle)
Amelanchier (serviceberry)
Amelasorbus (amelasorbus)
Aphanes (parsley piert)
Argentina (silverweed)
Aruncus (aruncus)
Cercocarus (mountain mahogany)
Chaenmeles (flowering quince)
Chamebatia (mountain misery)
Chamaebatiaria (fernbush)
Chamaerhodos (little rose)
Coleogyne (coleogyne)
Comarum (comarum)
Cotoneaster (cotoneaster)
Crataegus (hawthorn)
Cydonia (cydonia)
Dalibarda (dalibarda)
Dasiphora (shrubby cinquefoil)
Dryas (mountain avens)
Duchesnea (duchesnea)
Eriobotrya (loquat)
Exochorda (pearlbrush)
Fallugia (Apache plume)
Filipendula (queen)
Fragaria (strawberry)
Geum (avens)
Heteromeles (toyon)
Holodiscus (ocean spray)
Horkelia (horkelia)
Horkeliella (false horkelia)
Ivesia (mousetail)
Kelseya (kelseya)
Kerria (kerria)
Luetkea (luetkea)
Lyonothamnus (lyononthmnus)
Malacomeles (false serviceberry)
Malus (apple)
Mespilus (mespilus)
Neviusia (snow wreath)
Oemleria (oemleria)
Osteomeles (osteomeles)
Peraphyllum (peraphyllum)
Petrophyton (rock spiraea)
Photinia (chokeberry)
Physocarpus (ninebark)
Porteranthus (porteranthus)
Potentilla (cinquefoil)

(Continued)

Prunus (plum)
Pseudocydonia (Chinese-quince)
Purshia (bitterbrush)
Pyracantha (firethorn)
Pyrus (pear)
Quillaja (quillaja)
Rhodotypos (rhodotypos)
Rosa (rose)
Rubus (blackberry)
Sanguisorba (burnet)
Sibbaldia (sibbaldia)
Sibbaldiopsis (sibbaldiopsis)
Sorbaria (false spiraea)
Sorbus (mountain ash)
Spiraea (spiraea)
Stephanandra (stephanandra)
Vauquelinia (rosewood)
Waldsteinia (barren strawberry)

Crataegus species

Crataegus species (hawthorn, maybush, whitethorn) contain a variety of flavonoids, including rhamnosides, schaftosides, and spiraeosides. They have a positive inotropic effect on the heart by a mechanism different from that of cardiac glycosides, catecholamines, and the phosphodiesterase type III inhibitors (1) and are effective in mild heart failure (2).

Adverse effects
The main adverse effects of *Crataegus* are dizziness and vertigo (3).

Drug-drug interactions
It has been suggested that the flavonoids in *Crataegus* inhibit P glycoprotein function and might therefore interact with drugs that are substrates of P glycoprotein, such as digoxin. However, in a randomized, crossover trial in eight healthy volunteers *Crataegus* special extract WS 1442 (hawthorn leaves with flowers) had no effect on the pharmacokinetics of digoxin 0.25 mg/day for 10 days (4).

Prunus species

The raw pits or kernels of various *Prunus* species (such as apricot, bitter almond, choke cherry and peach) are promoted as health foods. However, they contain the cyanogenic glycoside, amygdalin, which yields hydrogen cyanide after ingestion.

During the late 1950s, and for about 20 years afterwards, laetrile (l-mandelonitrile-beta-glucuronoside), related to amygdalin, was touted as a cure for cancer (http://www.quackwatch.org/01QuackeryRelatedTopics/Cancer/laetrile.html).

However, in 1982 Arnold Relman, then Editor of the New England Journal of Medicine, pronounced that "Laetrile has had its day in court. The evidence, beyond reasonable doubt, is that it doesn't benefit patients with advanced cancer, and there is no reason to believe that it would be any more effective in the earlier stages of the disease····. The time has come to close the books." (5). Nevertheless, it is still being so touted.

Adverse effects

When ingested in sufficient quantities *Prunus* species cause cyanide poisoning. For instance, a total consumption of about 48 apricot kernels produced forceful vomiting, headache, flushing, heavy sweating, dizziness, and faintness before vomiting was induced in the emergency room, whereafter the symptoms rapidly subsided. In another case accidental poisoning was fatal (6).

- A 67-year-old woman with lymphoma presented with a neuromyopathy following treatment with laetrile. She had high blood and urinary thiocyanate and cyanide concentrations (7). Sural nerve biopsy specimen showed a mixed pattern of demyelination and axonal degeneration, the latter being prominent. Gastrocnemius muscle biopsy specimen showed a mixed pattern of denervation and myopathy with type II atrophy.

Besides the risk that a large dose can lead to acute cyanide poisoning, there is also the question whether continued ingestion of cyanogenic pits or kernels can cause chronic intoxication.

An outbreak of congenital malformations in swine has been retrospectively associated with the eating of the fruit, leaves, and bark of *Prunus serotina* (wild black cherry) (8). Prospective experimental evidence of teratogenicity was not available at that time, but amygdalin was later reported to be teratogenic in hamsters (9).

References

1. Joseph G, Zhao Y, Klaus W. Pharmakologisches Wirkprofil von Crataegus-Extrakt im Vergleich zu Epinephrin, Amrinon, Milrinon und Digoxin am isoliert perfundierten Meerschweinchenherzen. [Pharmacologic action profile of Crataegus extract in comparison to epinephrine, amirinone, milrinone and digoxin in the isolated perfused guinea pig heart.] Arzneimittelforschung 1995;45(12):1261–5.
2. Weihmayr T, Ernst E. Die therapeutische Wirksamkeit von Crataegus. [Therapeutic effectiveness of Crataegus.] Fortschr Med 1996;114(1–2):27–9.
3. Tauchert M. Efficacy and safety of crataegus extract WS 1442 in comparison with placebo in patients with chronic stable New York Heart Association class-III heart failure. Am Heart J 2002;143(5):910–5.
4. Tankanow R, Tamer HR, Streetman DS, Smith SG, Welton JL, Annesley T, Aaronson KD, Bleske BE. Interaction study between digoxin and a preparation of hawthorn (Crataegus oxyacantha). J Clin Pharmacol 2003;43(6):637–42.
5. Relman AS. Closing the books on Laetrile. N Engl J Med 1982;306(4):236.
6. Humbert JR, Tress JH, Braico KT. Fatal cyanide poisoning: accidental ingestion of amygdalin. JAMA 1977;238(6):482.
7. Kalyanaraman UP, Kalyanaraman K, Cullinan SA, McLean JM. Neuromyopathy of cyanide intoxication due to "laetrile" (amygdalin). A clinicopathologic study. Cancer 1983;51(11):2126–33.
8. Selby LA, Menges RW, Houser EC, Flatt RE, Case AA. Outbreak of swine malformations associated with the wild black cherry, Prunus serotina. Arch Environ Health 1971;22(4):496–501.
9. Willhite CC. Congenital malformations induced by laetrile. Science 1982;215(4539):1513–5.

Rubiaceae

General Information

The genera in the family of Rubiaceae (Table 50) include cinchona and coffee.

Table 50 The genera of Rubiaceae

Anthocephalus (anthocephalus)
Antirhea (quina)
Asperula (woodruff)
Bobea (ahakea)
Bouvardia (bouvardia)
Calycophyllum (calycophyllum)
Canthium (canthium)
Casasia (casasia)
Catesbaea (lilythorn)
Cephaelis (cephaelis)
Cephalanthus (buttonbush)
Chiococca (milkberry)
Chione (chione)
Cinchona (cinchona)
Coccocypselum (coccocypselum)
Coffea (coffee)
Coprosma (mirrorplant)
Crucianella (crucianella)
Cruciata (bedstraw)
Crusea (mountain saucerflower)
Diodia (buttonweed)
Erithalis (blacktorch)
Ernodea (ernodea)
Exostema (exostema)
Faramea (false coffee)
Galium (bedstraw)
Gardenia (gardenia)
Genipa (genipa)
Geophila (geophila)
Gonzalagunia (gonzalagunia)
Guettarda (guettarda)
Hamelia (hamelia)
Hedyotis (starviolet)
Hillia (hillia)
Hintonina
Houstonia (bluet)
Ixora (ixora)
Kelloggia (kelloggia)
Lasianthus (lasianthus)
Lucya (lucya)
Machaonia (machaonia)
Mitchella (mitchella)
Mitracarpus (girdlepod)
Morinda (morinda)
Neolaugeria (neolaugeria)
Neolamarckia (neolamarckia)
Neonauclea
Nertera (nertera)
Oldenlandia (oldenlandia)
Oldenlandiopsis (creeping-bluet)
Paederia (sewer vine)

(Continued)

Palicourea (cappel)
Pentas (pentas)
Pentodon (pentodon)
Phialanthus (phialanthus)
Pinckneya (pinckneya)
Psychotria (wild coffee)
Randia (indigoberry)
Richardia (Mexican clover)
Rondeletia (cordobancillo)
Rubia (rubia)
Sabicea (woodvine)
Schradera (schradera)
Scolosanthus (scolosanthus)
Serissa (snowrose)
Sherardia (sherardia)
Spermacoce (false buttonweed)
Strumpfia (strumpfia)
Timonius
Uncaria (uncaria)
Vangueria (vangueria)

Asperula odorata

Asperula odorata (sweet woodruff) is rich in coumarins.

Cephaelis ipecacuanha

The root of *Cephaelis ipecacuanha*, also known as *Psychotria ipecacuanha*, is the source of ipecacuanha, which contains the emetic alkaloids emetine and cephaeline. Ipecacuanha is covered in a separate monograph.

Hintonia latiflora

Hintonia latiflora (copalchi bark) has been advocated as a hypoglycemic agent, an effect that has been attributed to the neoflavonoid coutareagenin (1). Its use has been associated with liver toxicity but a causal relation could not be established (2).

Morinda citrifolia

Morinda citrifolia (noni) is one of the most important traditional Polynesian medicinal plants, which has a large range of therapeutic claims, including antibacterial, antiviral, antifungal, antihelminthic, antitumor, analgesic, hypotensive, anti-inflammatory, and immune enhancing effects (3,4).

Adverse effects
Electrolyte balance
The juice of *M. citrifolia* has been reported to cause hyperkalemia (5).

- A man with chronic renal insufficiency who followed dietary restriction of potassium developed a raised serum potassium concentration (5.8 mmol/l). He insisted that he had followed his dietary regimen as usual, except for taking noni juice, purchased from a health food store. He was treated with sodium polystyrene sulfonate and told to stop taking noni juice. At the next check-up his potassium was still raised; he said that he would never stop taking noni juice and that his physicians did not understand its healing power.

The potassium concentration in noni juice samples was 56 mmol/l.

Liver

Morinda citrifolia has been implicated in liver damage.

- A 45-year-old man developed very high liver transaminase activities and raised lactate dehydrogenase activity (6). His medical history was unremarkable and he took no regular medications. There was no evidence of viral hepatitis, Epstein–Barr virus or cytomegalovirus infection, autoimmune hepatitis, Budd–Chiari syndrome, hemochromatosis, or Wilson's disease. The patient said that he had been drinking the juice of M citrifolia for prophylactic reasons for 3 weeks. A liver biopsy confirmed acute hepatitis and there was an inflammatory infiltrate with numerous eosinophils in the portal tracts. Other causes were excluded and the histology suggested a causal role of the herb. He stopped taking noni and his transaminase activities normalized quickly and were within the reference ranges 1 month later.

Rubia tinctorum

The use of herbal medicines prepared from the root of *Rubia tinctorum* (madder) is no longer permitted in Germany. Root extracts have shown genotoxic effects in several test systems, which are attributed to the presence of the anthraquinone derivative lucidin. One of the other main components, alizarin primeveroside, is transformed into 1-hydroxyanthraquinone when given orally to the rat, in which this metabolite has carcinogenic activity (7).

Uncaria tomentosa

Uncaria tomentosa (cat's claw), an herb from the highlands of the Peruvian Amazon, contains the spiroindole alkaloids isopteropodine and rynchophylline; it has immunomodulatory properties and has been used to treat arthritis and inflammatory bowel disease (8).

Adverse effects

- A 59-year-old woman with mantle-cell lymphoma and no hepatic involvement took a range of unconventional medicines (9). During a routine check-up she had raised liver enzymes, and self-medication with cat's claw was deemed the most likely cause. Cat's claw was withdrawn and her liver tests normalized within 60 days.

References

1. Korec R, Heinz Sensch K, Zoukas T. Effects of the neoflavonoid coutareagenin, one of the antidiabetic active substances of Hintonia latiflora, on streptozotocin-induced diabetes mellitus in rats. Arzneimittelforschung 2000; 50(2):122–8.
2. In: Hansel R, Keller K, Rimpler H, Schneider G, editors. Hagers Handbuch der Pharmazeutischen Praxis. 5th ed.. Berlin: Springer-Verlag, 1993:1.
3. McClatchey W. From Polynesian healers to health food stores: changing perspectives of *Morinda citrifolia* (Rubiaceae). Integr Cancer Ther 2002;1(2):110–20.
4. Wang MY, West BJ, Jensen CJ, Nowicki D, Su C, Palu AK, Anderson G. *Morinda citrifolia* (noni): a literature review and recent advances in noni research. Acta Pharmacol Sin 2002;23(12):1127–41.
5. Mueller BA, Scott MK, Sowinski KM, Prag KA. Noni juice (*Morinda citrifolia*): hidden potential for hyperkalemia? Am J Kidney Dis 2000;35(2):310–2.
6. Millonig G, Stadlmann S, Vogel W. Herbal hepatotoxicity: acute hepatitis caused by a noni preparation (*Morinda citrifolia*). Eur J Gastroenterol Hepatol 2005;17(4):445–7.
7. De Smet PAGM, Stricker BHC. Meekrapwortel in Duitsland niet langer toegestaan. Pharm Weekbl 1993;128:503.
8. Steinberg PN. Una de gato: una hierba prodigiosa de la selva humeda Peruana. [Cat's claw: an herb from the Peruvian Amazon.] Sidahora 1995;35–6.
9. Gertz MA, Bauer BA. Caring (really) for patients who use alternative therapies for cancer. J Clin Oncol 2001;19(23):4346–9.

Rutaceae

General Information

The genera in the family of Rutaceae (Table 51) include citrus fruits (citrons, oranges, grapefruit, lemons, limes, tangerines, etc.), angostura, kumquat, pilocarpus, and rue.

Agathosma betulina

Agathosma betulina (buchu) contains diosmin (venosmine), and a synthetic form of diosmin is also available. It has been used to treat the pain and bleeding of hemorrhoids (1).

Adverse effects

Diosmin can rarely cause nausea and epigastric discomfort (SEDA-3, 181) and somnolence (2).

In 12 volunteers, diosmin 500 mg/day for 9 days significantly increased the AUC and C_{max} of metronidazole 800 mg without a change in t_{max}; this effect was attributed to inhibition of CYP3A4 (3).

Table 51   The genera of Rutaceae

Acronychia (acronychia)
Aegle (aegle)
Agathosma (agathosma)
Amyris (torchwood)
Boronia (boronia)
Casimiroa (sapote)
Choisya (Mexican orange)
Citrofortunella (citrofortunella)
Citroncirus (citroncirus)
Citrus (citrus)
Clausena (clausena)
Cneoridium (cneoridium)
Correa (Australian fuschia)
Cusparia (cusparia)
Dictamnus (dictamnus)
Eremocitrus (eremocitrus)
Esenbeckia (jopoy)
Flindersia (flindersia)
Fortunella (kumquat)
Galipea (galipea)
Glycosmis (glycosmis)
Helietta (helietta)
Limonia (limonia)
Melicope (melicope)
Microcitrus (microcitrus)
Murraya (murraya)
Phellodendron (cork tree)
Pilocarpus (pilocarpus)
Platydesma (platydesma)
Poncirus (poncirus)
Ptelea (hoptree)
Ravenia (ravenia)
Ruta (rue)
Severinia (severinia)
Thamnosma (desert rue)
Triphasia (triphasia)
Zanthoxylum (prickly ash)

Citrus aurantium

Volatile plant oils (often incorrectly termed "essential oils") are used in aromatherapy and are usually applied by gentle massage. The oil of *C. aurantium* (bergamot) is often used in this way. It has photosensitive and melanogenic properties and is potentially phototoxic and photomutagenic.

Adverse effects
Cardiovascular

Citrus aurantium (bitter orange) contains synephrine, which has ephedrine-like effects on the cardiovascular system.

- A 55-year-old woman suffered an acute lateral wall myocardial infarction after taking a herbal supplement containing bitter orange (Edita's Skinny Pill) 300 mg (4). She had no relevant past medical history and had been free of cardiovascular risk factors.
- A 52-year-old woman developed a tachydysrhythmia shortly after taking 500 mg of an extract of *Citrus aurantium* titrated to contain 6% synephrine (5). The extract was withdrawn and she remained in good health until she again started taking the same herbal supplement. She was admitted to hospital with a tachydysrhythmia and was again treated and discharged. She had not taken any other medication, except levothyroxine. After 6 months follow up she was in good health.

Skin

- Two patients had localized and disseminated bullous phototoxic skin reactions 48–72 hours after exposure to bergamot aromatherapy and ultraviolet light (6). One developed bullous skin lesions after exposure to aerosolized aromatherapy oil in a sauna.

Citrus paradisi (grapefruit)

Grapefruit juice is unusual in that it is not used therapeutically but can cause many drug interactions, since an unidentified constituent of grapefruit inhibits CYP3A4. The drugs most commonly involved are dihydropyridine calcium channel blockers and HMG-CoA reductase inhibitors (7).

Both whole grapefruit and grapefruit juice have been implicated, and the wording used on a warning label in New Zealand to alert patients to potential drug interactions with grapefruit is "Do not take grapefruit or grapefruit juice while being treated with this medicine" (8). Health Canada has advised the public not to take grapefruit or its juice (fresh or frozen) with certain drugs (9). Health Canada has also issued several communication documents to remind health professionals of possible interactions of grapefruit with therapeutic drugs and has worked with drug manufacturers whose products are adversely affected by grapefruit to ensure that the relevant information is printed on the product label.

Of the several hundred chemical entities in grapefruit juice (10), furanocoumarin, mainly bergamottin and 6,7-dihydroxybergamottin, are thought to be the components in grapefruit that are responsible for these interactions, although the flavonoid naringenin, which was originally suspected, cannot be excluded and may have a minor role (11).

Table 52   Drug–grapefruit juice interactions

Drug	Effect of grapefruit	Reference
Benzodiazepines	Pronounced nervous system effects	(14)
Ciclosporin	Inhibits metabolism and increases blood concentration	(15,16)
Coumarins	Interferes with the analysis of coumarins	(17)
Diazepam	Increases concentration	(18)
Felodipine	Enhances the effects and can cause increased blood pressure and heart rate, headache, flushing, and light-headedness	(19)
Lovastatin	Greatly increases serum concentration	(20)
Nifedipine	Increases plasma concentration	(21)
Nisoldipine	Increases plasma concentration	(22)
Quinidine	Delays absorption and maximal effect on QT interval	(23)
Saquinavir	Increases AUC and C_{max}	(11)
Terfenadine	Increases systemic availability, with cardiotoxicity at higher doses of grapefruit juice concentration	(24)
Triazolam	Increases plasma concentrations	(25)

The ability of grapefruit to increase the plasma concentrations of some drugs was accidentally discovered when grapefruit juice was used as a blinding agent in a drug interaction study of felodipine and alcohol (12). It was noticed that plasma concentrations of felodipine were much higher when the drug was taken with grapefruit juice than those previously reported for the dose of drug administered. In another study, (13) concurrent administration of grapefruit juice and felodipine increased the AUC, causing increased heart rate, and reduced diastolic blood pressure. Similarly, parallel administration of the juice with midazolam altered psychometric performance tests, while with nisoldipine or nitrendipine there was an increase in heart rate.

Table 52 lists some drugs whose effects are altered by grapefruit juice.

The Adverse Drug Reactions Advisory Committee (ADRAC) in Australia considers that "the safest course is to avoid grapefruit and its juice altogether when taking medicines that interact" (26).

Dictamnus dasycarpus

Dictamnus dasycarpus (densefruit pittany) contains rutaevin, terpene glycosides (dasycarpusides), sesquiterpene glycosides (dictamnosides), and phenolic glycosides. *D. dasycarpus* is a common ingredient of the complex traditional Chinese herbal medicines that have been associated with liver damage; however, a causative role remains to be established (27–30).

Adverse effects

The UK hepatologists have described two cases of severe liver damage, one fatal, attributed to Chinese herbal mixtures for minor complaints (31). *D. dasycarpus* was implicated in one. There have been other reports of hepatotoxicity due to *D. dasycarpus* (31).

Dermatitis has been attributed to *D. dasycarpus* (32).

The root bark of *D. dasycarpus* is mutagenic in bacteria, due not to dietary flavonoids, but to the furoquinoline alkaloids dictamnine and gamma-fagarine. The clinical relevance of this finding remains to be established.

Pilocarpus species

The leaves of *Pilocarpus* species, including *Pilocarpus jaborandi* (jaborandi) and *Pilocarpus racemosus* (aceitillo), contain pilocarpine or isopilocarpine as major alkaloids as well as a variety of minor alkaloids. Pilocarpine (see separate monograph) is an agonist at acetylcholine receptors, which enhances salivary flow, sweating, and gastrointestinal motility. It can interfere with asthma therapy and the effect of anticholinergic drugs.

Ruta graveolens

Ruta graveolens (rue) contains acridone alkaloids, such as furacridone and gravacridone, quinoline alkaloids, such as graveoline and graveolinine, the furanoquinoline dictamnine, coumarins, such as gravelliferone, isorutarin, rutacultin, rutaretin, and suberenone, and the furanocoumarins 5-methoxypsoralen (bergapten) and 8-methoxypsoralen (xanthotoxine). It has been used as an abortifacient and emmenagogue (33).

Adverse effects

The essential oil of rue can cause contact dermatitis and phototoxic reactions (34,35) and severe hepatic and renal toxicity. Therapeutic doses can lead to depression, sleep disorders, fatigue, dizziness, and cramps. The sap from fresh leaves can produce painful gastrointestinal irritation, fainting, sleepiness, a weak pulse, abortion, a swollen tongue, and a cool skin.

References

1. Diana G, Catanzaro M, Ferrara A, Ferrari P. Attivita della diosmina pura nel trattamento della malattia emorroidaria. [Activity of purified diosmin in the treatment of hemorrhoids.] Clin Ter 2000;151(5):341–4.
2. Jimenez Gomez R, Saldana Garrido D, Martin Arias LH, Carvajal Garcia Pando A. Somnolencia en el curso de un tratamiento con diosmina. [Somnolence during diosmin treatment.] Med Clin (Barc) 1991;97(5):198–9.
3. Rajnarayana K, Reddy MS, Krishna DR. Diosmin pretreatment affects bioavailability of metronidazole. Eur J Clin Pharmacol 2003;58(12):803–7.

4. Nykamp DL, Fackih MN, Compton AL. Possible association of acute lateral-wall myocardial infarction and bitter orange supplement. Ann Pharmacother 2004; 38: 812–6.

5. Firenzuoli F, Gori L, Galapai C. Adverse reaction to an adrenergic herbal extract (*Citrus aurantium*). Phytomedicine 2005; 12(3): 247–8.

6. Kaddu S, Kerl H, Wolf P. Accidental bullous phototoxic reactions to bergamot aromatherapy oil. J Am Acad Dermatol 2001;45(3):458–61.

7. Anonymous. Grapefruit juice. Specific report of drug interactions. WHO Pharmaceuticals Newslett 2003;1:5.

8. Anonymous. Grapefruit warning label: now official in some countries. Drugs Ther Perspect 1998;12:12–3.

9. Anonymous. Grapefruit juice. Potential for drug interactions. WHO Pharmaceuticals Newslett 2002;3:10–1.

10. Ranganna S, Govindarajan VS, Ramana KV. Citrus fruits—varieties, chemistry, technology, and quality evaluation. Part II. Chemistry, technology, and quality evaluation. A. Chemistry. Crit Rev Food Sci Nutr 1983;18(4):313–86.

11. Fuhr U. Drug interactions with grapefruit juice. Extent, probable mechanism and clinical relevance. Drug Saf 1998;18(4):251–72.

12. Bailey DG, Spence JD, Edgar B, Bayliff CD, Arnold JM. Ethanol enhances the hemodynamic effects of felodipine. Clin Invest Med 1989;12(6):357–62.

13. Rodvold KA, Meyer J. Drug-food interactions with grapefruit juice. Infect Med 1996;13:868–912.

14. Kupferschmidt HH, Ha HR, Ziegler WH, Meier PJ, Krahenbuhl S. Interaction between grapefruit juice and midazolam in humans. Clin Pharmacol Ther 1995;58(1):20–8.

15. Hollander AA, van Rooij J, Lentjes GW, Arbouw F, van Bree JB, Schoemaker RC, van Es LA, van der Woude FJ, Cohen AF. The effect of grapefruit juice on cyclosporine and prednisone metabolism in transplant patients. Clin Pharmacol Ther 1995;57(3):318–24.

16. Brunner LJ, Munar MY, Vallian J, Wolfson M, Stennett DJ, Meyer MM, Bennett WM. Interaction between cyclosporine and grapefruit juice requires long-term ingestion in stable renal transplant recipients. Pharmacotherapy 1998;18(1):23–9.

17. Runkel M, Tegtmeier M, Legrum W. Metabolic and analytical interactions of grapefruit juice and 1,2-benzopyrone (coumarin) in man. Eur J Clin Pharmacol 1996;50(3):225–30.

18. Ozdemir M, Aktan Y, Boydag BS, Cingi MI, Musmul A. Interaction between grapefruit juice and diazepam in humans. Eur J Drug Metab Pharmacokinet 1998;23(1):55–9.

19. Feldman EB. How grapefruit juice potentiates drug bioavailability. Nutr Rev 1997;55(11 Pt 1):398–400.

20. Kantola T, Kivisto KT, Neuvonen PJ. Grapefruit juice greatly increases serum concentrations of lovastatin and lovastatin acid. Clin Pharmacol Ther 1998;63(4):397–402.

21. Hashimoto Y, Kuroda T, Shimizu A, Hayakava M, Fukuzaki H, Morimoto S. Influence of grapefruit juice on plasma concentration of nifedipine. Jpn J Clin Pharmacol Ther 1996;27:599–606.

22. Azuma J, Yamamoto I, Wafase T, Orii Y, Tinigawa T, Terashima S, Yoshikawa K, Tanaka T, Kawano K. Effects of grapefruit juice on the pharmacokinetics of the calcium channel blockers nifedipine and nisoldipine. Curr Ther Res Clin Exp 1998;59:619–34.

23. Min DI, Ku YM, Geraets DR, Lee H. Effect of grapefruit juice on the pharmacokinetics and pharmacodynamics of quinidine in healthy volunteers. J Clin Pharmacol 1996;36(5):469–76.

24. Clifford CP, Adams DA, Murray S, Taylor GW, Wilkins MR, Boobis AR, Davies DS. The cardiac effects of terfenadine after inhibition of its metabolism by grapefruit juice. Eur J Clin Pharmacol 1997;52(4):311–5.

25. Hukkinen SK, Varhe A, Olkkola KT, Neuvonen PJ. Plasma concentrations of triazolam are increased by concomitant ingestion of grapefruit juice. Clin Pharmacol Ther 1995;58(2):127–31.

26. Anonymous. Grapefruit juice. Revised advice from ADRAC. WHO Pharm Newslett 2003; 3–6.

27. Pillans PI, Eade MN, Massey RJ. Herbal medicine and toxic hepatitis. NZ Med J 1994;107(988):432–3.

28. Kane JA, Kane SP, Jain S. Hepatitis induced by traditional Chinese herbs; possible toxic components. Gut 1995;36(1):146–7.

29. Vautier G, Spiller RC. Safety of complementary medicines should be monitored. BMJ 1995;311(7005):633.

30. Perharic L, Shaw D, Leon C, De Smet PAGM, Murray VSG. Liver damage associated with certain types of traditional Chinese medicines used for skin diseases. Hum Exp Toxicol, in press.

31. McRae CA, Agarwal K, Mutimer D, Bassendine MF. Hepatitis associated with Chinese herbs. Eur J Gastroenterol Hepatol 2002;14(5):559–62.

32. Stekhun FI, Kyrnakov BA. [Dermatitis caused by Dictamnus dasycarpus.]Vestn Dermatol Venerol 1962;36:67–70.

33. Conway GA, Slocumb JC. Plants used as abortifacients and emmenagogues by Spanish New Mexicans. J Ethnopharmacol 1979;1(3):241–61.

34. Wessner D, Hofmann H, Ring J. Phytophotodermatitis due to Ruta graveolens applied as protection against evil spells. Contact Dermatitis 1999;41(4):232.

35. Schempp CM, Schopf E, Simon JC. Dermatitis bullosa striata pratensis durch Ruta graveolens L. (Gartenraute). [Bullous phototoxic contact dermatitis caused by Ruta graveolens L. (garden rue), Rutaceae. Case report and review of literature.] Hautarzt 1999;50(6):432–4.

Salicaceae

General Information

The family of Salicaceae contains two genera:

1. *Populus* (cottonwood)
2. *Salix* (willow).

Salix species and salicylates

The barks of various species of *Salix* contain glycosides of saligenin (salicylalcohol), namely the simple O-glycoside salicin and more complex glycosides such as salicortin (1). When taken orally, these glycosides undergo intestinal transformation to saligenin, which is rapidly absorbed and converted by the liver to salicylic acid. When willow bark preparations are used according to current dosage recommendations, they will not provide sufficient salicylic acid to produce acute salicylate poisoning. However, the risk of hypersusceptibility reactions, such as skin reactions and bronchospasm, in sensitive individuals cannot be excluded.

Other natural sources of salicylates are listed in Table 53. Methyl salicylate is covered under *Gaultheria procumbens* in the monograph on the Ericaceae.

Anaphylaxis has been attributed to willow bark.

- A 25-year-old woman developed anaphylaxis after taking a herbal slimming aid containing willow bark, which is rich in salicylates (2). She had a history of salicylate allergy. She made a full recovery in intensive care.

References

1. Kammerer B, Kahlich R, Biegert C, Gleiter CH, Heide L. HPLC–MS/MS analysis of willow bark extracts contained in pharmaceutical preparations. Phytochem Anal 2005;16(6): 470–8.
2. Boullata JI, McDonnell PJ, Oliva CD. Anaphylactic reaction to a dietary supplement containing willow bark. Ann Pharmacother 2003; 37: 832–5.

Table 53 Some plant sources of salicylates

Salicylate	Plant species	Common name	Family
Methyl salicylate	*Gaultheria procumbens*	Wintergreen, Eastern teaberry	Ericaceae
	Spiraea alba	White meadowsweet	Rosaceae
Salicin	*Salix songorica*		Salicaceae
Salicylic acid	*Salix alba*	White willow	Salicaceae
	Burchardia multiflora	Milk maids	Colchicaceae
	Callicarpa integerrima		Verbenaceae
	Falcaria vulgaris	Sickleweed	Apiaceae
	Flemingia laevicarpa		Fabaceae
	Genista lucida		Fabaceae
	Populus pseudosimonii		Salicaceae
Saligenin	*Populus pseudosimonii*		Salicaceae

Sapindaceae

General Information

The genera in the family of Sapindaceae (Table 54) include akee and lychee.

Blighia sapida

Blighia sapida (akee) contains a large amount of a potent hypoglycemic amino acid, glycylglycylglycine, known as hypoglycin.

Adverse effects

Akee poisoning (Jamaican winter vomiting sickness or toxic hypoglycemic syndrome) is a potentially fatal illness. During an outbreak in 1991, symptoms included vomiting (77%), coma (25%), and seizures (24%) (1). In 29 African children, all of whom died in 2–48 hours, there were intense thirst (38%) and hypotonia (97%) and the effects were more severe in those with malnutrition.

Chronic poisoning can cause cholestatic jaundice (2).

- A 27-year-old Jamaican man developed jaundice, pruritus, intermittent diarrhea, and right upper quadrant abdominal pain after chronic ingestion of akee fruit. Liver biopsy showed centrilobular zonal necrosis and cholestasis. There was no evidence of another cause.

An anaphylactic reaction to akee has been described (3).

Paullinia cupana

Paullinia cupana (*guaran*á) contains the xanthines theobromine and theophylline as well as flavone glycosides. The crushed seeds of *P. cupana* are made into a dried paste called guaraná, from which a beverage is prepared. All of its effects are directly related to its high content of caffeine-like substances.

Adverse effects

US authors have reported two cases of ventricular extra beats associated with the intake of supplements containing guaraná (4). In both cases the supplements also contained multiple other ingredients, and causality was therefore not certain.

Table 54 The genera of Sapindaceae

Alectryon (alectryon)
Allophylus (allophylus)
Blighia (akee)
Cardiospermum (balloon vine)
Cupania (cupania)
Cupaniopsis (carrot wood)
Dimocarpus (dimocarpus)
Dodonaea (dodonaea)
Exothea (exothea)
Harpullia (harpullia)
Hypelate (hypelate)
Koelreuteria (koelreuteria)
Litchi (lychee)
Matayba (matayba)
Melicoccus (melicoccus)
Nephelium (nephelium)
Paullinia (bread and cheese)
Sapindus (soapberry)
Schleichera (schleichera)
Serjania (serjania)
Thouinia (thouinia)
Ungnadia (ungnadia)
Urvillea (urvillea)
Xanthoceras (xanthoceras)

References

1. Meda HA, Diallo B, Buchet JP, Lison D, Barennes H, Ouangre A, Sanou M, Cousens S, Tall F, Van de Perre P. Epidemic of fatal encephalopathy in preschool children in Burkina Faso and consumption of unripe ackee (Blighia sapida) fruit. Lancet 1999;353(9152):536–40.
2. Larson J, Vender R, Camuto P. Cholestatic jaundice due to ackee fruit poisoning. Am J Gastroenterol 1994;89(9): 1577–8.
3. Lebo DB, Ditto AM, Boxer MB, Grammer LC, Bonagura VR, Roberts M. Anaphylaxis to ackee fruit. J Allergy Clin Immunol 1996;98(5 Pt 1):997–8.
4. Baghkhani L, Jafari M. Cardiovascular adverse reactions associated with guaraná: is there a causal effect? J Herb Pharmacother 2002;2(1):57–61.

Selaginellaceae

General Information

The family of Selaginellaceae contains the single genus *Selaginella*.

Selaginella doederleinii

Selaginella doederleinii (spike moss) has been used as an alternative anticancer treatment.

Adverse effects

- A 52-year-old woman with cholangiocarcinoma developed severe bone marrow suppression after taking *S. doederleinii* daily for 2 weeks (1). She developed severe pancytopenia with skin ecchymoses and gum bleeding. A bone marrow smear and biopsy showed severe hypocellularity without malignant cell infiltration. One week after withdrawal her blood count became normal.

Reference

1. Pan KY, Lin JL, Chen JS. Severe reversible bone marrow suppression induced by Selaginella doederleinii. J Toxicol Clin Toxicol 2001;39(6):637–9.

Solanaceae

General Information

The genera in the family of Solanaceae (Table 55) include deadly nightshade, henbane, jimson weed, peppers, petunia, and tobacco.

Anisodus tanguticus

Anisodus tanguticus (Zangqie), a traditional Chinese herbal medicine, contains hyoscyamine and related toxic tropane alkaloids (1). Its active ingredient anisodamine has been used to treat snakebite (2).

Brugmansia species

Brugmansia candida (*Datura candida*, angel's trumpet) and *Brugmansia suaveolens* (*Datura suaveolens*, angel's tears) are ornamental flowers that have been used for hallucinogenic effects. Both contain tropane alkaloids,

Table 55 The genera of Solanaceae

Acnistus (acnistus)
Atropa (belladonna)
Bouchetia (bouchetia)
Browallia (browallia)
Brugmansia (brugmansia)
Brunfelsia (rain tree)
Calibrachoa (calibrachoa)
Capsicum (pepper)
Cestrum (jessamine)
Chamaesaracha (five eyes)
Datura (jimson weed)
Goetzea (goetzea)
Hunzikeria (hunzikeria)
Hyoscyamus (henbane)
Jaborosa (jaborosa)
Jaltomata (false holly)
Leucophysalis (leucophysalis)
Lycianthes (lycianthes)
Lycium (desert thorn)
Mandragora (mandrake)
Margaranthus (margaranthus)
Nectouxia (nectouxia)
Nicandra (nicandra)
Nicotiana (tobacco)
Nierembergia (cupflower)
Nothocestrum (aiea)
Oryctes (oryctes)
Petunia (petunia)
Physalis (ground cherry)
Quincula (quincula)
Salpichroa (salpichroa)
Salpiglossis (salpiglossis)
Schizanthus (schizanthus)
Scopolia
Solanum (nightshade)
Solandra (solandra)
Streptosolen (streptosolen)

such as hyoscine, hyoscyamine, meteloidine, and norhyoscine, which have anticholinergic properties.

Adverse effects
Anticholinergic toxicity can result from the alkaloids that *Brugmansia* species contain (3).

- A 53-year-old woman was admitted to hospital with vertigo, blurred vision, palpitation, mydriasis of the right eye, and tachycardia (120/minute) (4). She reported that she had been cutting leaves from an angel's trumpet when a drop of sap had entered her right eye.
- A 76-year-old man drank 3 teaspoons (15 ml) of a homemade wine made from *B. suaveolens* over 1 hour and about 1.5 hours later developed respiratory distress and weakness (5).

There have been reports of young men who had confusion and other signs of central nervous system involvement after taking angel's trumpet (6).

After ocular exposure to sap from *B. suaveolens*, seven patients developed unilateral mydriasis, at least three also had ipsilateral cycloplegia, and one developed transient tachycardia (7).

Capsicum annuum

Capsicum annuum (chili pepper) contains a variety of carotenoids, including capsanthin, capsorubin, beta-carotene, cryptoxanthin, lutein, phytofluene, and xanthophyll, and steroids, including capsicoside. One of the main constituents is capsaicin, which produces an intense burning sensation when it comes into contact with the skin, eyes, or mucous membranes and which gives peppers their burning taste. A hot shower or bath before topical application to the skin intensifies the burning sensation. Capsaicin is used internally for various conditions, including colic and for improving peripheral circulation, and externally for unbroken chilblains. A cream for topical application has been used to relieve the pain of postherpetic neuralgia and other pain syndromes.

Adverse effects
Sensitization to *C. annuum* has been reported in a patient who was allergic to latex (8).

Taken orally in regular high doses capsaicin can act as a carcinogen and could promote gastric cancer, but in low doses it seems to have anticarcinogenic activity (9).

Datura stramonium

Datura stramonium (Jimson weed) is a naturally occurring plant that is ingested to induce hallucinogenic effects. Toxicity after ingestion is due to an atropine-containing alkaloid that is present in all parts of the plant but is particularly concentrated in the seeds.

Adverse effects
Intoxication with *Datura* species is often reported (10–12). Eleven patients aged 13–21 years ingested large quantities of Jimson weed pods and seeds (13). The signs and

symptoms were classical of atropine poisoning. In milder cases there was asymptomatic mydriasis and tachycardia and in the more severely affected agitation, disorientation, and hallucinations. Nine of the 11 were admitted for observation. None died and none required pharmacological intervention with physostigmine to reverse the anticholinergic symptoms.

Lycium barbarum

Lycium barbarum (Chinese wolfberry) contains a variety of steroids. A tea prepared from it is used in China as a general tonic.

Adverse effects
An interaction of *L. barbarum* with warfarin has been described.

- A 61-year-old woman, who had taken warfarin for atrial fibrillation in weekly doses of 18–19 mg for years and had been completely stable, developed a raised INR after she consumed a tea made from Chinese wolfberry (14). Four days after drinking the tea (180 ml/day), she had an INR of 4.1. After withdrawal of the herbal tea her INR returned to within the target range and remained stable.

Although an extract of *L. barbarum* inhibited CYP2C9 in vitro, the effect was weak and the authors thought that another mechanism must have been involved.

Mandragora species

Mandragora (mandrake) species contain toxic tropane alkaloids, such as hyoscyamine and/or scopolamine.

Adverse effects
The tropane alkaloids in *Mandragora* species are powerful anticholinergic agents and can cause peripheral symptoms (for example blurred vision, dry mouth) and central effects (for example drowsiness, delirium) (15). They can potentiate the effects of synthetic drugs with similar pharmacological activity.

In 15 patients anticholinergic effects were due to poisoning by *Mandragora autumnalis* intermingled with leaves of chard (*Beta vulgaris*) and spinach (*Spinacia oleracea*) (16). The latency from the time of ingestion was 1–4 hours. There was no correlation between latency and severity. All had blurred vision and dryness of mouth, nine had difficult in micturition, nine dizziness, nine headache, eight vomiting, two difficulty in swallowing, and two abdominal pain. All had blushing, a reactive mydriasis, and tachycardia. The skin and mucosae were dry in 14. There was hyperactivity/hallucination in 14 and agitation/delirium in nine. One patient developed a florid psychotic episode. Prostigmine (2–6 mg) was given to 11 patients and physostigmine (0.5–2 mg) to six. The time to a definite response was variable (3–36 hours). The patients treated with physostigmine had better reversal of the psychoneurological symptoms.

In some cases of poisoning, formulations of ginseng may have been adulterated with *Mandragora officinarum* (17).

Anaphylaxis has been reported after the administration of a homeopathic preparation of mandragora (18).

Nicotiana tabacum

The leaves of *Nicotiana tabacum* (tobacco) contain the toxic alkaloid, nicotine as a major constituent and several other pyridine alkaloids as minor constituents. Transdermal nicotine is now widely used as an aid to smoking cessation. Nicotine is covered in a separate monograph.

Although tobacco enemas have been abandoned in official medicine because of their life-threatening toxicity, unorthodox self-medication with this agent has not completely died out. A case report in the 1970s described nausea and confusion, followed by hypotension and bradycardia, due to an enema apparently prepared from 5 to 10 cigarettes.

Green tobacco sickness is an occupational illness reported by tobacco workers worldwide (19). Among farm workers in shade tobacco fields in Connecticut 15% had diagnoses that could be attributed to possible green tobacco sickness (ICD-9) (20). Using a stricter case definition, the frequency fell to 4%. Non-smokers were significantly more likely than smokers to report symptoms, particularly isolated symptoms of headache and dizziness.

Diagnostic criteria for green tobacco sickness have not been established. The symptoms include dizziness or headache and nausea or vomiting, but can also include abdominal cramps, headache, prostration, difficulty in breathing, abdominal pain, and occasionally fluctuations in blood pressure or heart rate. Green tobacco sickness is normally a self-limiting condition, from which workers recover in 2 or 3 days. However, symptoms are sometimes severe enough to result in dehydration and a need for emergency medical care. Although the sickness is widely known among tobacco workers and has been reported in the major tobacco-growing states of the USA, the research literature is sparse.

In a prospective longitudinal study susceptibility factors for green tobacco sickness were picking tobacco, working in wet conditions, working later in the season, and not smoking or chewing tobacco (21).

Although early researchers argued that green tobacco sickness was caused by factors other than nicotine exposure (22), it has since been thought that dermal absorption of nicotine from plant surfaces results in the characteristic symptoms (23). Of 182 workers 25 had 31 episodes of green tobacco sickness. Among non-smokers, each incremental increase in the natural log of cotinine increased the odds of green tobacco sickness 2.11 times, adjusting for task and wet conditions (Arcury).

Urinary cotinine concentrations in 80 male tobacco-growing farmers were significantly higher than in 40 healthy men who did not handle wet tobacco leaves in Kelantan, Malaysia (24). Farmers with urinary cotinine concentrations of 50 ng/ml/m2 or above had eye symptoms more often than those with lower concentrations. Farmers who did not wear protective equipment had subjective symptoms more often than those who did.

Some of the symptoms were more common in organophosphate users than in non-users.

Scopolia species

Scopolia species (scopola) contain tropane alkaloids, such as atropine, hyoscine (scopolamine), hyoscyamine, and tropane. They can cause anticholinergic symptoms (25).

References

1. De Smet PA. Health risks of herbal remedies. Drug Saf 1995;13(2):81–93.
2. Li QB, Pan R, Wang GF, Tang SX. Anisodamine as an effective drug to treat snakebites. J Nat Toxins 1999;8(3):327–30.
3. Finlay P. Anticholinergic poisoning due to *Datura candida*. Trop Doct 1998;28(3):183–4.
4. Roemer HC, von Both H, Foellmann W, Golka K. Angel's trumpet and the eye. J R Soc Med 2000;93(6):319.
5. Smith EA, Meloan CE, Pickell JA, Oehme FW. Scopolamine poisoning from homemade "moon flower" wine. J Anal Toxicol 1991;15(4):216–9.
6. Greene GS, Patterson SG, Warner E. Ingestion of angel's trumpet: an increasingly common source of toxicity. South Med J 1996;89(4):365–9.
7. Havelius U, Asman P. Accidental mydriasis from exposure to angel's trumpet (*Datura suaveolens*). Acta Ophthalmol Scand 2002;80(3):332–5.
8. Gallo R, Cozzani E, Guarrera M. Sensitization to pepper (*Capsicum annuum*) in a latex-allergic patient. Contact Dermatitis 1997;37(1):36–7.
9. Surh YJ, Lee SS. Capsaicin in hot chili pepper: carcinogen, co-carcinogen or anticarcinogen? Food Chem Toxicol 1996;34(3):313–6.
10. Guharoy SR, Barajas M. Atropine intoxication from the ingestion and smoking of jimson weed (*Datura stramonium*). Vet Hum Toxicol 1991;33(6):588–9.
11. Coremans P, Lambrecht G, Schepens P, Vanwelden J, Verhaegen H. Anticholinergic intoxication with commercially available thorn apple tea. J Toxicol Clin Toxicol 1994;32(5):589–92.
12. Centers for Disease Control and Prevention (CDC). Jimson weed poisoning—Texas, New York, and California, 1994. MMWR Morb Mortal Wkly Rep 1995;44(3):41–4.
13. Tiongson J, Salen P. Mass ingestion of jimson weed by eleven teenagers. Del Med J 1998;70(11):471–6.
14. Lam AY, Elmer GW, Mohutsky MA. Possible interaction between warfarin and *Lycium barbarum* L. Ann Pharmacother 2001;35(10):1199–201.
15. Piccillo GA, Mondati EG, Moro PA. Six clinical cases of *Mandragora autumnalis* poisoning: diagnosis and treatment. Eur J Emerg Med 2002;9(4):342–7.
16. Jimenez-Mejias ME, Montano-Diaz M, Lopez Pardo F, Campos Jimenez E, Martin Cordero MC, Ayuso Gonzalez MJ, Gonzalez de la Puente. Intoxicacion atropinica por *Mandragora autumnalis*. Descripcion de quince casos. [Atropine poisoning by *Mandragora autumnalis*. A report of 15 cases.] Med Clin (Barc) 1990;95(18):689–92.
17. Chan TY. Anticholinergic poisoning due to Chinese herbal medicines. Vet Hum Toxicol 1995;37(2):156–7.
18. Helbling A, Brander KA, Pichler WJ, Muller UB. Anaphylactic shock after subcutaneous injection of mandragora D3, a homeopathic drug. J Allergy Clin Immunol 2000;106(5):989–90.
19. Centers for Disease Control and Prevention. Green tobacco sickness in tobacco harvesters—Kentucky, 1992. MMWR Morb Mortal Wkly Rep 1993; 42: 237–40.
20. Trape-Cardoso M, Bracker A, Grey M, Kaliszewski M, Oncken C, Ohannessian C, Barrera LV, Gould B. Shade tobacco and green tobacco sickness in Connecticut. J Occup Environ Med 2003; 45: 656–61.
21. Arcury TA, Quandt SA, Preisser JP, Bernett JT, Norton D, Wang J. High levels of transdermal nicotine exposure produce green tobacco sickness in Latino farm workers. Nicotine Tob Res 2003; 5: 315–21.
22. Weizenecker R, Deal WB. Tobacco croppers' sickness. J Flor Med Assoc 1970; 57: 13–4.
23. Hipke ME. Green tobacco sickness. South Med J 1993; 86: 989–92.
24. Onuki M, Yokoyama K, Kimura K, Sato H, Nordin RB, Naing L, Morita Y, Sakai T, Kobayashi Y, Araki S. Assessment of urinary cotinine as a marker of nicotine absorption from tobacco leaves: a study on tobacco farmers in Malaysia. J Occup Health 2003; 45: 140–5;erratum 270.
25. Cuculic M, Kalodera Z, Sindik J, Kvasic D, Petricic J. [Familial poisoning with the root of the plant, *Scopolia carniolica* Jacq.] Arh Hig Rada Toksikol 1988;39(3):345–8.

Sterculiaceae

General Information

The genera in the family of Sterculiaceae (Table 56) include cola, theobroma, and sterculia.

Table 56 The genera of Sterculiaceae

Ayenia (ayenia)
Brachychiton (brachychiton)
Chiranthodendron (chiranthodendron)
Cola (cola)
Firmiana (parasol tree)
Fremontodendron (flannel bush)
Guazuma (guazuma)
Helicteres (helicteres)
Heritiera
Hermannia (hermannia)
Kleinhovia (kleinhovia)
Melochia (melochia)
Pentapetes (pentapetes)
Sterculia (sterculia)
Theobroma (theobroma)
Waltheria (waltheria)

Sterculia species

The family of *Sterculia* plants (sterculia) yield a fiber that has bulk laxative effects.

Esophageal obstruction after ingestion of sterculia has been reported (1).

- A 91-year-old man presented with complete esophageal obstruction after taking a tablespoonful of sterculia granules (Normacol) without water. There was no predisposing esophageal disease. The severity of obstruction was such that endoscopic clearance was not possible, and the patient required gastrotomy and manual disimpaction of the lower esophagus.

Reference

1. Brown DC, Doughty JC, George WD. Surgical treatment of oesophageal obstruction after ingestion of a granular laxative. Postgrad Med J 1999;75(880):106.

Taxaceae

General Information

The family of Taxaceae contains two genera:

1. *Taxus* (yew)
2. *Torreya* (torreya).

Taxus species

Taxus (yew) trees contain alkaloids called taxanes that have anticancer activity and are the basis of the semisynthetic anticancer drugs paclitaxel and docetaxel which are based on taxol, from *Taxus brevifolia*.

Taxus baccata has been used as a means of suicide (1–3), successfully, because of cardiac dysrhythmias (4).

Taxoids from *Taxus cuspidata* (the Japanese yew) inhibit P glycoprotein and are candidates for reversing multidrug resistance in cancer cells (5).

Taxus celebica, which contains the flavonoid sciadopitysin, is traditionally used in China as an herbal treatment of diabetes mellitus.

Adverse effects
Hematologic
Thrombocytopenia has been attributed to the Chinese herbal mixture Jui, prepared from *T. cuspidata* (Japanese yew) (6).

- A 51-year-old Japanese woman developed gingival bleeding and petechiae. The only medication she had taken was Jui. She had thrombocytopenia and the causal relation was demonstrated by rechallenge. She recovered on withdrawal of the product.

Immunologic
Of 18 patients with seasonal allergic rhinitis, suffering mainly in April and May, five were sensitized to the pollen of *T. cuspidata* (7).

Drug overdose
In two cases the ingestion of a massive dose of *T. celebica* was followed by acute renal insufficiency; both patients initially presented with gastrointestinal upset and fever (8).

- A 40-year-old patient attempted suicide by drinking an extract made from 120 g of yew needles (*T. baccata*); recurrent episodes of ventricular tachycardia were successfully treated with high doses of lidocaine (9).
- Lethal intoxication with yew leaves (*T. baccata*) presented with dizziness 1 hour after ingestion, nausea, diffuse abdominal pain, unconsciousness, weak breathing, tachycardia, brief ventricular flutter followed by bradycardia, and finally death by respiratory arrest and cardiac standstill; the electrocardiogram showed atypical bundle branch block and absent P waves (10).
- A 14-year-old boy cut leaves from a yew tree (*T. baccata*), crushed them, and ingested them; he died soon afterwards (11). At autopsy pieces of partially crushed, partially preserved yew leaves were found in the stomach. The histological findings were non-specific.

Four prisoners drank a decoction of yew needles (*T. baccata*) (12). Two died in prison with cardiac arrest. One went into deep coma, and had several episodes of ventricular fibrillation, controlled by defibrillation; after return of consciousness his general condition deteriorated suddenly, he lost consciousness again, his circulation stopped, and he died on the fourth day. The other patient drank a much smaller amount of the decoction; he was conscious, had bradycardia requiring transient pacemaking, and had a mild ventricular dysrhythmia; he recovered after 10 days. In both cases there was excessive diuresis and severe hypokalemia and atropine was effective for a short time in the control of bradycardia.

Of 11 197 exposures to the berries of *Taxus* species children under 12 years of age were involved in 96% (under 6 years 93%; 6–12 years 3.7%) (13). When the final outcome of the exposure was documented ($n = 7269$), there were no adverse effects in 93% and minor effects in 7.0%. There were moderate (more pronounced, but not life-threatening) effects in 30 individuals and major (life-threatening) effects in four. There were no deaths. Decontamination therapy had no impact on outcome compared with no therapy. When symptoms occurred after exposure to *Taxus*, the most frequent were gastrointestinal (66%), followed by skin reactions (8.3%), nervous system effects (6.0%), and cardiovascular effects (6.0%).

References

1. Stebbing J, Simmons HL, Hepple J. Deliberate self-harm using yew leaves (*Taxus baccata*). Br J Clin Pract 1995;49(2):101.
2. Janssen J, Peltenburg H. Een klassieke wijze van zelfdoding: met *Taxus baccata*. [A classical method of committing suicide: with *Taxus baccata*.] Ned Tijdschr Geneeskd 1985;129(13):603–5.
3. Frohne D, Pribilla O. Todliche Vergiftung mit *Taxus baccata*. [Fatal poisoning with *Taxus baccata*.] Arch Toxikol 1965;21(3):150–62.
4. Willaert W, Claessens P, Vankelecom B, Vanderheyden M. Intoxication with *Taxus baccata*: cardiac arrhythmias following yew leaves ingestion. Pacing Clin Electrophysiol 2002;25(4 Pt 1):511–2.
5. Kobayashi J, Shigemori H, Hosoyama H, Chen Z, Akiyama S, Naito M, Tsuruo T. Multidrug resistance reversal activity of taxoids from *Taxus cuspidata* in KB-C2 and 2780AD cells. Jpn J Cancer Res 2000;91(6):638–42.
6. Azuno Y, Yaga K, Sasayama T, Kimoto K. Thrombocytopenia induced by Jui, a traditional Chinese herbal medicine. Lancet 1999;354(9175):304–5.
7. Maguchi S, Fukuda S. *Taxus cuspidata* (Japanese yew) pollen nasal allergy. Auris Nasus Larynx 2001;28(Suppl):S43–7.
8. Lin JL, Ho YS. Flavonoid-induced acute nephropathy. Am J Kidney Dis 1994;23(3):433–40.
9. von Dach B, Streuli RA. Lidocainbehandlung einer Vergiftung mit Eibennadeln (*Taxus baccata* L.). [Lidocaine treatment of poisoning with yew needles (*Taxus baccata* L.).] Schweiz Med Wochenschr 1988;118(30):1113–6.
10. Schulte T. Todliche Vergiftung mit Eibennadeln (*Taxus baccata*). [Lethal intoxication with leaves of the yew tree (*Taxus baccata*).] Arch Toxikol 1975;34(2):153–8.

11. Wehner F, Gawatz O. Suizidale Eibenintoxikationen—von Casar bis heute—oder: Suizidanleitung im Internet. [Suicidal yew poisoning—from Caesar to today—or suicide instructions on the internet.] Arch Kriminol 2003;211(1–2):19–26.

12. Feldman R, Chrobak J, Liberek Z, Szajewski J. Cztery przypadki zatrucia wywarem z igiel cisu (*Taxus baccata*).

[4 cases of poisoning with the extract of yew (*Taxus baccata*) needles.] Pol Arch Med Wewn 1988;79(1):26–9.

13. Krenzelok EP, Jacobsen TD, Aronis J. Is the yew really poisonous to you? J Toxicol Clin Toxicol 1998;36(3):219–23.

Theaceae

General Information

The genera in the family of Theaceae (Table 57) include camellia.

Camellia sinensis

Camellia sinensis (green tea) contains caffeine and antioxidant polyphenols. It has been touted as being useful in a wide variety of conditions, including cancer prevention, mostly on relatively slim epidemiological evidence (1), cardiovascular disorders, and AIDS.

Table 57   The genera of Theaceae

Camellia (camellia)
Cleyera (cleyera)
Eurya (eurya)
Franklinia (Franklin tree)
Gordonia (gordonia)
Laplacea (laplacea)
Stewartia (stewartia)
Ternstroemia (ternstroemia)

Adverse effects

The main adverse effects of green tea are those of caffeine, with tremulousness and insomnia and withdrawal symptoms (headache, drowsiness, and fatigue).

Liver

The French and Spanish Advisory Boards have suspended the marketing authorization of Exolise, an ethanolic extract of green tea (*Camelia sinensis*), because of several reports of hepatic disorders (2). There have been 13 reports of hepatic disorders (nine in France and four in Spain). All were women, aged 27-69 years, with a time to onset of 9 days to 5 months. Five of the patients had not taken any other medications. Viral serology was negative in eight cases. Rechallenge was positive in eight cases.

References

1. Bushman JL. Green tea and cancer in humans: a review of the literature. Nutr Cancer 1998;31(3):151–9.
2. Anonymous. *Camelia sinensis*. Ethanolic extract products withdrawn due to hepatoxicity WHO Pharm Newslett 2003; 3:1.

Urticaceae

General Information

The genera in the family of Urticaceae (Table 58) include various types of nettle.

Urtica dioica

Urtica dioica (stinging nettle) has been used to treat rheumatism and benign prostatic hyperplasia.

Table 58 The genera of Urticaceae

Boehmeria (false nettle)
Cypholophus (lopleaf)
Elatostema
Hesperocnide (stinging nettle)
Laportea (laportea)
Neraudia (ma'oloa)
Parietaria (pellitory)
Pilea (clearweed)
Pipturus (pipturus)
Pouzolzia (pouzolzia)
Rousselia (rousselia)
Soleirolia (soleirolia)
Touchardia (touchardia)
Urera (urera)
Urtica (nettle)

Adverse effects

Gastrointestinal

A secondary source states that tea prepared from stinging nettle leaves can occasionally cause gastric irritation, skin reactions, edema, and oliguria, but primary references are not provided.

Skin

The blister-raising properties of topical stinging nettle extracts are well known; there is no definitive treatment (1).

A positive patch test reaction to *U. dioica* has been obtained in a patient who had developed edematous gingivostomatitis following the regular use of stinging nettle tea as a tonic; the patient also had positive reactions to chamomile (*Anthemis nobilis*) and its allergens (sesquiterpene lactones) (2).

References

1. Anderson BE, Miller CJ, Adams DR. Stinging nettle dermatitis. Am J Contact Dermat 2003;14(1):44–6.
2. Bossuyt L, Dooms-Goossens A. Contact sensitivity to nettles and camomile in "alternative" remedies. Contact Dermatitis 1994;31(2):131–2.

Valerianaceae

General Information

The family of Valerianaceae contains four genera:

1. *Centranthus* (centranthus)
2. *Plectritis* (seablush)
3. *Valeriana* (valerian)
4. *Valerianella* (cornsalad).

Valeriana species

Valeriana (all-heal, amantilla, heliotrope, valerian), 400–900 mg 30–60 minutes before bedtime, is a traditional herbal sleep remedy (1), but although it has subjective effects on sleep objective sleep measures have been inconsistently affected (2). The roots of members of the *Valeriana* (valerian) species contain valepotriates, which have alkylating properties. Valtrate/isovaltrate and dihydrovaltrate are mutagenic in bacterial test systems in the presence of a metabolic activator, and their degradation products baldrinal (from valtrate) and homobaldrinal (from isovaltrate) are mutagenic even without metabolic activation. These latter compounds also have direct genotoxic effects. As far as is known, decomposition of dihydrovaltrate does not yield baldrinals.

The amounts of valepotriates and baldrinals in valerian extracts depend on the botanical species: root extracts of *Valeriana officinalis* contain up to 0.9% of valepotriates, compared with 2–4% and 5–7% of valepotriates in root extracts of *Valeriana wallichii* and *Valeriana mexicana* respectively.

Another relevant variable is the dosage form: when a herbal tea is prepared by hot extraction from valerian root, up to 60% of the valepotriates remain in the root material and only 0.1% can be recovered from the tea. A freshly prepared tincture contains 11% of the valepotriates originally found in the root material. Storage at room temperature rapidly reduces this to 3.7% after 1 week and 0% after 3 weeks. In view of this rapid degradation, it is not surprising that commercially available tincture samples yield baldrinals.

Valerian-containing tablets and capsules provide up to 1 mg of baldrinals per piece.

Valepotriates are poorly absorbed from the gastrointestinal tract, but 2% is degraded in vivo to baldrinals after oral administration of valtrate/isovaltrate to mice. In other words, a tablet with 50 mg of valepotriates may add 1 mg of baldrinals to the amount of baldrinals, which are already present before ingestion. In contrast to the valepotriates, the degradation product homobaldrinal is absorbed fairly well after oral administration to mice. As much as 71% of the administered dose can be recovered from the urine in the form of baldrinal glucuronide. Since no unchanged homobaldrinal can be demonstrated in body fluids or liver samples after oral administration, the compound appears to undergo substantial first-pass metabolism. As this glucuronidation leads to loss of its mutagenic properties, the primary target organs that are at risk from valepotriates and baldrinals are the gastrointestinal tract and the liver (3). However, the toxicological significance of all these data is still unclear, since the carcinogenic potential of valerian formulations has not been evaluated.

Adverse effects
Liver

- A 13-year-old child with fulminant seronegative liver failure required liver transplantation (4). A liver biopsy showed non-specific necrosis of more than 90% of hepatocytes. Possible exogenous toxic factors were ruled out and it was thought that the most likely cause of the liver damage was self-medication with Euphytose, a herbal mixture of *V. officinalis*, *Ballota nigra*, *Crataegus oxyacantha*, *Passiflora incarnata*, and *Cola nitida*.

The authors suggested that of these medicinal plants valerian was the most likely to have caused the liver damage. However, it could just as easily have been due to one of the other constituents or some other cause entirely.

Drug withdrawal

Delirium has been attributed to valerian withdrawal (5).

- A 58-year-old man who had regularly taken excessive doses of valerian root extract for many years was given naloxone postoperatively and developed extreme tremulousness and worsening ventilation. His condition deteriorated and he became delirious. He was treated with midazolam, lorazepam, and finally reducing doses of clonazepam and made an uneventful recovery.

The authors suggested that the anxiolytic action of valerian is similar to that of benzodiazepines and that the patient had been suffering from a valerian withdrawal syndrome.

Drug overdose

In one case of overdose with about 20 times the recommended therapeutic dose, there were only mild symptoms, all of which resolved within 24 hours (6).

In 23 patients who took an overdose of "Sleep-Qik" (valerian dry extract 75 mg, mean dose 2.5 g, hyoscine hydrobromide 0.25 mg, and cyproheptadine hydrochloride 2 mg) the main problems were central nervous system depression and anticholinergic poisoning (7). There was no evidence of liver damage, subclinical, acute, or delayed. It is unlikely that valerian contributed to the adverse effects in these cases.

References

1. Hadley S, Petry JJ. Valerian. Am Fam Physician 2003;67(8):1755–8.
2. Pallesen S, Bjorvatn B, Nordhus IH, Skjerve A. Valeriana som sovemiddel. [Valerian as a sleeping aid?.] Tidsskr Nor Laegeforen 2002;122(30):2857–9.

3. Dieckmann H. Untersuchungen zur Pharmakokinetik, Metabolismus und Toxikologie von Baldrinalen. Inaugural Dissertation, Free University, Berlin, 1988.

4. Bagheri H, Broue P, Lacroix I, Larrey D, Olives JP, Vaysse P, Ghisolfi J, Montastruc JL. Fulminant hepatic failure after herbal medicine ingestion in children. Therapie 1998;53(1):82–3.

5. Garges HP, Varia I, Doraiswamy PM. Cardiac complications and delirium associated with valerian root withdrawal. JAMA 1998;280(18):1566–7.

6. Willey LB, Mady SP, Cobaugh DJ, Wax PM. Valerian overdose: a case report. Vet Hum. Toxicol 1995;37(4):364–5.

7. Chan TY, Tang CH, Critchley JA. Poisoning due to an over-the-counter hypnotic, Sleep-Qik (hyoscine, cyproheptadine, valerian). Postgrad Med J 1995;71(834):227–8.

Verbenaceae

General Information

The genera in the family of Verbenaceae (Table 59) include mangrove and vervain.

Table 59 The genera of Verbenaceae

Aegiphila (spiritweed)
Aloysia (bee brush)
Avicennia (mangrove)
Bouchea (bouchea)
Callicarpa (beauty berry)
Caryopteris (caryopteris)
Citharexylum (fiddlewood)
Clerodendrum (glory bower)
Congea (congea)
Cornutia (cornutia)
Duranta (duranta)
Faradaya
Glandularia (mock vervain)
Gmelina (gmelina)
Holmskioldia (holmskioldia)
Lantana (lantana)
Lippia (lippia)
Nashia (nashia)
Petitia (petitia)
Petrea (petrea)
Phryma (phryma)
Phyla (fog fruit)
Premna (premna)
Priva (priva)
Stachytarpheta (porterweed)
Stylodon (stylodon)
Tamonea (tamonea)
Tectona (tectona)
Tetraclea (tetraclea)
Verbena (vervain)
Vitex (chaste tree)

Vitex agnus-castus

Vitex agnus-castus (chaste tree, hemp tree, monk's pepper) is a plant with estrogen-like activities, used for a variety of gynecological problems, particularly on the European continent, particularly premenstrual symptoms (1).

Adverse effects

Vitex agnus-castus can cause ovarian hyperstimulation and increase the risk of miscarriage.

A systematic review of all clinical reports of adverse events in clinical trials, post-marketing surveillance studies, surveys, spontaneous reporting schemes, and to manufacturers and herbalist organizations has suggested that the adverse events are mild and reversible (2). The most frequent adverse events are nausea, headache, gastrointestinal disturbances, menstrual disorders, acne, pruritus, and erythematous rashes. No drug-drug interactions were reported.

Endocrine

- A woman took *V. agnus-castus* during an unstimulated cycle while undergoing in vitro fertilization (3). She had considerable derangements of gonadotrophin and ovarian hormone concentrations.

References

1. Wuttke W, Jarry H, Christoffel V, Spengler B, Seidlova-Wuttke D. Chaste tree (*Vitex agnus-castus*)—pharmacology and clinical indications. Phytomedicine 2003;10(4):348–57.
2. Daniele C, Thompson Coon J, Pittler MH, Ernst E. *Vitex agnus castus*: a systematic review of adverse events. Drug Saf 2005; 28(4): 319–32.
3. Cahill DJ, Fox R, Wardle PG, Harlow CR. Multiple follicular development associated with herbal medicine. Hum Reprod 1994;9(8):1469–70.

Viscaceae

General Information

The family of Viscaceae contains four genera of mistletoe:

1. *Arceuthobium* (dwarf mistletoe)
2. *Korthalsella* (korthal mistletoe)
3. *Phoradendron* (American mistletoe)
4. *Viscum* (mistletoe).

Phoradendron flavescens

In 14 cases of accidental exposure to *Phoradendron fla-vescens* (*Phoradendron serotinum*, American mistletoe) there were no symptoms, but there was one death from intentional ingestion of an unknown amount of an elixir brewed from the berries (1). In 92 patients aged 4 months to 42 years (median 2 years), 14 were symptomatic, 11 related to mistletoe exposure. All the symptomatic cases had onset of symptoms within 6 hours. The symptoms included gastrointestinal upsets ($n = 6$), mild drowsiness ($n = 2$), eye irritation ($n = 1$), ataxia ($n = 1$), and seizures ($n = 1$). Treatment included gastrointestinal decontamination in 54 patients, ocular irrigation in one, and an intravenous benzodiazepine in one; decontamination did not affect the outcome. The amount ingested ranged from one berry or leaf to more than 20 berries or five leaves. Eight of ten patients who had taken at least five berries were symptom-free. Of 11 patients who took only leaves (range 1–5 leaves), three had gastrointestinal upsets. There were no cardiovascular effects in any case.

Viscum album

The stems and leaves of *Viscum album* (all heal, bird lime, devil's fuge, golden bough, mistletoe) contain alkaloids, viscotoxins, and lectins. While the toxicity of the alkaloids remains to be assessed, the viscotoxins and lectins have been found to be very poisonous in animals when given parenterally. Mistletoe is a widely used alternative treatment for cancer, without convincing evidence of efficacy (2).

Adverse effects
Liver
Mistletoe is sometimes assumed to have hepatotoxic potential, based on a case report of hepatitis due to an herbal combination product claimed to have had mistletoe as one of its ingredients (3). However, the allegation that mistletoe was the probable cause of the illness has been rightly criticized, inter alia because the botanical material was not authenticated. The incriminated product also contained skullcap, which is hepatotoxic.

Immunologic
Parenteral administration of *V. album* can cause serious allergic reactions (4), and anaphylactic reactions have been described (5).

Tumorigenicity

- A 73-year-old man with a 5-year history of centrocytic non-Hodgkin's lymphoma presented with subcutaneous nodules in the abdominal wall at the sites where he had previously received subcutaneous injections of mistletoe (6). The nodules turned out to be infiltrations by the centrocytic lymphoma. The patient died 6 weeks later of bilateral pneumonia. The authors hypothesized that mistletoe has a growth-promoting action on lymphoma cells, mediated by high local concentrations of interleukin-6 liberated from the skin by mistletoe lectins.

References

1. Spiller HA, Willias DB, Gorman SE, Sanftleban J. Retrospective study of mistletoe ingestion. J Toxicol Clin Toxicol 1996;34(4):405–8.
2. Kleijnen J, Knipschild P. Mistletoe treatment for cancer. Review of controlled trials in humans. Phytomedicine 1994;1:255–60.
3. Harvey J, Colin-Jones DG. Mistletoe hepatitis. BMJ (Clin Res Ed) 1981;282(6259):186–7.
4. Pichler WJ, Angeli R. Allergie auf Mistelextrakt. [An allergy to mistletoe extract.] Dtsch Med Wochenschr 1991;116(35):1333–4.
5. Hutt N, Kopferschmitt-Kubler M, Cabalion J, Purohit A, Alt M, Pauli G. Anaphylactic reactions after therapeutic injection of mistletoe (*Viscum album* L.) Allergol Immunopathol (Madr) 2001;29(5):201–3.
6. Hagenah W, Dorges I, Gafumbegete E, Wagner T. Subkutane Manifestationen eines zentrozytischen Non-Hodgkin-Lymphoms an Injektionsstellen eines Mistelpräparats. [Subcutaneous manifestations of a centrocytic non-Hodgkin lymphoma at the injection site of a mistletoe preparation.] Dtsch Med Wochenschr 1998;123(34–35):1001–4.

Zingiberaceae

General Information

The genera in the family of Zingiberaceae (Table 60) include cardamom and ginger.

Curcuma longa

Because of its bright yellow color, this plant (turmeric) is sometimes used as a food coloring agent; in Ayurvedic medicine turmeric has a long history as a medicine for a wide range of conditions for both topical and internal use. A systematic review of all human trials found only six studies (1). The dosage of curcumin was 1125-2500 mg/day. Serious adverse events were not reported, and the authors conclude that curcumin "has been demonstrated to be safe".

Zingiber officinale

Zingiber officinale (ginger) contains a variety of compounds, including diarylheptanoids and the phenol gingerol. It has been used to treat motion sickness and other forms of nausea and vomiting, and may have some efficacy (2). However, definitive clinical evidence is only available for pregnancy-related nausea and vomiting (3, 4). Its anti-inflammatory properties (5) have not been well studied clinically. Pharmacokinetic data are only available for 6-gingerol and zingiberene. Preclinical safety data have not ruled out potential adverse effects, and therapy should be carefully monitored, especially during consumption of ginger over longer periods.

Adverse effects

In a systematic review of 24 randomized controlled trials in 1073 patients ginger had no beneficial effects in postoperative nausea and vomiting (6). Of 777 patients in 15 studies, 3.3% had mild adverse effects, mainly mild *gastrointestinal symptoms* and *sleepiness*, not requiring specific treatment. There was one serious adverse event, *abortion* in the 12th week of gestation, but that could not clearly be attributed to the ginger.

Table 60 The genera of Zingiberaceae

Aframomum (aframomum)
Alpinia (alpinia)
Amomum (cardamom)
Boesenbergia (boesenbergia)
Curcuma (curcuma)
Elettaria (elettaria)
Etlingera (waxflower)
Hedychium (garland-lily)
Hitchenia (hitchenia)
Kaempferia (kaempferia)
Renealmia (renealmia)
Zingiber (ginger)

Skin

Of about 1000 patients with occupational skin diseases, five had occupational allergic contact dermatitis from spices (7). They were chefs, or workers in kitchens, coffee rooms, and restaurants. In all cases the dermatitis affected the hands. The causative spices were garlic, cinnamon, ginger, allspice, and clove. The same patients had positive patch test reactions to carrot, lettuce, and tomato.

Among 55 patients with suspected contact dermatitis, skin patch tests that were positive at concentrations of both 10% and 25% were most common with ginger ($n = 7$), nutmeg ($n = 5$), and oregano ($n = 4$); other spices produced no responses or one positive response (8). Positive reactions at only one concentration were more likely at 25%: nutmeg ($n = 5$), ginger and cayenne ($n = 4$), curry, cumin, and cinnamon ($n = 3$), turmeric, coriander, and sage ($n = 2$), oregano ($n = 1$), and basil and clove ($n = 0$).

Pregnancy

Because *Zingiber officinale* (ginger) is often used for morning sickness, it is important to define its safety in pregnancy. The Canadian Motherisk Program monitored 187 pregnancies of women taking ginger during early pregnancy (9). There were 181 live births, two stillbirths, three spontaneous abortions, one elective termination, and three malformations. Compared with non-ginger taking mothers, these data were not significantly different. The authors therefore argued that ginger does not appear to increase the rates of major malformations. However, the sample size may well have been too small to draw such a conclusion with certainty.

Ginger and vitamin B6 have been compared in the treatment of nausea and vomiting in pregnancy (10). They were equally well tolerated but belching was more common in those who took ginger. The live birth rate was significantly lower in those who took vitamin B6. The safety of ginger in pregnancy has been confirmed in a meta-analysis of six trials (11).

Drug interactions

Ginger may potentiate the effects of oral anticoagulants (12).

- A 76-year-old white woman taking long-term phenprocoumon, with an INR in the target range, began using ginger products (13). Several weeks later, her INR rose to 10 and she had epistaxis. The INR returned to the target range after ginger was stopped and vitamin K1 was given.

References

1. Chainani-Wu N. Safety and anti-inflammatory activity of curcumin: a component of tumeric (*Curcuma longa*). J Altern Complement Med 2003; 9: 161–8.
2. Ernst E, Pittler MH. Efficacy of ginger for nausea and vomiting: a systematic review of randomized clinical trials. Br J Anaesth 2000;84(3):367–71.

3. Chrubasik S, Pittler MH, Roufogalis BD. *Zingiberis rhizoma*: a comprehensive review on the ginger effect and efficacy profiles. Phytomedicine 2005;12(9):684–701.

4. Boone SA, Shields KM. Treating pregnancy-related nausea and vomiting with ginger. Ann Pharmacother 2005; 39(10): 1710–3.

5. Grzanna R, Lindmark L, Frondoza CG. Ginger—an herbal medicinal product with broad anti-inflammatory actions. J Med Food 2005;8(2):125–32.

6. Betz O, Kranke P, Geldner G, Wulf H, Eberhart LH. Ist Ingwer ein klinisch relevantes Antiemetikum? Eine systematische ubersicht randomisierter kontrollierter Studien. Forsch Komplementarmed Klass Naturheilkd 2005; 12(1): 14–23.

7. Kanerva L, Estlander T, Jolanki R. Occupational allergic contact dermatitis from spices. Contact Dermatitis 1996; 35(3):157–62.

8. Futrell JM, Rietschel RL. Spice allergy evaluated by results of patch tests. Cutis 1993;52(5):288–90.

9. Portnoi G, Chng LA, Karimi-Tabesh L, Koren G, Tan MP, Einarson A. Prospective comparative study of the safety and effectiveness of ginger for the treatment of nausea and vomiting in pregnancy. Am J Obstet Gynecol 2003; 189: 1374–7.

10. Smith C, Crowther C, Hotham N, McMillan V. Ginger was as effective as vitamin B6 in improving symptoms of nausea and vomiting in early pregnancy. Obstet Gynecol 2004; 103: 639–45.

11. Borrelli F, Capasso R, Aviello G, Pittler MH, Izzo AA. Effectiveness and safety of ginger in the treatment of pregnancy-induced nausea and vomiting. Obstet Gynecol 2005; 105(4): 849–56.

12. Heck AM, DeWitt BA, Lukes AL. Potential interactions between alternative therapies and warfarin. Am J Health Syst Pharm 2000;57(13):1221–7.

13. Kruth P, Brosi E, Fux R, Morike K, Gleiter CH. Ginger-associated overanticoagulation by phenprocoumon. Ann Pharmacother 2004;38(2):257–60.

Zygophyllaceae

General Information

The family of Zygophyllaceae (Table 61) contains nine genera.

Larrea tridentata

The name chaparral (*Larrea tridentata,* creosote bush) is also sometimes used for other desert plants such as *Larrea mexicana, Larrea glutinosa, Larrea nitida,* and *Larrea caneifolia,* which are all widely used, for instance to cure colds, diarrhea, urinary tract infections, rheumatism, and skin problems, or to reduce body weight.

The major phenolic component of *Larrea tridentata* is a catechol lignan called nordihydroguaiaretic acid. It causes lymphatic and renal lesions when given chronically in high doses to rodents.

Adverse effects
Liver

There have been several reports of hepatotoxicity attributed to herbal medicines containing *L. tridentata* leaves (1,2).

- A 22-year-old, previously healthy Finnish woman developed toxic hepatitis after taking chaparral tablets (3). She recovered after dechallenge but again developed toxic hepatitis after re-starting the herbal remedy against the advice of her doctors. Her liver function normalized 6 months after this re-challenge but she still had hepatic fibrosis.

Of 18 reports of illnesses associated with the ingestion of chaparral, there was evidence of hepatotoxicity in 13 cases (4). The presentation was characterized by jaundice with a marked increase in serum liver enzymes at 3–52

Table 61 The genera of Zygophyllaceae

Balanites (balanites)
Bulnesia (lignum vitae)
Fagonia (fagon bush)
Guajacum (lignum vitae)
Kallstroemia (caltrop)
Larrea (creosote bush)
Peganum (peganum)
Tribulus (punturevine)
Zygophyllum (bean caper)

weeks after ingestion, and it resolved 1–17 weeks after withdrawal. The predominant pattern of liver damage was cholestatic; in four cases there was progression to cirrhosis and in two there was acute fulminant liver failure that required liver transplantation.

Chaparral-induced hepatotoxicity in 16 published cases has been summarized in the light of a further report (5).

- A 27-year-old Hispanic man presented with nausea and vomiting, diarrhea, and upper abdominal pain 12 months after starting to take chaparral capsules. A liver biopsy showed hepatocellular injury with necrosis and periportal inflammation. His liver function stabilized after withdrawal of chaparral.

However, in four patients who were given topical chaparral tincture there was no evidence of liver damage (6).

Skin

Contact dermatitis has been attributed to *L. tridentata* (7).

A cystic renal cell carcinoma and acquired renal cystic disease associated with consumption of chaparral tea has been reported (8).

References

1. Gordon DW, Rosenthal G, Hart J, Sirota R, Baker AL. Chaparral ingestion. The broadening spectrum of liver injury caused by herbal medications. JAMA 1995;273(6):489–90.
2. Batchelor WB, Heathcote J, Wanless IR. Chaparral-induced hepatic injury. Am J Gastroenterol 1995;90(5):831–3.
3. Kauma H, Koskela R, Mäkisalo H, Autio-Harmainen H, Lehtola J, Höckerstedt K. Toxic acute hepatitis and hepatic fibrosis after consumption of chaparral tablets. Scand J Gastroenterol 2004;39:1168–71.
4. Sheikh NM, Philen RM, Love LA. Chaparral-associated hepatotoxicity. Arch Intern Med 1997;157(8):913–9.
5. Grant KL, Boyer LV, Erdman BE. Chaparral-induced hepatotoxicity. Integrative Med 1998;1:83–7.
6. Heron S, Yarnell E. The safety of low-dose Larrea tridentata (DC) Coville (creosote bush or chaparral): a retrospective clinical study. J Altern Complement Med 2001;7(2):175–85.
7. Shasky DR. Contact dermatitis from Larrea tridentata (creosote bush). J Am Acad Dermatol 1986;15(2 Pt 1):302.
8. Smith AY, Feddersen RM, Gardner KD Jr, Davis CJ Jr. Cystic renal cell carcinoma and acquired renal cystic disease associated with consumption of chaparral tea: a case report. J Urol 1994;152(6 Pt 1):2089–91.

Compounds found in Herbal products

This section contains the following compounds that are found in herbal products or are directly derived from herbal sources:

Amygdalin
Anticholinergic alkaloids
Cannabinoids
Cardiac glycosides
Coumarin
Ephedrine and pseudoephedrine

Haematitium
HMG coenzyme-A reductase inhibitors
Ipecacuanha, emetine, and dihydroemetine
Limonene
Nicotine and nicotine replacement therapy
Phytoestrogens
Pyrrolizidine alkaloids
Sparteine
Vitamin E
Yohimbine

Amygdalin

Amygdalin and laetrile (a synthetic form of amygdalin) have been used in the treatment of cancer.

Drug-drug interactions

An interaction of amygdalin with vitamin C has been suggested.

- A 68-year-old woman with cancer became comatose, with a reduced Glasgow Coma Score, seizures, and severe lactic acidosis, requiring intubation and ventilation shortly after taking amygdalin 3 g (1). She also took vitamin C 4800 mg/day. She responded rapidly to hydroxocobalamin. The adverse drug reaction was rated probable on the Naranjo probability scale.

Vitamin C increases the in vitro conversion of amygdalin to cyanide and reduces body stores of cysteine, which detoxifies cyanide; the authors suggested that this was a plausible explanation for this adverse event.

Reference

1. Bromley J, Hughes BG, Leong DC, Buckley NA. Life-threatening interaction between complementary medicines: cyanide toxicity following ingestion of amygdalin and vitamin C. Ann Pharmacother 2005;39(9):1566–9.

Anticholinergic Alkaloids

Tropane alkaloids, such as hyoscyamine and/or scopolamine, occur in the solanaceous plants *Atropa belladonna*, *Datura stramonium*, *Hyoscyamus niger*, and *Mandragora officinarum* (1). These alkaloids are powerful anticholinergic agents and can elicit peripheral symptoms (for example blurred vision, dry mouth) (2–6) as well as central effects (for example drowsiness, delirium). They can potentiate the effects of anticholinergic medicaments.

References

1. Müller JL. Love potions and the ointment of witches: historical aspects of the nightshade alkaloids. J Toxicol Clin Toxicol 1998;36(6):617–27.

2. Ceha LJ, Presperin C, Young E, Allswede M, Erickson T. Anticholinergic toxicity from nightshade berry poisoning responsive to physostigmine. J Emerg Med 1997;15(1):65–9.

3. Hatziisaak T, Weber A, Zeng ZL, Wang Z. [*Scopolica carniolica* Jacq. tea.] Schweiz Rundsch Med Prax 1998;87(49):1705–8.

4. Parisi P, Francia A. A female with central anticholinergic syndrome responsive to neostigmine. Pediatr Neurol 2000;23(2):185–7.

5. Piccillo GA, Mondati EG, Moro PA. Six clinical cases of *Mandragora autumnalis* poisoning: diagnosis and treatment. Eur J Emerg Med 2002;9(4):342–7.

6. Piccillo GA, Miele L, Mondati E, Moro PA, Musco A, Forgione A, Gasbarrini G, Grieco A. A Anticholinergic syndrome due to 'Devil's herb': when risks come from the ancient time. Int J Clin Pract 2006;60(4):492–4.

Cannabinoids

General Information

Cannabis is the abbreviated name for the hemp plant *Cannabis sativa*. The common names for cannabis include marijuana, grass, and weed. Other names for cannabis refer to particular strains; they include bhang and ganja. The most potent forms of cannabis come from the flowering tops of the plants or from the dried resinous exudate of the leaves, and are referred to as hashish or hash. *Cannabis sativa* contains more than 450 substances and only a few of the main active cannabinoids have been evaluated.

Cannabis is the most commonly used illicit drug. In 2001, 83 million Americans and 37% of those aged 12 and older had tried marijuana (1).The long history of marijuana use both as a recreational drug and as a herbal medicine for centuries has been reviewed (2).

In different Western countries the possible therapeutic use of cannabinoids as antiemetics in patients with cancer or in patients with multiple sclerosis has been debated, because of the prohibition of cannabis, and has polarized opinion about the seriousness of its adverse effects (3,4).

Pharmacology

The primary active component of cannabis is Δ9-tetrahydrocannabinol (THC), which is responsible for the greater part of the pharmacological effects of the cannabis complex. Δ8-THC is also active. However, the cannabis plant contains more than 400 chemicals, of which some 60 are chemically related to Δ9-THC, and it is evident that the exact proportions in which these are present can vary considerably, depending on the way in which the material has been harvested and prepared. In man, Δ9-THC is rapidly converted to 11-hydroxy-Δ9-THC (5), a metabolite that is active in the central nervous system. A specific receptor for the cannabinols has been identified; it is a member of the G-protein-linked family of receptors (6). The cannabinoid receptor is linked to the inhibitory G-protein, which is linked to adenyl cyclase in an inhibitory fashion (7). The cannabinoid receptor is found in highest concentrations in the basal ganglia, the hippocampus, and the cerebellum, with lower concentrations in the cerebral cortex.

When cannabis is smoked, usually in a cigarette with tobacco, the euphoric and relaxant effects occur within minutes, reach a maximum in about 30 minutes, and last up to 4 hours. Some of the motor and cognitive effects can persist for 5–12 hours. Cannabis can also be taken orally, in foods such as cakes (for example "space cake") or sweetmeats (for example hashish fudge) (8).

Many variables affect the psychoactive properties of cannabis, including the potency of the cannabis used, the route of administration, the smoking technique, the dose, the setting, the user's past experience, the user's expectations, and the user's biological vulnerability to the effects of the drug.

Animal and in vitro toxicology

Δ9-tetrahydrocannabinol, the active component in herbal cannabis, is very safe. Laboratory animals (rats, mice, dogs, monkeys) can tolerate doses of up to 1000 mg/kg, equivalent to some 5000 times the human intoxicant dose. Despite the widespread illicit use of cannabis, there are very few, if any, instances of deaths from overdose (9).

Long-term toxicology studies with THC were carried out by the National Institute of Mental Health in the late 1960s (10). These included a 90-day study with a 30-day recovery period in both rats and monkeys and involved not only Δ9-THC but also Δ8-THC and a crude extract of marijuana. Doses of cannabis or cannabinoids in the range 50–500 mg/kg caused reduced food intake and lower body weight. All three substances initially depressed behavior, but later the animals became more active and were irritable or aggressive. At the end of the study the weights of the ovaries, uterus, prostate, and spleen were reduced and the weight of the adrenal glands was increased. The behavioral and organ changes were similar in monkeys, but less severe than those seen in rats. Further studies were carried out to assess the damage that might be done to the developing fetus by exposure to cannabis or cannabinoids during pregnancy. Treatment of pregnant rabbits with THC at doses up to 5 mg/kg had no effect on birth weight and did not cause any abnormalities in the offspring (10).

A similarly detailed toxicology study was carried out with THC by the National Institute of Environmental Health Sciences in the USA, in response to a request from the National Cancer Institute (11). Rats and mice were given THC up to 500 mg/kg five times a week for 13 weeks; some were followed for a period of recovery over 9 weeks. By the end of the study more than half of the rats treated with the highest dose (500 mg/kg) had died, but all of the remaining animals appeared to be healthy, although in both species the higher doses caused lethargy and increased aggressiveness. The THC-treated animals ate less food and their body weights were consequently significantly lower than those of untreated controls at the end of the treatment period, but returned to normal during recovery. During this period the animals were sensitive to touch and some had convulsions. There was a trend towards reduced uterine and testicular weights.

In further studies rats were treated with doses of THC up to 50 mg/kg and mice with up to 500 mg/kg 5 times a week for 2 years in a standard carcinogenicity test (11). After 2 years, more treated animals had survived than controls, probably because the treated animals ate less and had lower body weights. The treated animals also had a significantly lower incidence of the various cancers normally seen in aged rodents in testes, pancreas, pituitary gland, mammary glands, liver, and uterus. Although there was an increased incidence of precancerous changes in the thyroid gland in both species and in the mouse ovary after one dose (125 mg/kg), these changes were not dose-related. The conclusion was that there was "no evidence of carcinogenic activity of THC at doses up to 50 mg/kg." This was also supported by the failure to detect any genetic toxicity in other tests designed to identify drugs capable of causing chromosomal damage. For example, THC was negative in the so-called "Ames test," in which bacteria are exposed to very high concentrations of a drug to see whether it causes mutations. In another test, hamster ovary cells were exposed to high concentrations of the drug in tissue culture; there were no effects on cell division that might suggest chromosomal damage.

By any standards, THC must be considered to be very safe, both acutely and during long-term exposure. This probably partly reflects the fact that cannabinoid receptors are virtually absent from those regions at the base of the brain that are responsible for such vital functions as breathing and blood pressure control. The available animal data are more than adequate to justify its approval as a human medicine, and indeed it has been approved by the FDA for certain limited therapeutic indications (generic name = dronabinol) (9).

Respiratory

There have been several attempts to address this question by exposing laboratory animals to cannabis smoke. After such exposure on a daily basis for periods of up to 30 months, extensive damage has been observed in the lungs of rats (12), dogs (13), and monkeys (14), but it is very difficult to extrapolate these findings to man, as it is difficult or impossible to imitate human exposure to cannabis smoke in any animal model.

Nervous system

Animal studies on neurotoxicity have yielded conflicting results. Treatment of rats with high doses of THC given orally for 3 months (15) or subcutaneously for 8 months (16) produced neural damage in the hippocampal CA3 zone, with shrunken neurons, reduced synaptic density, and loss of cells. But in perhaps the most severe test of all, rats and mice treated on 5 days each week for 2 years had no histopathological changes in the brain, even after 50 mg/kg/day (rats) or 250 mg/kg/day (mice) (11). Although claims were made that exposure of a small number of rhesus monkeys to cannabis smoke led to ultrastructural changes in the septum and hippocampus (17,18), subsequent larger-scale studies failed to show any cannabis-induced histopathology in monkey brain (19).

Studies of the effects of cannabinoids on neurons in vitro have also yielded inconsistent results. Exposure of rat cortical neurons to THC shortened their survival: twice as many cells were dead after exposure to THC 5 μmol/l for 2 hours than in control cultures (20). Concentrations of THC as low as 0.1 μmol/l had a significant effect. The effects of THC were accompanied by release of cytochrome c, activation of caspase-3, and DNA fragmentation, suggesting an apoptotic mechanism. All of the effects of THC could be blocked by the antagonist AM-251 or by pertussis toxin, suggesting that they were mediated through CB1 receptors. Toxic effects of THC have also been reported in hippocampal neurons in culture, with 50% cell death after exposure to THC 10 μmol/l for 2 hours or 1 μmol/l for 5 days (21). The antagonist rimonabant blocked these effects, but pertussis toxin did not. The authors proposed a toxic mechanism involving arachidonic acid release and the formation of free radicals. On the other hand, other authors have failed to observe any damage in rat cortical neurons exposed for up to 15 days to THC 1 mmol/l, although they found that this concentration killed rat C6 glioma cells, human astrocytoma U373MG cells, and mouse neuroblastoma N18TG12 cells (22). In a remarkable study, injection of THC into solid tumors of C6 glioma in rodent brain led to increased survival times, and there was complete eradication of the tumors in 20–35% of the treated animals (23). A stable analogue of anandamide also produced a drastic reduction in the tumor volume of a rat thyroid epithelial cell line transformed by K-ras oncogene, implanted in nude mice (24). The antiproliferative effect of cannabinoids has suggested a potential use for such drugs in cancer treatment (25).

Some authors have reported neuroprotective actions of cannabinoids. WIN55,212-2 reduced cerebral damage in rat hippocampus or cerebral cortex after global ischemia or focal ischemia in vivo (26). The endocannabinoid 2AG protected against damage elicited by closed head injury in mouse brain, and the protective effects were blocked by rimonabant (27). THC had a similar effect in vivo in protecting against damage elicited by ouabain (28). Rat hippocampal neurons in tissue culture were protected against glutamate-mediated damage by low concentrations of WIN55,212-2 or CP-55940, and these effects were mediated through CB1 receptors (29). But not all of these effects seem to require mediation by cannabinoid receptors. The protective effects of WIN55,212-2 did not require either CB1 or CB2 cannabinoid receptors in cortical neurons exposed to hypoxia (26), and there were similar findings for the protective actions of anandamide and 2-AG in cortical neuronal cultures (30). Both THC and cannabidiol, which is not active at cannabinoid receptors, protected rat cortical neurons against glutamate toxicity (31) and these effects were also independent of CB1 receptors. The authors suggested that the protective effects of THC might be due to the antioxidant properties of these polyphenolic molecules, which have redox potentials higher than those of known antioxidants (for example ascorbic acid).

Pregnancy

In animals, THC can cause spontaneous abortion, low birth weight, and physical deformities (32). However, these were only seen after treatment with extremely high doses of THC (50–150 times higher than human doses), and only in rodents and not in monkeys.

Tolerance and dependence

Many animal studies have shown that tolerance develops to most of the behavioral and physiological effects of THC (33). Dependence on cannabinoids in animals is clearly observable, because of the availability of CB_1 receptor antagonists, which can be used to precipitate withdrawal. Thus, a behavioral withdrawal syndrome was precipitated by rimonabant in rats treated for only 4 days with THC in doses as low as 0.5–4.0 mg/kg/day (34). The syndrome included scratching, face rubbing, licking, wet dog shakes, arched back, and ptosis, many of the signs that are seen in rats undergoing opiate withdrawal. Similar withdrawal signs occurred when rats treated chronically with the synthetic cannabinoid CP-55940 were given rimonabant (35). Rimonabant-induced withdrawal after 2 weeks of treatment of rats with the cannabinoid HU-120 was accompanied by a marked increase in release of the stress-related neuropeptide corticotropin-releasing factor in the amygdala, a result that also occurred in animals undergoing heroin withdrawal (36). An electrophysiological study showed that precipitated withdrawal was also associated with reduced firing of dopamine neurons in the ventral tegmental area of rat brain (37).

These data clearly show that chronic administration of cannabinoids leads to adaptive changes in the brain, some of which are similar to those seen with other drugs of dependence. The ability of THC to cause selective release of dopamine from the nucleus accumbens (38) also suggests some similarity between THC and other drugs in this category.

Furthermore, although many earlier attempts to obtain reliable self-administration behavior with THC were unsuccessful (33), some success has been obtained recently. Squirrel monkeys were trained to self-administer low doses of THC (2 μg/kg per injection), but only after the animals had first been trained to self-administer cocaine (39). THC is difficult to administer intravenously, but these authors succeeded, perhaps in part because they used doses comparable to those to which human cannabis users are exposed, and because the potent synthetic cannabinoids are far more water-soluble than THC, which makes intravenous administration easier. Mice could be trained to self-administer intravenous WIN55212-2, but CB_1 receptor knockout animals could not (40).

Another way of demonstrating the rewarding effects of drugs in animals is the conditioned place preference paradigm, in which an animal learns to approach an environment in which it has previously received a rewarding stimulus. Rats had a positive THC place preference after doses as low as 1 mg/kg (41).

Some studies have suggested that there may be links between the development of dependence to cannabinoids and to opiates (42). Some of the behavioral signs of rimonabant-induced withdrawal in THC-treated rats can be mimicked by the opiate antagonist naloxone (43). Conversely, the withdrawal syndrome precipitated by naloxone in morphine-dependent mice can be partly relieved by THC (44) or endocannabinoids (45). Rats treated chronically with the cannabinoid WIN55212-2 became sensitized to the behavioral effects of heroin (46). Such interactions can also be demonstrated acutely. Synergy between cannabinoids and opiate analgesics has been described above. THC also facilitated the antinociceptive effects of RB 101, an inhibitor of enkephalin inactivation, and acute administration of THC caused increased release of Met-enkephalin into microdialysis probes placed into the rat nucleus accumbens (47).

The availability of receptor knockout animals has also helped to illustrate cannabinoid–opioid interactions. CB_1 receptor knockout mice had greatly reduced morphine self-administration behavior and less severe naloxone-induced withdrawal signs than wild type animals, although the antinociceptive actions of morphine were unaffected in the knockout animals (40). The rimonabant–precipitated withdrawal syndrome in THC-treated mice was significantly attenuated in animals with knockout of the pro-enkephalin gene (48). Knockout of the μ opioid (OP_3) receptor also reduced rimonabant-induced withdrawal signs in THC-treated mice, and there was an attenuated naloxone withdrawal syndrome in morphine-dependent CB_1 knockout mice (49,50).

These findings clearly point to interactions between the endogenous cannabinoid and opioid systems in the CNS, although the neuronal circuitry involved is unknown. Whether this is relevant to the so-called "gateway" theory is unclear. In the US National Household Survey of Drug Abuse, respondents aged 22 years or over who had started to use cannabis before the age of 21 years were 24 times more likely than non-cannabis users to begin using hard drugs (51). However, in the same survey the proportion of cannabis users who progressed to heroin or cocaine use was very small (2% or less). Mathematical modeling using the Monte Carlo method suggested that the association between cannabis use and hard drug use need not be causal, but could relate to some common predisposing factor, for example "drug-use propensity" (52).

Tumorigenicity

THC does not appear to be carcinogenic, but there is plenty of evidence that the tar derived from cannabis smoke is. Bacteria exposed to cannabis tar develop mutations in the standard Ames test for carcinogenicity (53), and hamster lung cells in tissue culture develop accelerated malignant transformations within 3–6 months of exposure to tobacco or cannabis smoke (54).

General adverse effects

A review has summarized the evidence related to the adverse effects of acute and chronic use of cannabis (55). The effects of acute usage include anxiety, impaired attention, and increased risk of psychotic symptoms. Probable risks of chronic cannabis consumption include bronchitis and subtle impairments of attention and memory.

The adverse effects of cannabis can be considered under two main headings, reflecting psychoactive and autonomic effects, in addition to which there are direct toxic effects. The most frequently reported psychoactive effects include enhanced sensory perception (for example a heightened appreciation of color and sound). Cannabis intoxication commonly heightens the user's sensitivity to other external stimuli as well, but subjectively slows the appreciation of time. In high doses, users may also experience depersonalization and derealization. Various forms of psychomotor performance, including driving, are significantly impaired for 8–12 hours after using cannabis. The most serious possible consequence of cannabis use is a road accident if a user drives while intoxicated.

Adverse reactions have been reported at relatively low doses and principally affect the psyche, leading to anxiety states, panic reactions, restlessness, hallucinations, fear, confusion, and rarely toxic psychosis. These effects appear to be reversible (56). Ingestion of cake with cannabis by people who seldom use or have never used cannabis before can result in mental changes, including confusion, anxiety, loss of logical thinking, fits of laughter, hallucinations, hypertension, and/or paranoid psychosis, which can last as long as 8 hours.

The autonomic effects of cannabis lead to tachycardia, peripheral vasodilatation, conjunctival congestion, hyperthermia, bronchodilatation, dry mouth, nystagmus, tremor, ataxia, hypotension, nausea, and vomiting, that is a spectrum of effects that closely resembles the consequences of overdosage with anticholinergic agents. Some individuals have sleep disturbances. Increased appetite and dry mouth are other common effects of cannabis intoxication.

Hypersensitivity reactions are rare, but a few have been reported after inhalation. Delayed hypersensitivity reactions, particularly affecting vascular tissue, have been recorded with chronic systemic administration. Tumor-inducing effects are difficult to attribute to cannabis alone. Animal studies have shown neoplastic pulmonary lesions superimposed on chronic inflammation, but such pathology may be primarily associated with the "tar" produced by burning marijuana. The most serious potential adverse effects of cannabis use come from the inhalation of the same carcinogenic hydrocarbons that are present in tobacco, and some data suggest that heavy cannabis users are at risk of chronic respiratory diseases and lung cancer.

The effects of oral cannabinoids (dronabinol or *Cannabis sativa* plant extract) in relieving pain and muscle spasticity have been studied in 16 patients with multiple sclerosis (mean age 46 years, mean duration of disease 15 years) in a double-blind, placebo-controlled, crossover study (57). The initial dose was 2.5 mg bd, increasing to 5 mg bd after 2 weeks if the dose was well tolerated. The plant extract was more likely to cause adverse events; five patients had increased spasticity and one rated an adverse event of acute psychosis as severe. All physical measures were in the reference ranges. There were no significant differences in any measure of efficacy score that would indicate a therapeutic benefit of cannabinoids. This study is the largest and longest of its kind, but the authors acknowledged some possible shortcomings. The route of administration could affect subjective ratings, since the gastrointestinal tract is a much slower and more inefficient route than the lungs. Another possibility is that the dose was too small to have the desired therapeutic effects.

Organs and Systems

Observational studies

In an open trial the safety, tolerability, dose range, and efficacy of the whole-plant extracts of *Cannabis sativa* were evaluated in 15 patients with advanced multiple sclerosis and refractory lower urinary tract symptoms

(58). The patients took extracts containing delta-9-tetrahydrocannabinol (THC) and cannabidiol (CBD; 2.5 mg per spray) for 8 weeks followed by THC only for a further 8 weeks. Urinary urgency, the number and volume of incontinence episodes, frequency, and nocturia all reduced significantly after treatment with both extracts. Patients' self assessments of pain, spasticity, and quality of sleep improved significantly, and the improvement in pain continued for up to a median of 35 weeks. Most of the patients had symptoms of intoxication, such as mild drowsiness, disorientation, and altered time perception, during the dose titration period. Three had single short-lived hallucinations that did not occur when the dose was reduced. All complained of a worsening of dry mouth that was already present from other treatments and two complained of mouth soreness at the site of drug administration.

Of 220 patients with multiple sclerosis in Halifax, Canada 72 (36%) reported ever having used cannabis (59). Ever use of cannabis for medicinal purposes was associated with male sex, the use of tobacco, and recreational use of cannabis. Of the 34 medicinal cannabis users, 10 reported mild, eight moderate and one strong adverse effects; none reported severe adverse effects. The most common adverse effects were feeling "high" (n = 24), drowsiness (20), dry mouth (14), paranoia (3), anxiety (3), and palpitation (3).

Placebo-controlled studies

Cannabis has been used to treat many medical conditions, especially those involving pain and inflammation. Many studies with improved designs and larger sample sizes are providing preliminary data of efficacy and safety in conditions such as multiple sclerosis and chronic pain syndromes.

In a parallel group, double-blind, randomized, placebo-controlled study undertaken at three sites in 160 patients with multiple sclerosis a cannabis-based medicinal extract containing equal amounts of delta-9-tetrahydrocannabinol (THC) and cannabidiol (CBD) at doses of 2.5–120 mg of each daily in divided doses for 6 weeks, spasticity scores were significantly improved by cannabis (60). However, when the changes in symptoms were measured using the Primary Symptoms Scale, there were no significant differences between cannabis and placebo. The main adverse events were dizziness (33%), local discomfort at the site of application (26%), fatigue (15%), disturbance in attention (8.8%), disorientation (7.5%), a feeling of intoxication (5%), and mouth ulcers (5%).

In a randomized, double-blind, placebo-controlled, crossover trial the effect of the synthetic delta-9-tetrahydrocannabinol dronabinol on central neuropathic pain was evaluated in 24 patients with multiple sclerosis (61). Oral dronabinol reduced central pain. Adverse events were reported by 96% of the patients compared with 46% during placebo treatment. They were more common during the first week of treatment. The most common adverse events during dronabinol treatment were dizziness (58%), tiredness (42%), headache (25%), myalgia (25%), and muscle weakness (13%). There was increased tolerance to the adverse effects over the course of treatment and with dosage adjustments.

Three cannabis-based medicinal extracts in sublingual form recently became available for use against pain. In a randomized, double-blind, placebo-controlled, crossover study for 12 weeks in 34 patients with chronic neuropathic pain THC extracts were effective in symptom control (62). Drowsiness and euphoria/dysphoria were common in the first 2 weeks. Dizziness was less of a problem. Anxiety and panic were infrequent but occurred during the run-in period. Dry mouth was the most common complaint.

Cardiovascular

Marijuana has several effects on the cardiovascular system, and can increase resting heart rate and supine blood pressure and cause postural hypotension. It is associated with an increase in myocardial oxygen demand and a decrease in oxygen supply. Peripheral vasodilatation, with increased blood flow, orthostatic hypotension, and tachycardia, can occur with normal recreational doses of cannabis. High doses of THC taken intravenously have often been associated with ventricular extra beats, a shortened PR interval, and reduced T wave amplitude, to which tolerance readily develops and which are reversible on withdrawal. While the other cardiovascular effects tend to decrease in chronic smokers, the degree of tachycardia continues to be exaggerated with exercise, as shown by bicycle ergometry.

Marijuana use is most popular among young adults (18–25 years old). However, with a generation of post-1960s smokers growing older, the use of marijuana in the age group that is prone to coronary artery disease has increased. The cardiovascular effects may present a risk to those with cardiovascular disorders, but in adults with normal cardiovascular function there is no evidence of permanent damage associated with marijuana (56,63,64), and it is not known whether marijuana can precipitate myocardial infarction, although mixed use of tobacco and cannabis make the evaluation of the effects of cannabis very difficult.

Postural syncope after marijuana use has been studied in 29 marijuana-experienced volunteers, using transcranial Doppler to measure cerebral blood velocity in the middle cerebral artery in response to postural changes (65). They were required to abstain from marijuana and other drugs for 2 weeks before the assessment, as confirmed by urine drug screening. They were then given marijuana, tetrahydrocannabinol, or placebo and lying and standing measurements were made. When marijuana or tetrahydrocannabinol was administered, 48% reported a dizziness rating of three or four and had significant falls in standing cerebral blood velocity, mean arterial blood pressure, and systolic blood pressure. Eight subjects were so dizzy that they had to be supported. The authors suggested that marijuana interferes with the protective mechanisms that maintain standing blood pressure and cerebral blood velocity. All but one of the subjects who took marijuana or tetrahydrocannabinol reported some degree of dizziness. Women tended to be dizzier. As the

postural dizziness was significant and unrelated to plasma concentrations of tetrahydrocannabinol or other indices, the authors raised concerns about marijuana use in those who are medically compromised or elderly.

Popliteal artery entrapment occurred in a patient with distal necrosis and cannabis-related arteritis, two rare or exceptional disorders that have never been described in association (66).

- A 19-year-old man developed necrosis in the distal third right toe, with loss of the popliteal and foot pulses. Arteriography showed posterior popliteal artery compression in the right leg and unusually poor distal vascularization in both legs. An MRI scan did not show a cyst and failed to identify the type of compression and the causal agent. Surgery showed that the patient had type III entrapment. Surprisingly, the pain failed to regress and the loss of distal pulses persisted despite a perfect result on the postoperative MRA scan. The patient then admitted consuming cannabis 10 times a day for 4 years, which suggested a Buerger-type arteritis related to cannabis consumption. A 21-day course of intravenous vasodilators caused the leg pain to disappear and the toe necrosis to regress. An MRA scan confirmed permanent occlusion of three arteries on the right side of the leg and the peroneal artery on the left side. Capillaroscopy excluded Buerger's disease.

The authors suggested that popliteal artery entrapment in a young patient with non-specific symptoms should raise the suspicion of a cannabis-related lesion. Their review of literature suggested that this condition affects young patient and that complications secondary to popliteal artery entrapment did not occur in those who were under 38 years age.

Ischemic heart disease

Investigators in the Determinants of Myocardial Infarction Onset Study recently reported that smoking marijuana is a rare trigger of acute myocardial infarction (67). Interviews of 3882 patients (1258 women) were conducted on an average of 4 days after infarction. Reported use of marijuana in the hour preceding the first symptoms of myocardial infarction was compared with use in matched controls. Among the patients, 124 reported smoking marijuana in the previous year, 37 within 24 hours, and 9 within 1 hour of cardiac symptoms. The risk of myocardial infarction was increased 4.8 times over baseline in the 60 minutes after marijuana use and then fell rapidly. The authors emphasized that in a majority of cases, the mechanism that triggered the onset of myocardial infarction involved a ruptured atherosclerotic plaque secondary to hemodynamic stress. It was not clear whether marijuana has direct or indirect hemodynamic effects sufficient to cause plaque rupture.

- Two young men, aged 18 and 30 years, developed retrosternal pain with shortness of breath, attributed to acute coronary syndrome (68). Each had smoked marijuana and tobacco and admitted to intravenous drug use. Urine toxicology was positive for tetrahydrocannabinol.

Aspartate transaminase and creatine kinase activities and troponin-I and C-reactive protein concentrations were raised. Echocardiography in the first patient showed hypokinesia of the posterior and inferior walls and in the second hypokinesia of the basal segment of the anterolateral wall. Coronary angiography showed normal coronary anatomy with coronary artery spasm. Genetic testing for three common genetic polymorphisms predisposing to acute coronary syndrome was negative.

The authors suggested that marijuana had increased the blood carboxyhemoglobin concentration, leading to reduced oxygen transport capacity, increased oxygen demand, and reduced oxygen supply.

Two other cases have been reported (69).

- A 48-year-old man, a chronic user of cannabis who had had coronary artery bypass grafting 10 years before and recurrent angina over the past 18 months, developed chest pain. An electrocardiogram showed intermittent resting ST segment changes and coronary angiography showed that of the three previous grafts, only one was still patent. There was also sub-total occlusion of a stent in the left main stem. After 24 hours he had a cardiac arrest while smoking cannabis and had multiple episodes of ventricular fibrillation, requiring both electrical and pharmacological cardioversion. He then underwent urgent percutaneous coronary intervention which involved stenting of his left main stem. He eventually stabilized and recovered for discharge 11 weeks later.
- A 22-year-old man had two episodes of tight central chest pain with shortness of breath after smoking cannabis. He had been a regular marijuana smoker since his mid-teens and had used more potent and larger amounts during the previous 2 weeks. An electrocardiogram showed ST segment elevation in leads V1-5, with reciprocal ST segment depression in the inferior limb leads. A provisional diagnosis of acute myocardial infarction was made. Thrombolysis was performed, but the electrocardiographic changes continued to evolve. Angiography showed an atheromatous plaque in the left anterior descending artery which was dilated and stented. There was early diffuse disease in the cardiac vessels.

The authors suggested that in the first case ventricular fibrillation had been caused by increased myocardial oxygen demand in the presence of long-standing coronary artery disease. In the second case, they speculated that chronic cannabis use may have contributed to the unexpectedly severe coronary artery disease in a young patient with few risk factors.

Coronary no-flow and ventricular tachycardia after habitual marijuana use has been reported (70).

Cardiac dysrhythmias

Paroxysmal atrial fibrillation has been reported in two cases after marijuana use (71).

- A healthy 32-year-old doctor, who smoked marijuana 1–2 times a month, had paroxysmal tachycardia for several months. An electrocardiogram was normal and a Holter recording showed sinus rhythm with isolated supraventricular extra beats. He was treated with propranolol. He later secretly smoked marijuana while undergoing another Holter recording, which showed numerous episodes of paroxysmal atrial tachycardia and atrial fibrillation lasting up to 2 minutes. He abstained from marijuana for 12 months and maintained stable sinus rhythm.
- A 24-year-old woman briefly lost consciousness and had nausea and vomiting several minutes after smoking marijuana. She had hyporeflexia, atrial fibrillation (maximum 140/minute with a pulse deficit), and a blood pressure of 130/80 mmHg. Echocardiography was unremarkable. Within 12 hours, after metoprolol, propafenone, and intravenous hydration with electrolytes, sinus rhythm was restored.

The authors discussed the possibility that Δ9-THC, the active ingredient of marijuana, can cause intra-atrial re-entry by several mechanisms and thereby precipitate atrial fibrillation.

Sustained atrial fibrillation has also been attributed to marijuana (72).

- A 14-year-old African-American man with no cardiac history had palpitation and dizziness, resulting in a fall, within 1 hour of smoking marijuana. After vomiting several times he had a new sensation of skipped heartbeats. The only remarkable finding was a flow murmur. The electrocardiogram showed atrial fibrillation. Echocardiography was normal. Serum and urine toxicology showed cannabis. He was given digoxin, and about 12 hours later his cardiac rhythm converted to sinus rhythm. Digoxin was withdrawn. He abstained from marijuana over the next year and was symptom free.

The authors noted that marijuana's catecholaminergic properties can affect autonomic control, vasomotor reflexes, and conduction-enhancement of perinodal fibers in cardiac muscle, and thus lead to an event such as this.

- A 34-year-old man developed palpitation, shortness of breath, and chest pain. He had smoked a quarter to a half an ounce of marijuana per week and had taken it 3 hours before the incident. He had ventricular tachycardia at a rate of 200/minute with a right bundle branch block pattern. Electrical cardioversion restored sinus rhythm. Angiography showed a significant reduction in left anterior descending coronary artery flow rate, which was normalized by intra-arterial verapamil 200 micrograms.

The authors thought that marijuana may have enhanced triggered activity in the Purkinje fibers along with a reduction in coronary blood flow, perhaps through coronary spasm.

In terms of its potential for inducing cardiac dysrhythmias, cannabis is most likely to cause palpitation due to a dose-related sinus tachycardia. Other reported dysrhythmias include sinus bradycardia, second-degree atrioventricular block, and atrial fibrillation. Also reported are ventricular extra beats and other reversible electrocardiographic changes. Supraventricular tachycardia after the use of cannabis has been reported (73).

- A 35-year-old woman with a 1-month history of headaches was found to be hypertensive, with a blood pressure of 179/119 mmHg. She smoked 20 cigarettes a day and used cannabis infrequently. Her family history included hypertension. Electrocardiography suggested left ventricular hypertrophy but echocardiography was unremarkable. She was given amlodipine 10 mg/day and the blood pressure improved. While in the hospital, she smoked marijuana and about 30 minutes later developed palpitation, chest pain, and shortness of breath. The blood pressure was 233/120 mmHg and the pulse rate 150/minute. Electrocardiography showed atrial flutter with 2:1 atrioventricular block. Cardiac troponin was normal at 12 hours. Urine toxicology was positive for cannabis only. Two weeks later, while she was taking amlodipine 10 mg/day and atenolol 25 mg/day, her blood pressure was 117/85 mmHg.

The authors reviewed the biphasic effect of marijuana on the autonomic nervous system. At low to moderate doses it causes increased sympathetic activity, producing a tachycardia and increase in cardiac output; blood pressure therefore increases. At high doses it causes increased parasympathetic activity, leading to bradycardia and hypotension. They thought that this patient most probably had adrenergic atrial flutter.

Arteritis

A case of progressive arteritis associated with cannabis use has been reported (74).

- A 38-year-old Afro-Caribbean man was admitted after 3 months of severe constant ischemic pain and numbness affecting the right foot. The pain was worse at night. He also had intermittent claudication after walking 100 yards. He had a chronic history of smoking cannabis about 1 ounce/day, mixed with tobacco in the early years of usage. However, at the time of admission, he had not used tobacco in any form for over 10 years. He had patchy necrosis and ulceration of the toes and impalpable pulses in the right foot. The serum cotinine concentrations were consistent with those found in non-smokers of tobacco. Angiography of his leg was highly suggestive of Buerger's disease (thromboangiitis obliterans).

Remarkably, this patient, despite having abstained from tobacco for more than 10 years, developed a progressive arteritis leading to ischemic changes. While arterial pathology with cannabis has been reported before, it has been difficult to dissociate the effects of other drugs.

Respiratory

Acute inhalation of marijuana or THC causes bronchodilatation, but with chronic use resistance in the bronchioles increases (75,76). Prolonged use of cannabis by inhalation can cause chronic inflammatory changes in the bronchial tree, in part related to the inhalants that accompany the smoke. In some cases attacks of asthma and glottal and uvular angioedema can occur. Reduced respiratory gas exchange has been reported in long-term smokers, and under experimental conditions THC can depress respiratory function slightly and act as a respiratory irritant. In fact, chronic marijuana cigarette smoking and chronic tobacco cigarette smoking produce very similar changes, but these occur after smoking fewer cigarettes when marijuana is smoked, compared with tobacco-smoking. With marijuana inhalation, when a filter is never used, inhalation is deeper and the smoke is held in the lungs for longer than when smoking commercially produced tobacco-based cigarettes (77). There is therefore a greater build-up of carbon monoxide, reduction in carboxyhemoglobin saturation, and alveolar cellular irritation with depression of macrophages (SEDA-13, 25). Pneumothorax, pneumopericardium, and pneumomediastinum have been reported when positive pulmonary pressure is applied or a Valsalva maneuver used, as often happens (78,79).

Cannabis smoking can cause pneumothorax (80).

- A 23-year-old man who had smoked cannabis regularly for about 10 years developed severe respiratory distress. He had bilateral pneumothoraces with complete collapse of the left lung.

No obvious reason for the problem was found and the authors suggested that coughing while breath-holding during cannabis inhalation had caused the problem.

The term "Bong lung" is used to refer to a histological change that occurs in the lungs of chronic cannabis smokers (81). Patients with cannabis-induced recurrent pneumothorax often undergo resection of bullae. In Australia, the histopathology of resected lung was examined in 10 cannabis smokers, 5 heavy tobacco smokers, and 5 non-smokers. All marijuana smokers had irregular emphysema with cystic blebs and bullae in the lung apices. There was also massive accumulation of intra-alveolar pigmented histiocytes or "smoker's macrophages" throughout the pulmonary parenchyma, but sparing of the peribronchioles, similar to desquamative interstitial pneumonia.

Four men, who smoked both tobacco and marijuana, developed large, multiple, bilateral, peripheral bullae at their lung apices, with normal parenchymal tissue elsewhere (82). Three patients with large bullae in the upper lung lobe have been reported (83). All had been heavy marijuana smokers over 10–24 years. However, they all had at least nine pack-years of cigarette exposure and so marijuana may not have been the only cause of their lung bullae. Nevertheless, the authors recommended that all those who present with upper lung bullae should be screened for cannabis use.

While Δ9-THC may not contribute directly to lung bullae, it is possible that the respiratory dynamics of smoking the drug explains it. Typically, a draw on a marijuana joint has, on average, a depth of inspiration that is one-third greater, a volume two-thirds greater, and a breath-holding time four times longer than a draw on a cigarette. The marijuana joint lacks a filter tip, and the practice of smoking "leads to a fourfold greater delivery of tar and a five times greater increase in carboxyhemoglobin per cigarette smoked" (75). Smoking three to four joints of marijuana per day is reported to produce a symptom profile and damage to the respiratory airways similar to that caused by smoking 20 tobacco cigarettes daily.

Nervous system

Propriospinal myoclonus has been reported after cannabis use (84).

- A 25-year-old woman developed spinal myoclonus 18 months after having experienced acute-onset repetitive involuntary flexion and extension spasms of her trunk immediately after smoking cannabis. The jerks, which lasted 2–5 seconds, involved the trunk, neck, and to a lesser extent the limbs. The attacks occurred in clusters lasting up to 2 weeks and she was asymptomatic for 2-3 months between clusters. The myoclonus was not present during sleep. During a bout of jerks, myoclonus would occur every few minutes and continue for up to 9 hours, with associated fatigue and back pain. Neurological examination showed repetitive flexion jerks of the trunk with no other abnormal signs. An electroencephalogram, an MRI scan of the head and spine, and a full-length myelogram were all normal. Multi-channel surface electromyography with parallel frontal electroencephalography showed propriospinal myoclonus of mid-thoracic origin.

There have been no previous reports of propriospinal myoclonus precipitated by marijuana. The etiology was not clear but may have involved cannabinoid receptors located in the brain and spinal cord as well as the peripheral nervous system.

Occipital stroke has been reported after cannabis use (85).

- A 37-year-old Albanian man had an uneventful medical history except that he smoked 20 cigarettes/day and marijuana joints regularly for 10 years. In the previous 6 months he had increased his marijuana smoking to 2-3 joints/week from 1-2 joints/month. He suddenly developed left-sided hemiparesis, left-sided hemihypesthesia, and recurrent double vision 15 minutes after having smoked a joint containing about 250 mg marijuana. Most of the symptoms disappeared within 1 hour after onset. An MRI scan showed an area of impaired diffusion, 2 cm in diameter, in the right occipital area subcortically. He responded well to acetylsalicylic acid with dipyridamole and atorvastatin and was discharged 3 days later with blurred vision when looking to the left. There was no cardiac source of embolism. Other causes of stroke were carefully excluded, and it could only be attributed to the use of marijuana.

The authors, based on previous reports of the vasogenic effects of marijuana, suggested that this event may have been related to increased concentrations of catecholamine and carboxyhemoglobin, and diminishing cerebral autoregulatory capacity.

Transient global amnesia, an amnesia of sudden onset regarding events in the present and recent past, typically occurs in elderly people. Transient global amnesia following accidental marijuana ingestion has been reported in a young boy (86).

- A 6-year-old boy accidentally became intoxicated with marijuana after eating cookies laced with marijuana. He developed retentive memory deficits of sudden onset, later diagnosed as transient global amnesia. He was anxious and had a tachycardia, fine tremors in the upper and lower limbs, and an ataxic gait. His CSF was unremarkable. He had cannabinoids in his urine. His memory returned to normal after 14 hours. His mother admitted baking marijuana cookies and leaving them out on the kitchen table. Up to 12 months later he had no memory of the episode.

This is the first case of transient global amnesia from marijuana in a 6-year-old. With increased use of marijuana in society, children can sometimes be exposed to marijuana inadvertently.

A review of the evidence has suggested that, particularly with high doses, cannabis users are 3–7 times more likely to cause motor accidents than non-drug users (87).

Marijuana can interact with the neurotransmitter dopamine, and the effects of marijuana on the brain in schizophrenia have been studied by single photon emission computerized tomography (SPECT) (88).

A 38-year-old man with schizophrenia secretively smoked marijuana during a neuroimaging study. A comparison of two sets of images, before and after marijuana inhalation, showed a 20% reduction in the striatal dopamine D_2 receptor binding ratio, suggestive of increased synaptic dopaminergic activity.

On the basis of this in vivo SPECT study, the authors speculated that marijuana may interact with dopaminergic systems in brain reward pathways.

Consistent with this, extrapyramidal effects have been reported in a patient taking neuroleptic drugs who smoked cannabis (89).

- A 20-year-old man with no previous movement disorders, who had smoked marijuana for 4 years was given risperidone 9 mg/day and clorazepate 10–20 mg/day for paranoid schizophrenia. After 4 weeks he started using marijuana again and had least two episodes of cervical and jaw dystonia with oculogyric crises, for which intramuscular biperiden was effective. He acknowledged heavy marijuana use before each episode. He was then given oral biperiden 2–4 mg/day and risperidone was replaced by olanzapine 30 mg/day. He again started smoking marijuana and had similar episodes of dystonia and oculogyric crises. No other causes of secondary extrapyramidal disorders were found.

The authors suggested a causal association between use of marijuana and extrapyramidal disorders. The research literature contains evidence that the endogenous cannabinoid system plays a role in basal ganglia transmission circuitry, possibly by interfering with dopamine reuptake. Furthermore, central cannabinoid receptors are located in two areas that regulate motor activity, the lateral globus pallidus and substantia nigra (90A).

Cerebrovascular disease due to marijuana is infrequent.

- A 36-year-old man developed acute aphasia followed by a convulsive seizure a few hours later after heavy consumption of hashish and 3–4 alcoholic beverages at a party (91). He had no previous vascular risk factors. His blood pressure was 120/80 mmHg. An MRI scan showed two ischemic infarcts, one in the left temporal lobe and one in the right parietal lobe. Magnetic resonance angiography of the head and neck showed narrowing of the distal temporal branches of the left middle cerebral artery without involvement of its proximal segment. There was no evidence of diffuse atherosclerotic disease. There was tetrahydrocannabinol in the urine. Electroencephalography and transesophageal echocardiography were normal. He was given ticlopidine. A year later, he had a second episode of aphasia and right hemiparesis immediately after smoking marijuana. His blood pressure was 140/80 mmHg. An MRI scan showed acute left and right frontal cortical infarctions. He had a third stroke 18 months later, when he developed auditory agnosia after heavy use of hashish and 3–4 drinks of alcohol. On this occasion he was normotensive. An MRI scan showed acute infarcts in the right posterior temporal lobe and lower parietal lobe.

The authors discussed the importance of the close temporal relation between the use of cannabis and alcohol and the episodes of stroke. The mechanism was unclear. However, they speculated that a vasculopathy, either toxic or immune inflammatory, was the most likely mechanism.

In another case a cannabis smoker had recurrent transient ischemic attacks.

- A 50-year-old male cigarette smoker with hypertension had episodes of transient right-sided hemisensory loss lasting only a few minutes (92). Coughing while smoking marijuana was the apparent trigger. Electroencephalography was negative. An MRI scan of the brain showed chain-like low-flow infarctions in the white matter of the left parietal subcortex. Duplex sonography and digital subtraction angiography showed a subocclusive stenosis of the left internal carotid artery. Blood flow to the left middle cerebral artery was reduced and delayed and there was steal by the right middle cerebral artery. Endarterectomy of the left internal carotid resolved the symptoms.

The authors suggested that marijuana may increase the risk of reduced cerebral perfusion. Coughing may have contributed by reduced flow velocity within arteries supplying the brain and by a sudden increase in intracranial pressure.

Sensory systems

Eyes

No consistent effects of cannabinoids on the eyes have been reported, apart from a reduction in intraocular pressure (93). The initial reduction in intraocular pressure is followed by a rebound increase associated with increased prostaglandin concentrations.

However, bilateral angle-closure glaucoma has been reported after combined consumption of ecstasy and cannabis (94).

- A 29-year-old woman developed severe headaches, blurred vision, and malaise. Her visual acuity was <20/400 in the right eye and 20/40 in the left eye. Intraocular pressures were raised at 38 and 40 mmHg. Slit lamp examination showed bilateral conjunctival hyperemia, corneal edema, and shallow anterior chambers. Gonioscopy showed bilateral circular closed angles. The pupils were mid-dilated and non-reactive to light. The optic nerve heads in both eyes had slightly enlarged cups. She admitted to recreational use of ecstasy and marijuana before this ophthalmic crisis and also 2 years earlier, when she had had an episode of ophthalmic migraine with headache and transient blurred vision. Ophthalmic examination showed a narrow anterior chamber angle in both eyes.

The authors suggested that the bilateral angle-closure glaucoma had been precipitated by a combined mydriatic effect of ecstasy and cannabis.

Ears

The effect of THC, 7.5 mg and 15 mg, on auditory functioning has been investigated in eight men in a double-blind, randomized, placebo-controlled, crossover trial (95). Blood concentrations of THC were measured for up to 48 hours after ingestion, and audiometric tests were carried out at 2 hours. There were no significant differences across treatments, suggesting that cannabis does not affect the basic unit of auditory perception.

Psychological, psychiatric

The psychological effects of cannabis vary with personal and social factors. However, some guidance to the essential effects of the drug can be derived from investigations with THC and marijuana in non-user volunteers. Blood concentrations of THC over 75 µg/ml under these conditions are associated with euphoria, and somewhat higher concentrations with dissociation of events and memory and impairment of psychomotor tasks lasting over 24 hours (56).

Through random urine testing of draftees to the Italian army, 133 marijuana users were identified, tested, and interviewed (96). Among these marijuana users, 83% of those with cannabis dependence, 46% with cannabis abuse, and 29% of occasional users had at least one DSM-IIIR psychiatric diagnosis. With greater cannabis use, the risk of associated psychiatric disabilities tended to increase progressively.

Occasional and regular users can suffer panic attacks, paranoia, hallucinations, or feelings of unreality (depersonalization and derealization).

In a critical English-language literature review of the cannabis research done during the 10 years from 1994 to 2004 the relation between the rate of cannabis use, behavioral problems, and mental disorders in young people was explored (43). Although there are shortcomings in the studies done in this area, the data suggest that early and heavy use of cannabis has negative effects on psychosocial functioning and psychopathology. Although infrequent use causes few mental health or behavioral problems, cannabis is not necessarily harmless. Accumulating evidence suggests that regular marijuana use during adolescence may have effects, whether biological, psychological, or social, that are different from those in later life. Most recent data challenge the notion that marijuana relieves psychotic or depressive symptoms.

Psychosis

The causal relation between cannabis abuse and schizophrenia is controversial. Cannabis abuse, and particularly heavy abuse, can exacerbate symptoms of schizophrenia and can be considered as a risk factor eliciting relapse in schizophrenia (97). Chronic cannabis use can precipitate schizophrenia in vulnerable individuals (98).

Neurotrophins, such as nerve growth factor and brain-derived neurotrophic factor (BDNF), are implicated in neuronal development, growth, plasticity, and maintenance of function. Neurodevelopment is impaired in schizophrenia and vulnerable schizophrenic brains may be more sensitive to toxic influences. Thus, cannabis may be more neurotoxic to schizophrenic brains than to non-schizophrenic brains when used chronically. In 157 drug-naïve first-episode schizophrenic patients there were significantly raised BDNF serum concentrations by up to 34% in patients with chronic cannabis abuse or multiple substance abuse before the onset of the disease (99). Thus, raised BDNF serum concentrations are not related to schizophrenia and /or substance abuse itself but may reflect cannabis-related idiosyncratic damage to the schizophrenic brain. Disease onset was 5.2 years earlier in the cannabis-consuming group.

The effects of marijuana on psychotic symptoms and cognitive deficits in schizophrenia have been studied in 13 medicated stable patients with schizophrenia and 13 healthy subjects in a double-blind, placebo-controlled randomized study using 2.5 and 5 mg of intravenous delta-9-tetrahydrocannabinol (THC, 100). Tetrahydrocannabinol transiently worsened cognitive deficits, perceptual alterations, and a range of positive and negative symptoms in those with schizophrenia. There were no positive effects. These results suggest a role for cannabinoid receptors in the pathophysiology of schizophrenia.

Four cases in which psychosis developed after relatively small amounts of marijuana were smoked for the first time have been reported (101). All required hospitalization and neuroleptic drug treatment. Each had a mother with manic disorder and two had psychotic features. The authors noted that marijuana is a dopamine receptor agonist, and mania may be associated with excessive dopaminergic neurotransmission. The use of marijuana may precipitate psychosis or mania in subjects who are genetically vulnerable to major mental illness.

Marijuana abuse and its possible associated risks in reinforcing further use, causing dependence, and producing withdrawal symptoms among adolescents with conduct symptoms and substance use disorders has been investigated in 165 men and 64 women selected and then interviewed from a group of 255 consecutive admissions to a university-based adolescent substance abuse treatment program (102). All had DSM-IIIR substance dependence, 82% had conduct disorder, 18% had major depression, and 15% had attention-deficit/hyperactivity disorder. Most (79%) met the criteria for cannabis dependence. Two-thirds of the cannabis-dependent individuals admitted serious drug-related problems and reported associated drug withdrawal symptoms according to the Comprehensive Addiction Severity Index in adolescents (CASI). For the majority, progression from first to regular cannabis use was as rapid as tobacco progression and more rapid than that of alcohol.

Memory

Long-term heavy use of cannabis impairs mental performance, causes defects in memory (especially short-term memory), and leads to impairment of memory, attention, and organization and integration of complex information (103). Adolescents with pre-existing disabilities in learning and cognition have experienced serious aggravation of their problem from regular use of cannabis (104).

Memory involves two components: an initial delay-independent discrimination or "encoding" and a second delay-dependent discrimination or "recall" of information. In five subjects tetrahydrocannabinol acutely impaired delay-dependent discrimination but not delay-independent discrimination (105). In other words, smoking marijuana increased the rates of forgetting but did not alter initial discriminability.

Cognitive effects

It has been reported that long-term marijuana altered the electroencephalogram during abstinence (106). In 29 individuals who met DSM-III R criteria for marijuana dependence or abuse and 21 drug-free controls, electroencephalograms were recorded for 3 minutes (107). Marijuana abusers had significantly lower log power for the theta and alpha1 bands during abstinence compared with controls. The authors also observed increased cerebrovascular resistance using transcranial Doppler sonography in an overlapping sample of marijuana abusers. They proposed that this combination of electroencephalographic findings and changes in cerebral blood flow may explain cognitive deficits reported in chronic marijuana users.

Delta-9-tetrahydrocannabinol (THC) activates cannabinoid receptors in frontal cortex and hippocampus. Electroencephalograms were obtained from 10 subjects who performed cognitive tasks before and after smoking marijuana or a placebo, to examine the effects on performance and neurophysiology signals of cognitive functions (108). Marijuana increased heart rate and reduced global theta band electroencephalographic power, consistent with increased autonomic arousal. Responses in working memory tasks were slower and less accurate after smoking marijuana, and were accompanied by reduced alpha-band electroencephalographic reactivity in response to increased task difficulty. Marijuana disrupted both sustained and transient attention processes, resulting in impaired memory task performance. In the episodic memory task, marijuana use was associated with an increased tendency to identify distracter words erroneously as having been previously studied. In both tasks, marijuana attenuated stimulus-locked event-related potentials (ERP). In subjects most affected by marijuana, a pronounced ERP difference between previously studied words and new distracter words was also reduced, suggesting disruption of neural mechanisms underlying memory for recent episodes.

The effect of regular marijuana on binocular depth perception has been examined using the Binocular Depth Inversion Illusion (BDII) to identify individuals with an impairment of "top down" processing in perceptual networks in 10 regular users of marijuana and 10 healthy, non-cannabis-using controls (109). The subjects had consumed marijuana at least every other day for a full year. The results suggested that regular marijuana users had significantly higher scores on the BDII, which implies subtle neurocognitive impairment that affects the sensory system involved in correcting ambiguous perceptions. The authors proposed that these impairments are similar to those seen in individuals with schizophrenia, and that cannabis use could be an independent risk factor for the development of schizophrenia.

In a Vietnamese study of 54 monozygotic male twin pairs who were discordant for regular marijuana use and who had not used any other illicit drug regularly the marijuana users significantly differed from the non-users on the general intelligence domain; however, within that domain only the performance of the block design subtests of Wechsler Adult Intelligence Scale-Revised reached statistical significance (110). The marijuana users had not used it for at least 1 year, and a mean of almost 20 years had passed since the last time marijuana had been used regularly. There were no marked long-term residual effects of marijuana use on cognitive abilities.

The neurocognitive effects of marijuana have been studied in 113 young adults (111). Marijuana users, identified by self-reporting and urinalysis, were categorized as light users (<5 joints per week) or heavy users (5 or more joints per week) and current users or former users, the latter having used the drug regularly in the past (1 or more joint per week) but not for at least 3 months. IQ, memory, processing speed, vocabulary, attention, and abstract reasoning were assessed. Current regular heavy users performed significantly worse than non-users beyond the acute intoxication period. Memory, both immediate and delayed, was most strongly affected. However, after 3 months abstinence, there were no residual effects of marijuana, even among those who had formerly had heavy use.

The effects of chronic marijuana smoking on human brain function and cognition have been further investigated (75). Normalized regional brain blood flow and regional brain metabolism, measured using PET scanning with ^{15}O, were compared in 17 frequent marijuana users and 12 non-users. Testing was performed after at least 26 hours of monitored abstinence in all subjects. Marijuana users had hypoactivity or reduced brain blood flow in a large region of the posterior cerebellum compared with controls. This is consistent with what was reported in the only previous PET study of chronic marijuana use (112). The cerebellum is hypothesized to have input to aspects of cognition, specifically timing, the processing of sensory information, and attention and prediction of real-time events. Users often report that marijuana smoking is followed by alterations in the sense of time and less efficient cognitive processing.

The safety and possible benefits of long-term marijuana use have been studied in four seriously ill patients in the Missoula Chronic Clinical Cannabis Use Study with a quality-controlled sample of marijuana (113). They were evaluated using an extensive neurocognitive battery.

- A 62-year-old woman with congenital cataracts smoked marijuana illicitly for 12 years (current use 3–4 g/day smoked and 3–4 g/day orally). She had mild-to-moderate difficulty with attention and concentration and minimal-to-mild difficulty with acquisition and storage of very complex new verbal material. Her executive functioning was not affected.
- A 50-year-old man with hereditary osteo-onychodysplasia had smoked marijuana since 1974 to alleviate muscle spasms and pain (current use 7 g/day of 3.75% THC). He had mild-to-moderate impairment of attention and concentration and reduced ability to acquire new verbal material. He scored poorly on the California Verbal Learning Test (CVLT), a measure of short-term memory recall, and had difficulty with motor tasks.
- A 48-year-old man with multiple congenital cartilaginous exostoses had smoked marijuana since the late 1970s (current use 9 g/day of 2.7% THC). His neurocognitive scores suggest mild difficulty in sustaining attention and a minimal-to-mild deficit in the acquisition of new verbal material.
- A 45-year-old woman with multiple sclerosis had smoked cannabis since 1990 to control pain and muscle spasms (current use marijuana cigarettes containing 3.5% THC 10/day). She had impairment of concentration, learning, and memory efficiency. Her ability to acquire new verbal information was also impaired.

The authors attributed these cognitive deficits not to marijuana use but rather to the patients' illnesses, arguing that it is difficult for patients with painful debilitating diseases to concentrate on neurocognitive tasks. Any abnormalities in MRI imaging and electroencephalography were attributed to age-related brain deterioration. There were no significant abnormalities of respiratory function, apart from a "slight downward trend in FEV_1 and FEV_1/FVC ratios, and perhaps an increase in FVC" in three patients, interpretation of these findings being complicated by concomitant tobacco smoking. One patient had mild polycythemia and a raised white cell count. None had abnormal endocrine tests. This was a comprehensive study of the long-term effects of cannabis, but concomitant illnesses and use of tobacco made the results difficult to interpret.

Concerns have been raised about the possible adverse effects of acute as well as chronic medicinal and recreational use of cannabis on cognition and the body (114). The author, while acknowledging the therapeutic role of cannabinoids in the management of pain and other conditions, expressed concern that in recent years the prevalence of recreational cannabis use (especially in the young) and the potency of the available products have markedly increased in the UK.

An unusual account of transient amnesia after marijuana use has been reported from Europe (115).

- A 40-year-old healthy man with a long history of cannabis use was hospitalized with an acute memory disturbance after smoking for several hours a strong type of marijuana called "superskunk." After smoking, he had difficulty recollecting recent events and would ask the same questions repeatedly. While his routine laboratory results were within the reference ranges, his urine and blood toxic screens had very high concentrations of cannabinoids (and no other drugs). He was alert and oriented to his name, address, date, and place of birth, but could not recall his marital status, whether he had children, or the nature of his job. He was disoriented in time. He performed normally in tests of general cognitive functioning (for example Raven's matrices, word fluency, Rey's complex figurecopy) and short-term memory (for example digit span, verbal cues), but showed severe impairment in verbal and non-verbal long-term components of anterograde memory tests. He had a severe retrograde memory defect mainly affecting autobiographical memory, with a temporal gradient such that remote facts were preserved. These memory impairments lasted 4 days and then rapidly improved, leaving amnesia for the acute episode. Electroencephalography during the amnestic episode was normal, except for brief trains of irregular slow activity in the frontal areas bilaterally. A SPECT scan of his brain was normal. A week later, repeat neuropsychological examination showed normal memory and a normal electroencephalogram and MRI scan of the brain with enhancement. One year later, he had stopped using marijuana and had no further amnestic episodes.

The authors found similarities between the memory disorder seen here and transient global amnesia (see above), which consists of anterograde amnesia and a variably graded retrograde amnesia. The authors stated that although memory impairment has been reported with marijuana before, it has never involved retrieval of already learned material. They wondered if the memory impairment was due to marijuana-induced changes in cerebral blood flow and ischemia through vasospasm. However, their SPECT data did not support this theory. They considered the possibility that cannabinoid receptors, which are dense in the hippocampus, could have been occupied by marijuana, resulting in such memory loss. They cautioned that the effects of marijuana on memory may be more severe than previously thought.

Behavior

The use of marijuana is related to risky behaviors that may result in other drug use, high-risk sexual activity, risky car driving, traffic accidents, and crime. The acute effects of marijuana on human risk taking has been investigated in a laboratory setting in 10 adults who were given three doses of active marijuana cigarettes (half placebo and half 1.77%, 1.77%, and 3.58 % tetrahydrocannabinol) and placebo cigarettes (116). There were measurable changes in risky decisions after marijuana. Tetrahydrocannabinol 3.58% increased selection of the risky response option and also caused shifts in trial-by-trial response probabilities, suggesting altered sensitivity to both reinforced and losing risky outcomes. The authors suggested that the effect on risk taking was possibly seen only at the 3.58% dose because it created a requisite level of impairment to disrupt inhibitory processes in the mesolimbic-prefrontal cortical network.

Endocrine

In animals (particularly monkeys), cannabis depresses ovarian and testicular function. In man, chronic use has been associated with reduced serum FSH and LH concentrations in a few people, often accompanied by reduced serum testosterone, oligospermia, reduced sperm motility, and gynecomastia (117). There is no evidence of impairment of male fertility; no studies have been carried out on female fertility. There is evidence of slightly shortened gestation periods in chronic users (118). There are variable non-specific effects on serum prolactin and growth hormone and a rise in plasma cortisol concentrations has been recorded in one study.

Hematologic

Of the hematological changes very occasionally noted, polycythemia appears to be secondary to reduced pulmonary oxygen exchange (see the Respiratory section in this monograph).

Gastrointestinal

Although cannabis has been used as an antiemetic, in 19 patients it was associated with cyclical hyperemesis, and in seven cases withdrawal was followed by the disappearance of symptoms; in three cases there was a positive rechallenge (119).

Immunologic

Tetrahydrocannabinol depresses lymphocyte and macrophage activity in cell cultures, while in rats in vivo it directly suppresses natural killer cell activity and impairs T lymphocyte transformation by phytohemagglutinin in concentrations of cannabinoids achievable with the usual doses (120). Variable results have been obtained in man in tests of circulating T cells and hormonal immunity (121).

In animals and man, chronic use often suppresses the immune system's response to inhaled bacterial or fungal material. In this connection it is relevant to note that a contaminant mould (*Aspergillus*) found in cannabis can predispose immunocompromised cannabis smokers to infection. It has been suggested that baking the cannabis (at 300°F for 15 minutes) before smoking will kill the fungus and reduce the potential risk (122).

The effects of oral cannabinoids on immune functioning were studied in 16 patients with multiple sclerosis in a crossover study of dronabinol, *Cannabis sativa* plant extract, or placebo for 4 weeks (123). There was a modest increase in pro-inflammatory cytokine tumor necrosis factor alfa during cannabis plant extract treatment in all the subjects. Those with high adverse event scores (n = 7) had significant increases in pro-inflammatory plasma cytokine IL-12p40 while taking the plant extract; this was not the case with tetrahydrocannabinol. Other cytokines were not affected. Tumor necrosis factor alfa and IL-12p40 are known to worsen the course of multiple sclerosis (124). These results are interesting because they suggest immunoactivation by cannabinoids in patients with multiple sclerosis, rather than immunosuppression, as previously reported with the plant extract (125). More studies are needed, because these pro-inflammatory effects could have negative influence on the course of the disease.

The effects of marijuana on immune function have been reviewed (126). The studies suggest that marijuana affects immune cell function of T and B lymphocytes, natural killer cells, and macrophages. In addition, cannabis appears to modulate host resistance, especially the secondary immune response to various infectious agents, both viral and bacterial. Lastly, marijuana may also affect the cytokine network, influencing the production and function of acute-phase and immune cytokines and modulating network cells, such as macrophages and T helper cells. Under some conditions, marijuana may be immunomodulatory and promote disease.

A severe allergic reaction after intravenous marijuana has been reported (127).

- A 25-year-old man with intermittent metamfetamine use developed facial edema, pruritus, and dyspnea 45 minutes after injecting a mixture of crushed marijuana leaves and heated water. He was anxious, and had tachypnea, respiratory stridor, wheezing, edema of the face and oral mucosa, and truncal urticaria. There was mild pre-renal uremia and urine toxicology was positive for metamfetamine and marijuana. Skin testing was not done. With appropriate medical intervention there was resolution of symptoms within a day.

The authors noted that marijuana may have contaminants, including *Aspergillus*, *Salmonella*, herbicides, and mercury, which can trigger allergic reactions.

Long-Term Effects

Drug tolerance

Tolerance develops with heavy chronic use in individuals who report problems in controlling their use and who continue to use cannabis despite adverse personal consequences (55).

Drug withdrawal

Withdrawal symptoms occur after chronic heavy use (55). Abrupt withdrawal of high-level use of cannabinoids causes irritability, restlessness, and insomnia, with a rebound increase in REM sleep, tremor, and anorexia lasting up to a week (128–130). Occasional use does not appear to be associated with major consequences.

Tumorigenicity

Three different associations of cannabinoids with cancer have been discussed (131). Firstly, there is a possible direct carcinogenic effect. In in vitro studies and in mice tetrahydrocannabinol alone does not seem to be carcinogenic or mutagenic. However, cannabis smoke is both carcinogenic and mutagenic and contains similar carcinogens to those in tobacco smoke. Cannabis is possibly linked to digestive and respiratory system cancers. Case reports support this association but epidemiological cohort studies and case-control studies have provided conflicting evidence. Secondly, there is conflicting evidence on the beneficial effects of tetrahydrocannabinol and other cannabinoids in patients with cancer. In some in vitro and in vivo studies, tetrahydrocannabinol and synthetic cannabinoids had antineoplastic effects, but in others tetrahydrocannabinol had a negative effect on the immune system. No anticancer effects of tetrahydrocannabinol in humans have so far been reported. Thirdly, cannabis may palliate some of the symptoms and adverse effects of cancer. Cannabis may improve appetite, reduce nausea and vomiting, and alleviate moderate neuropathic pain in patients with cancer. The authors defined the challenge for the medical use of cannabinoids as the development of safe, effective, and therapeutic methods of using it that are devoid of the adverse psychoactive effects. Lastly, they discussed the possible associations between cannabis smoking and tumors of the prostate and brain, noting the need for larger, controlled studies.

Second-Generation Effects

Pregnancy

Behavioral anomalies have been identified in the offspring of monkeys and women exposed to cannabis during pregnancy (132,133). These include reduced visual responses, increased auditory responses, and reduced quietude. Most of the effects resolved within 4–5 weeks postpartum and there were no abnormalities at 1 year.

Teratogenicity

In animals, THC crosses the placenta and is excreted in breast milk. There is conflicting evidence concerning teratogenicity in animals, but no definitive evidence in man. However, there have been many anecdotal reports of abnormalities. Although these were without consistent characteristics, the descriptions would readily fit the fetal alcohol syndrome (134–137) and clinical evaluation of the use of cannabis during pregnancy is complicated by the frequent concomitant use of alcohol and tobacco.

Fetotoxicity

The effect of maternal and prenatal marijuana exposure on offspring from birth to adolescence is being investigated (138). The Ottawa Prenatal Prospective Study (OPPS), a longitudinal project begun in 1978, recently reported its findings in 146 low-risk, middle-class children aged 9–12 years. Their performances on neurobehavioral tasks that focus on visuoperceptual abilities (ranging from basic skills to those requiring integration and cognitive manipulation of such skills) were analysed. Performance outcomes were different in children with prenatal exposure to cigarette smoking and those with prenatal exposure to marijuana. Maternal cigarette smoking affected fundamental visuoperceptual functioning. Prenatal marijuana use had a negative effect on performance in visual problem-solving, which requires integration, analysis, and synthesis. In a second prospective study, the effects of prenatal marijuana exposure and child behavior problems were studied in 763 subjects aged 10 years (139). Prenatal maternal marijuana exposure was associated with increased hyperactivity, impulsivity, and inattention in the children. There was also increased delinquency and externalizing problems. The authors suggested a possible pathway between prenatal marijuana exposure and delinquency, which may be mediated by the effects of marijuana exposure on symptoms of inattention.

The effects of prenatal marijuana exposure on cognitive functioning have been examined in 145 children aged 13-16 years (140). The age breakdown was 45 13-year-olds, 36 14-year-olds, 51 15-year-olds, and 13 16-year-olds. These groups were further classified by maternal marijuana use: less than six joints per week (n = 120) and six or more joints per week (n = 25). A standard neurocognitive test battery was administered, and two of the tests differed significantly between non-users/light users and heavy users. On the Abstract Designs test the children of heavy users had significantly slower response times. Children in the heavy user group also scored significantly lower on the Peabody Spelling test. These results suggest a dose-related effect of prenatal marijuana exposure on cognition. The two tests that differed across groups depend, to a lesser degree, on cognitive manipulation or comprehension and, to a greater degree, on visual memory, analysis, and integration. Unlike cigarette exposure, marijuana does not seem to affect overall intelligence. It is therefore possible that heavy marijuana use during pregnancy causes subtle deficits in visual analysis.

Cannabis is the illicit drug that is most commonly used by young women and they are not likely to withdraw until the early stages of pregnancy. The effects of early maternal marijuana use on fetal growth have been reported in pregnant women who elected voluntary saline-induced abortion at mid-gestation (weeks 17–22, 141). Marijuana (n = 44) and non-marijuana exposed fetuses (n = 95) were compared and adjusted for maternal alcohol and cigarette use. Both fetal foot length and body weight were significantly reduced by marijuana. Fetal growth impairment was greatest in the group with moderate, regular exposure to about 3–6 joints/week and not in those with heavy maternal marijuana use. There was no significant effect on fetal body length and head circumference due to prenatal marijuana exposure.

Susceptibility Factors

Age

Children

In young children, accidental ingestion leads to the rapid onset of drowsiness, hypotonia, dilated pupils, and coma. Fortunately, gradual recovery occurs spontaneously, barring accidents. Passive inhalation of marijuana in infants can have serious consequences.

- A 9-month-old girl presented with extreme lethargy and a modified Glasgow coma scale of 10, after having been exposed to cigarette and cannabis smoke at the home of her teenage sister's friend (142). The physical examination and laboratory results were unremarkable. Cannabinoids were detected in a urine screen.

While chronic adult users can display apathy and impaired concentration, these effects are possibly in part associated with other factors. No permanent organic brain damage has been demonstrated (142,143).

HIV infection

The use of cannabinoids has been studied in 62 patients with HIV-1 infection (144). Cannabinoids and HIV are of interest because there is the chance of an interaction between tetrahydrocannabinol and antiretroviral therapy. Tetrahydrocannabinol inhibits the metabolism of other drugs (145, 146) and cannabinoids are broken down by the same cytochrome P-450 enzymes that metabolize HIV protease inhibitors. The subjects were randomly assigned to marijuana, dronabinol (synthetic delta-9-tetrahydro-cannabinol), or placebo, given three times a day, 1 hour before meals. The amounts of HIV RNA in the blood did not increase significantly over the course of the study and there were no significant effects on CD4+ or CD8+ cell counts. However, there was significant weight gain in both cannabinoid groups compared with placebo. Although this study was of very short duration, the results suggested that either oral or smoked marijuana may be safe for individuals with HIV-1.

Other features of the patient

People with pre-existing coronary artery disease may have an increased incidence of attacks of angina (147). In individuals who are vulnerable to schizophrenia, cannabis can precipitate psychoses or aggravate schizophrenia. The control of epilepsy may be impaired. Users undergoing anesthesia may react unexpectedly and may have enhanced nervous system depression. Because of impairment of judgement and psychomotor performance, users should not drive or operate machinery for at least 24 hours after administration.

Drug–Drug Interactions

Alcohol

Additive psychoactive effects sought by users may be achieved by combinations of cannabis and alcohol, but at the same time the ability of THC to induce microsomal enzymes will increase the rate of metabolism of alcohol and so reduce the additive effects (128).

Anticholinergic drugs

The anticholinergic effects of cannabis (128) may result in interactions with other drugs with anticholinergic effects, such as some antidysrhythmic drugs.

Barbiturates, short-acting

Additive psychoactive effects sought by users may be achieved by combinations of cannabis and short-acting barbiturates, but at the same time the ability of THC to induce microsomal enzymes will increase the rate of metabolism of barbiturates and so reduce the additive effects (128).

Disulfiram

Concurrent administration with disulfiram is associated with hypomania (148).

Lysergic acid diethylamide

"Flashbacks," or the return of hallucinogenic effects, occur in almost a quarter of those who have used LSD, particularly if they have also used other CNS stimulants, such as alcohol or marijuana. They can experience distortions of perception of objects, space, or time, which intrude without warning into reality, resulting in delusions, panic, and unusual images. A "trailing phenomenon" has also been reported, in which the visual perception of objects is reduced to a series of interrupted pictures rather than a constant view. The frequency of these events may slowly abate over several years, but in a significant number their incidence later increases (149,150).

Psychotropic drugs

Cannabis alters the effects of psychotropic drugs, such as opioids, anticholinergic drugs, and antidepressants, although variably and unpredictably (128).

Sildenafil

Myocardial infarction has been attributed to the combination of cannabis and sildenafil.

- A 41-year-old man developed chest tightness radiating down both arms (151). He had taken sildenafil and cannabis recreationally the night before. His vital signs were normal and he had no signs of heart failure. However, electrocardiography showed an inferior evolving non-Q-wave myocardial infarct and his creatine kinase activity was raised (431 U/l).

Cannabis inhibits CYP3A4, which is primarily responsible for the metabolism of sildenafil, increased concentrations of which may have caused this cardiac event.

References

1. Rey JM, Martin A, Krabman P. Is the party over? Cannabis and juvenile psychiatric disorder: the past 10 years. J Am Acad Child Adolesc Psychiatry 2004; 43(10):1194–205.
2. Goodin D. Marijuana and multiple sclerosis. Lancet Neurol 2004;3(2):79–80.
3. Caswell A. Marijuana as medicine. Med J Aust 1992; 156(7):497–8.
4. Voelker R. Medical marijuana: a trial of science and politics. JAMA 1994;271(21):1645–8.
5. Woody GE, MacFadden W. Cannabis related disorders. In: Kaplan HI, Sadock B, editors. 6th ed.Comprehensive Textbook of Psychiatry vol. 1. Baltimore: Williams & Wilkins, 1995:810–7.
6. Herkenham M. Cannabinoid receptor localization in brain: relationship to motor and reward systems. Ann NY Acad Sci 1992;654:19–32.
7. Childers SR, Fleming L, Konkoy C, Marckel D, Pacheco M, Sexton T, Ward S. Opioid and cannabinoid receptor inhibition of adenylyl cyclase in brain. Ann NY Acad Sci 1992;654:33–51.
8. Aronson J. When I use a word···: Sloe gin. BMJ 1997;314:1106.
9. Iversen LL. The Science of MarijuanaNew York: Oxford University Press;. 2000.
10. Braude MC. Toxicology of cannabinoids. In: Paton WM, Crown J, editors. Cannabis and its Derivatives. Oxford: Oxford University Press, 1972:89–99.
11. Chan PC, Sills RC, Braun AG, Haseman JK, Bucher JR. Toxicity and carcinogenicity of delta 9-tetrahydrocannabinol in Fischer rats and B6C3F1 mice. Fundam Appl Toxicol 1996;30(1):109–17.
12. Fleischman RW, Baker JR, Rosenkrantz H. Pulmonary pathologic changes in rats exposed to marihuana smoke for 1 year. Toxicol Appl Pharmacol 1979;47(3):557–66.
13. Roy PE, Magnan-Lapointe F, Huy ND, Boutet M. Chronic inhalation of marijuana and tobacco in dogs: pulmonary pathology. Res Commun Chem Pathol Pharmacol 1976;14(2):305–17.
14. Fligiel SE, Beals TF, Tashkin DP, Paule MG, Scallet AC, Ali SF, Bailey JR, Slikker W Jr. Marijuana exposure and pulmonary alterations in primates. Pharmacol Biochem Behav 1991;40(3):637–42.
15. Scallet AC, Uemura E, Andrews A, Ali SF, McMillan DE, Paule MG, Brown RM, Slikker W Jr. Morphometric studies of the rat hippocampus following chronic delta-9-tetrahydrocannabinol (THC). Brain Res 1987;436(1):193–8.
16. Landfield PW, Cadwallader LB, Vinsant S. Quantitative changes in hippocampal structure following long-term exposure to delta 9-tetrahydrocannabinol: possible mediation by glucocorticoid systems. Brain Res 1988;443(1–2):47–62.
17. Harper JW, Heath RG, Myers WA. Effects of *Cannabis sativa* on ultrastructure of the synapse in monkey brain. J Neurosci Res 1977;3(2):87–93.
18. Heath RG, Fitzjarrell AT, Fontana CJ, Garey RE. *Cannabis sativa*: effects on brain function and ultrastructure in rhesus monkeys. Biol Psychiatry 1980;15(5): 657–90.

19. Scallet AC. Neurotoxicology of cannabis and THC: a review of chronic exposure studies in animals. Pharmacol Biochem Behav 1991;40(3):671–6.

20. Downer E, Boland B, Fogarty M, Campbell V. Delta 9-tetrahydrocannabinol induces the apoptotic pathway in cultured cortical neurones via activation of the CB₁ receptor. Neuroreport 2001;12(18):3973–8.

21. Chan GC, Hinds TR, Impey S, Storm DR. Hippocampal neurotoxicity of Δ9-tetrahydrocannabinol. J Neurosci 1998;18(14):5322–32.

22. Sanchez C, Galve-Roperh I, Canova C, Brachet P, Guzman M. Δ9-tetrahydrocannabinol induces apoptosis in C6 glioma cells. FEBS Lett 1998;436(1):6–10.

23. Galve-Roperh I, Sanchez C, Cortes ML, del Pulgar TG, Izquierdo M, Guzman M. Anti-tumoral action of cannabinoids: involvement of sustained ceramide accumulation and extracellular signal-regulated kinase activation. Nat Med 2000;6(3):313–9.

24. Bifulco M, Laezza C, Portella G, Vitale M, Orlando P, De Petrocellis L, Di Marzo V. Control by the endogenous cannabinoid system of ras oncogene-dependent tumor growth. FASEB J 2001;15(14):2745–7.

25. Guzman M, Sanchez C, Galve-Roperh I. Control of the cell survival/death decision by cannabinoids. J Mol Med 2001;78(11):613–25.

26. Nagayama T, Sinor AD, Simon RP, Chen J, Graham SH, Jin K, Greenberg DA. Cannabinoids and neuroprotection in global and focal cerebral ischemia and in neuronal cultures. J Neurosci 1999;19(8):2987–95.

27. Panikashvili D, Simeonidou C, Ben-Shabat S, Hanus L, Breuer A, Mechoulam R, Shohami E. An endogenous cannabinoid (2-AG) is neuroprotective after brain injury. Nature 2001;413(6855):527–31.

28. van der Stelt M, Veldhuis WB, Bar PR, Veldink GA, Vliegenthart JF, Nicolay K. Neuroprotection by Δ9-tetrahydrocannabinol, the main active compound in marijuana, against ouabain-induced in vivo excitotoxicity. J Neurosci 2001;21(17):6475–9.

29. Shen M, Thayer SA. Cannabinoid receptor agonists protect cultured rat hippocampal neurons from excitotoxicity. Mol Pharmacol 1998;54(3):459–62.

30. Sinor AD, Irvin SM, Greenberg DA. Endocannabinoids protect cerebral cortical neurons from in vitro ischemia in rats. Neurosci Lett 2000;278(3):157–60.

31. Hampson AJ, Grimaldi M, Axelrod J, Wink D. Cannabidiol and (-)Δ9-tetrahydrocannabinol are neuroprotective antioxidants. Proc Natl Acad Sci USA 1998;95(14):8268–73.

32. Zimmer L, Morgan JP. Marijuana Myths, Marijuana FactsNew York: Lindesmith Centre;. 1997.

33. Pertwee RG. Tolerance to and dependence on psychotropic cannabinoids. In: Pratt J, editor. The Biological Basis of Drug Tolerance. London: Academic Press, 1991:232–65.

34. Aceto MD, Scates SM, Lowe JA, Martin BR. Dependence on delta 9-tetrahydrocannabinol: studies on precipitated and abrupt withdrawal. J Pharmacol Exp Ther 1996;278(3):1290–5.

35. Rubino T, Patrini G, Massi P, Fuzio D, Vigano D, Giagnoni G, Parolaro D. Cannabinoid-precipitated withdrawal: a time-course study of the behavioral aspect and its correlation with cannabinoid receptors and G protein expression. J Pharmacol Exp Ther 1998;285(2):813–9.

36. Rodriguez de Fonseca F, Carrera MR, Navarro M, Koob GF, Weiss F. Activation of corticotropin-releasing factor in the limbic system during cannabinoid withdrawal. Science 1997;276(5321):2050–4.

37. Diana M, Melis M, Muntoni AL, Gessa GL. Mesolimbic dopaminergic decline after cannabinoid withdrawal. Proc Natl Acad Sci USA 1998;95(17):10269–73.

38. Tanda G, Pontieri FE, Di Chiara G. Cannabinoid and heroin activation of mesolimbic dopamine transmission by a common mu1 opioid receptor mechanism. Science 1997;276(5321):2048–50.

39. Tanda G, Munzar P, Goldberg SR. Self-administration behavior is maintained by the psychoactive ingredient of marijuana in squirrel monkeys. Nat Neurosci 2000;3(11):1073–4.

40. Ledent C, Valverde O, Cossu G, Petitet F, Aubert JF, Beslot F, Bohme GA, Imperato A, Pedrazzini T, Roques BP, Vassart G, Fratta W, Parmentier M. Unresponsiveness to cannabinoids and reduced addictive effects of opiates in CB₁ receptor knockout mice. Science 1999;283(5400):401–4.

41. Lepore M, Vorel SR, Lowinson J, Gardner EL. Conditioned place preference induced by delta 9-tetrahydrocannabinol: comparison with cocaine, morphine, and food reward. Life Sci 1995;56(23–24):2073–80.

42. Manzanares J, Corchero J, Romero J, Fernandez-Ruiz JJ, Ramos JA, Fuentes JA. Pharmacological and biochemical interactions between opioids and cannabinoids. Trends Pharmacol Sci 1999;20(7):287–94.

43. Kaymakcalan S, Ayhan IH, Tulunay FC. Naloxone-induced or postwithdrawal abstinence signs in Δ9-tetrahydrocannabinol-tolerant rats. Psychopharmacology (Berl) 1977;55(3):243–9.

44. Hine B, Friedman E, Torrelio M, Gershon S. Morphine-dependent rats: blockade of precipitated abstinence by tetrahydrocannabinol. Science 1975;187(4175):443–5.

45. Yamaguchi T, Hagiwara Y, Tanaka H, Sugiura T, Waku K, Shoyama Y, Watanabe S, Yamamoto T. Endogenous cannabinoid, 2-arachidonoylglycerol, attenuates naloxone-precipitated withdrawal signs in morphine-dependent mice. Brain Res 2001;909(1–2):121–6.

46. Pontieri FE, Monnazzi P, Scontrini A, Buttarelli FR, Patacchioli FR. Behavioral sensitization to heroin by cannabinoid pretreatment in the rat. Eur J Pharmacol 2001;421(3):R1–3.

47. Valverde O, Maldonado R, Valjent E, Zimmer AM, Zimmer A. Cannabinoid withdrawal syndrome is reduced in pre-proenkephalin knock-out mice. J Neurosci 2000;20(24):9284–9.

48. Valverde O, Noble F, Beslot F, Dauge V, Fournie-Zaluski MC, Roques BP. Δ9-tetrahydrocannabinol releases and facilitates the effects of endogenous enkephalins: reduction in morphine withdrawal syndrome without change in rewarding effect. Eur J Neurosci 2001;13(9):1816–24.

49. Lichtman AH, Fisher J, Martin BR. Precipitated cannabinoid withdrawal is reversed by Δ(9)-tetrahydrocannabinol or clonidine. Pharmacol Biochem Behav 2001;69(1–2):181–8.

50. Lichtman AH, Sheikh SM, Loh HH, Martin BR. Opioid and cannabinoid modulation of precipitated withdrawal in Δ(9)-tetrahydrocannabinol and morphine-dependent mice. J Pharmacol Exp Ther 2001;298(3):1007–14.

51. US Department of Health and Human Services. National Household Survey on Drug Abuse, 1982–94. Computer Files (ICPSR Version)Ann Arbor, MI: Inter-University Consortium for Political Social;. 1999.

52. Morral AR, McCaffrey DF, Paddock SM. Reassessing the marijuana gateway effect. Addiction 2002;97(12): 1493–504.

53. Wehner FC, van Rensburg SJ, Thiel PG. Mutagenicity of marijuana and Transkei tobacco smoke condensates in the Salmonella/microsome assay. Mutat Res 1980;77(2): 135–42.

54. Leuchtenberger C, Leuchtenberger R. Cytological and cytochemical studies of the effects of fresh marihuana smoke on growth and DNA metabolism of animal and human lung cultures. In: Braude MC, Szara S, editors. The Pharmacology of Marijuana. New York: Raven Press, 1976:595–612.

55. Hall W, Solowij N. Adverse effects of cannabis. Lancet 1998;352(9140):1611–6.

56. Institute of Medicine. Marijuana and HealthWashington, DC: National Academy Press;. 1982.

57. Killestein J, Hoogervorst EL, Reif M, Kalkers NF, Van Loenen AC, Staats PG, Gorter RW, Uitdehaag BM, Polman CH. Safety, tolerability, and efficacy of orally administered cannabinoids in MS. Neurology 2002;58(9):1404–7.

58. Brady CM, DasGupta R, Dalton C, Wiseman OJ, Berkley KJ, Fowler CJ. An open-label pilot study of cannabis-based extracts for bladder dysfunction in advanced multiple sclerosis. Mult Scler 2004;10(4):425–33.

59. Clark AJ, Ware MA, Yazer E, Murray TJ, Lynch ME. Patterns of cannabis use among patients with multiple sclerosis. Neurology 2004;62(11):2098–100.

60. Wade DT, Makela P, Robson P, House H, Bateman C. Do cannabis-based medicinal extracts have general or specific effects on symptoms in multiple sclerosis? A double-blind, randomized, placebo-controlled study on 160 patients. Mult Scler 2004;10(4):434–41.

61. Svendsen KB, Jensen TS, Bach FW. Does the cannabinoid dronabinol reduce central pain in multiple sclerosis? Randomised double blind placebo controlled crossover trial. BMJ 2004;329(7460):253.

62. Notcutt W, Price M, Miller R, Newport S, Phillips C, Simmons S, Sansom C. Initial experiences with medicinal extracts of cannabis for chronic pain: results from 34 'N of 1' studies. Anaesthesia 2004;59(5):440–52.

63. Avakian EV, Horvath SM, Michael ED, Jacobs S. Effect of marihuana on cardiorespiratory responses to submaximal exercise. Clin Pharmacol Ther 1979;26(6):777–81.

64. Relman AS. Marijuana and health. N Engl J Med 1982;306(10):603–5.

65. Mathew RJ, Wilson WH, Davis R. Postural syncope after marijuana: a transcranial Doppler study of the hemodynamics. Pharmacol Biochem Behav 2003;75:309–18.

66. Ducasse E, Chevalier J, Dasnoy D, Speziale F, Fiorani P, Puppinck P. Popliteal artery entrapment associated with cannabis arteritis. Eur J Vasc Endovasc Surg 2004;27(3):327–32.

67. Mittleman MA, Lewis RA, Maclure M, Sherwood JB, Muller JE. Triggering myocardial infarction by marijuana. Circulation 2001;103(23):2805–9.

68. Papp E, Czopf L, Habon T, Halmosi R, Horvath B, Marton Z, Tahin T, Komocsi A, Horvath I, Melegh B, Toth K. Drug-induced myocardial infarction in young patients. Int J Cardiol 2005;98:169–70.

69. Lindsay AC, Foale RA, Warren O, Henry JA. Cannabis as a precipitant of cardiovascular emergencies. Int J Cardiol 2005;104:230–2.

70. Rezkalla SH, Sharma P, Kloner RA. Coronary no-flow and ventricular tachycardia associated with habitual marijuana use. Ann Emerg Med 2003;42:365–9.

71. Kosior DA, Filipiak KJ, Stolarz P, Opolski G. Paroxysmal atrial fibrillation following marijuana intoxication: a two-case report of possible association. Int J Cardiol 2001;78(2):183–4.

72. Singh GK. Atrial fibrillation associated with marijuana use. Pediatr Cardiol 2000;21(3):284.

73. Fisher BAC, Ghuran A, Vadamalai V, Antonios TF. Cardiovascular complications induced by cannabis smoking: a case report and review of the literature. Emerg Med J 2005;22:679–80.

74. Schneider HJ, Jha S, Burnand KG. Progressive arteritis associated with cannabis use. Eur J Vasc Endovasc Surg 1999;18(4):366–7.

75. Wu TC, Tashkin DP, Djahed B, Rose JE. Pulmonary hazards of smoking marijuana as compared with tobacco. N Engl J Med 1988;318(6):347–51.

76. Tashkin DP, Calvarese BM, Simmons MS, Shapiro BJ. Respiratory status of seventy-four habitual marijuana smokers. Chest 1980;78(5):699–706.

77. Tashkin DP. Pulmonary complications of smoked substance abuse. West J Med 1990;152(5):525–30.

78. Douglass RE, Levison MA. Pneumothorax in drug abusers. An urban epidemic? Am Surg 1986;52(7):377–80.

79. Tashkin DP, Coulson AH, Clark VA, Simmons M, Bourque LB, Duann S, Spivey GH, Gong H. Respiratory symptoms and lung function in habitual heavy smokers of marijuana alone, smokers of marijuana and tobacco, smokers of tobacco alone, and nonsmokers. Am Rev Respir Dis 1987;135(1):209–16.

80. Goodyear K, Laws D, Turner J. Bilateral spontaneous pneumothorax in a cannabis smoker. J Roy Soc Med 2004;97:435–6.

81. Gill A. Bong lung: regular smokers of cannabis show relatively distinctive histologic changes that predispose to pneumothorax. Am J Surg Pathol 2005;29:980–1.

82. Johnson MK, Smith RP, Morrison D, Laszlo G, White RJ. Large lung bullae in marijuana smokers. Thorax 2000;55(4):340–2.

83. Thompson CS, White RJ. Lung bullae and marijuana. Thorax 2002;57(6):563.

84. Lozsadi DA, Forster A, Fletcher NA. Cannabis-induced propriospinal myoclonus. Mov Disord 2004;19(6):708–9.

85. Finsterer J, Christian P, Wolfgang K. Occipital stroke shortly after cannabis consumption. Clin Neurol Neurosurg 2004;106(4):305–8.

86. Shukla PC, Moore UB. Marijuana-induced transient global amnesia. South Med J 2004;97(8):782–4.

87. Ramaekers JG, Berghaus G, van Laar M, Drummer OH. Dose related risk of motor vehicle crashes after cannabis use. Drug Alcohol Depend 2004;73:109–19.

88. Voruganti LN, Slomka P, Zabel P, Mattar A, Awad AG. Cannabis induced dopamine release: an in-vivo SPECT study. Psychiatry Res 2001;107(3):173–7.

89. Altable CR, Urrutia AR, Martinez MIC. Cannabis-induced extrapyramidalism in a patient on neuroleptic treatment. J Clin Psychopharmacol 2005;25:91–2.

90. Arsenault L, Cannon M, Witton J. Causal association between cannabis and psychosis: examination of the evidence. Br J Psychiatry 2004;184:110–7.

91. Mateo I, Pinedo A, Gomez-Beldarrain M, Basterretxea JM, Garcia-Monco JC. Recurrent stroke associated with cannabis use. J Neurol Neurosurg Psychiatry 2005;76:435–7.

92. Haubrich, Diehl R, Donges M, Schiefer J, Loos M, Kosinski C. Recurrent transient ischemic attacks in a cannabis smoker. J Neurol 2005;252:369–70.

93. Dawson WW, Jimenez-Antillon CF, Perez JM, Zeskind JA. Marijuana and vision—after ten years' use in Costa Rica. Invest Ophthalmol Vis Sci 1977;16(8):689–99.

94. Trittibach P, Frueh BE, Goldblum D. Bilateral angle-closure glaucoma after combined consumption of "ecstasy" and marijuana. Am J Emerg Med 2005;23:813–4.

95. Mulheran M, Middleton P, Henry JA. The acute effects of tetrahydrocannabinol on auditory threshold and frequency resolution in human subjects. Hum Exp Toxicol 2002;21(6):289–92.

96. Troisi A, Pasini A, Saracco M, Spalletta G. Psychiatric symptoms in male cannabis users not using other illicit drugs. Addiction 1998;93(4):487–92.

97. Linszen DH, Dingemans PM, Lenior ME. Cannabis abuse and the course of recent-onset schizophrenic disorders. Arch Gen Psychiatry 1994;51(4):273–9.

98. Andreasson S, Allebeck P, Engstrom A, Rydberg U. Cannabis and schizophrenia. A longitudinal study of Swedish conscripts. Lancet 1987;2(8574):1483–6.

99. Jockers-Scherubl MC, Danker-Hopfe H, Mahlberg R, Selig F, Rentzsch J, Schurer F, Lang UE, Hellweg R. Brain-derived neurotrophic factor serum concentrations are increased in drug-naive schizophrenic patients with chronic cannabis abuse and multiple substance abuse. Neurosci Lett 2004;371(1):79–83.

100. D'Souza DC, Abi-Saab WM, Madonick S, Forselius-Bielen K, Doersch A, Braley G, Gueorguieva R, Cooper TB, Krystal JH. Delta-9-tetrahydrocannabinol effects in schizophrenia: implications for cognition, psychosis and addiction. Biol Psychiatry 2005;57:594–608.

101. Bowers MB Jr. Family history and early psychotogenic response to marijuana. J Clin Psychiatry 1998;59(4):198–9.

102. Crowley TJ, Macdonald MJ, Whitmore EA, Mikulich SK. Cannabis dependence, withdrawal, and reinforcing effects among adolescents with conduct symptoms and substance use disorders. Drug Alcohol Depend 1998;50(1): 27–37.

103. Solowij N. Cannabis and Cognitive Functioning Cambridge: Cambridge University Press;. 1998.

104. Schwartz RH, Gruenewald PJ, Klitzner M, Fedio P. Short-term memory impairment in cannabis-dependent adolescents. Am J Dis Child 1989;143(10):1214–9.

105. Lane SD, Cherek DR, Lieving LM, Tcheremissine OV. Marijuana effects on human forgetting functions. J Exp Analysis Behav 2005;83:67–83.

106. Struve FA, Straumanis JJ, Patrick G, Leavitt J, Manno JE, Manno BR. Topographic quantitative EEG sequelae of chronic marihuana use: a replication using medically and psychiatrically screened normal subjects. Drug Alcohol Depend 1999;56:167–79.

107. Herning RI, Better W, Tate K, Cadet JL. EEG deficits in chronic marijuana abusers during monitored abstinence: preliminary findings. Ann N Y Acad Sci 2003;993:75–8;discussion 79-81.

108. Ilan AB, Smith ME, Gevins A. Effects of marijuana on neurophysiological signals of working and episodic memory. Psychopharmacology(Berl) 2004;176(2):214–22.

109. Semple DM, Ramsden F, McIntosh AM. Reduced binocular depth inversion in regular cannabis users. Pharmacol Biochem Behav 2003;75:789–93.

110. Lyons MJ, Bar JL, Panizzon MS, Toomey R, Eisen S, Xian H, Tsuang MT. Neuropsychological consequences of regular marijuana use: a twin study. Psychol Med 2004;34(7):1239–50.

111. Fried PA, Watkinson B, Gray R. Neurocognitive consequences of marihuana—comparison with pre-drug performance. Neurotoxicol Teratol 2005;27:231–9.

112. Volkow ND, Gillespie H, Mullani N, Tancredi L, Grant C, Valentine A, Hollister L. Brain glucose metabolism in chronic marijuana users at baseline and during marijuana intoxication. Psychiatry Res 1996;67(1):29–38.

113. Russo E, Mathre ML, Byrne A, Velin R, Bach PJ, Sanchez-Ramos J, Kirlin KA. Chronic cannabis use in the compassionate investigational new drug program: an examination of benefits and adverse effects of legal clinical cannabis. J Cannabis Ther 2002;2:3–57.

114. Ashton CH. Adverse effects of cannabis and cannabinoids. Br J Anaesth 1999;83(4):637–49.

115. Stracciari A, Guarino M, Crespi C, Pazzaglia P. Transient amnesia triggered by acute marijuana intoxication. Eur J Neurol 1999;6(4):521–3.

116. Lane SD, Cherek DR, Tcheremissine OV, Lieving LM, Pietras CJ. Acute marijuana effects on human risk taking. Neuropsychopharmacology 2005;30:800–9.

117. Kolodny RC, Masters WH, Kolodner RM, Toro G. Depression of plasma testosterone levels after chronic intensive marihuana use. N Engl J Med 1974;290(16):872–4.

118. Fried PA, Watkinson B, Willan A. Marijuana use during pregnancy and decreased length of gestation. Am J Obstet Gynecol 1984;150(1):23–7.

119. Allen JH, de Moore GM, Heddle R, Twartz JC. Cannabinoid hyperemesis: cyclical hyperemesis in association with chronic cannabis abuse. Gut 2004;53:1566–70.

120. Klein TW, Newton C, Friedman H. Inhibition of natural killer cell function by marijuana components. J Toxicol Environ Health 1987;20(4):321–32.

121. Pillai R, Nair BS, Watson RR. AIDS, drugs of abuse and the immune system: a complex immunotoxicological network. Arch Toxicol 1991;65(8):609–17.

122. Levitz SM, Diamond RD. Aspergillosis and marijuana. Ann Intern Med 1991;115(7):578–9.

123. Killestein J, Hoogervorst EL, Reif M, Blauw B, Smits M, Uitdehaag BM, Nagelkerken L, Polman CH. Immunomodulatory effects of orally administered cannabinoids in multiple sclerosis. J Neuroimmunol 2003;137:140–3.

124. Huang YM, Liu X, Steffensen K, Sanna A, Arru G, Fois ML, Rosati G, Sotgiu S, Link H. Immunological heterogeneity of multiple sclerosis in Sardinia and Sweden. Mult Scler 2005;11:16–23.

125. Van Boxel-Dezaire AH, Hoff SC, Van Ooosten BW, Verweij CL, Drager AM, Ader HJ, Van Houwelingen JC, Barkhof F, Polman CH. Nagelkerken Decreased interleukin-10 and increased interleukin-12p40 mRNA are associated with disease activity and characterize different disease stages in multiple sclerosis. Ann Neurol 1999;45:695–703.

126. Klein TW, Friedman H, Specter S. Marijuana, immunity and infection. J Neuroimmunol 1998;83(1–2):102–15.

127. Perez JA Jr. Allergic reaction associated with intravenous marijuana use. J Emerg Med 2000;18(2):260–1.

128. Jones RT. Cannabis and health. Annu Rev Med 1983;34:247–58.

129. Carney MW, Bacelle L, Robinson B. Psychosis after cannabis abuse. BMJ (Clin Res Ed) 1984;288(6423):1047.

130. Liakos A, Boulougouris JC, Stefanis C. Psychophysiologic effects of acute cannabis smoking in long-term users. Ann NY Acad Sci 1976;282:375–86.

131. Hall W, Christie M, Currow D. Cannabinoids and cancer: causation, remediation and palliation. Lancet Oncol 2005;6:35–42.

132. Fried PA. Marihuana use by pregnant women: neurobehavioral effects in neonates. Drug Alcohol Depend 1980;6(6):415–24.

133. Abel EL. Prenatal exposure to cannabis: a critical review of effects on growth, development, and behavior. Behav Neural Biol 1980;29(2):137–56.

134. Qazi QH, Mariano E, Milman DH, Beller E, Crombleholme W. Abnormalities in offspring associated

with prenatal marihuana exposure. Dev Pharmacol Ther 1985;8(2):141–8.

135. Fried PA. Marihuana use by pregnant women and effects on offspring: an update. Neurobehav Toxicol Teratol 1982;4(4):451–4.

136. Greenland S, Staisch KJ, Brown N, Gross SJ. Effects of marijuana on human pregnancy, labor, and delivery. Neurobehav Toxicol Teratol 1982;4(4):447–50.

137. Nahas G, Frick HC. Developmental effects of cannabis. Neurotoxicology 1986;7(2):381–95.

138. Fried PA, Watkinson B. Visuoperceptual functioning differs in 9- to 12-year olds prenatally exposed to cigarettes and marihuana. Neurotoxicol Teratol 2000;22(1):11–20.

139. Goldschmidt L, Day NL, Richardson GA. Effects of prenatal marijuana exposure on child behavior problems at age 10. Neurotoxicol Teratol 2000;22(3):325–36.

140. Fried PA, Watkinson B, Gray R. Differential effects on cognitive functioning in 13- to 16-year-olds prenatally exposed to cigarettes and marihuana. Neurotoxicol Teratol 2003;25:427–36.

141. Hurd YL, Wang X, Anderson V, Beck O, Minkoff H, Dow-Edwards D. Marijuana impairs growth in mid-gestation fetuses. Neurotoxicol Teratol 2005;27:221–9.

142. Wert RC, Raulin ML. The chronic cerebral effects of cannabis use. II. Psychological findings and conclusions. Int J Addict 1986;21(6):629–42.

143. Wert RC, Raulin ML. The chronic cerebral effects of cannabis use. I. Methodological issues and neurological findings. Int J Addict 1986;21(6):605–28.

144. Abrams DI, Hilton JF, Leiser RJ, Shade SB, Elbeik TA, Aweeka FT, Benowitz NL, Bredt BM, Kosel B, Aberg JA, Deeks SG, Mitchell TF, Mulligan K, Bacchetti P, McCune JM, Schambelan M. Short-term effects of cannabinoids in patients with HIV-1 infection: a randomized, placebo-controlled clinical trial. Ann Intern Med 2003;139:258–66.

145. Benowitz NL, Nguyen TL, Jones RT, Herning RI, Bachman J. Metabolic and psychophysiologic studies of cannabidiol-hexobarbital interaction. Clin Pharmacol Ther 1980;28:115–20.

146. Benowitz NL, Jones RT. Effects of delta-9-tetrahydrocannabinol on drug distribution and metabolism. Antipyrine, pentobarbital, and ethanol. Clin Pharmacol Ther 1977;22:259–68.

147. Aronow WS, Cassidy J. Effect of marihuana and placebo-marihuana smoking on angina pectoris. N Engl J Med 1974;291(2):65–7.

148. Lacoursiere RB, Swatek R. Adverse interaction between disulfiram and marijuana: a case report. Am J Psychiatry 1983;140(2):243–4.

149. Watson SJ. Hallucinogens and other psychotomimetics: biological mechanisms. In: Barchas JD, Berger PA, Cioranello RD, Elliot GR, editors. Psychopharmacology from Theory to Practise. New York: Oxford University Press, 1977:1.

150. Strassman RJ. Adverse reactions to psychedelic drugs. A review of the literature. J Nerv Ment Dis 1984;172(10):577–95.

151. McLeod AL, McKenna CJ, Northridge DB. Myocardial infarction following the combined recreational use of Viagra and cannabis. Clin Cardiol 2002;25(3):133–4.

Cardiac glycosides

Numerous plants worldwide contain cardiac glycosides that have been used both therapeutically as herbal formulations and for the purposes of self-poisoning. Details are given in Table 61.

- A 59-year-old man developed third-degree atrioventricular block after using an extract of *Nerium oleander* transdermally to treat psoriasis (1). A fatality due to drinking a herbal tea prepared from *N. oleander* leaves, erroneously believed to be eucalyptus leaves, has been reported (2).

Poisoning from ingestion of the seeds of *Thevetia peruviana* (yellow oleander) can be treated with oral multiple-dose activated charcoal (3).

In a Turkish case, the ingestion of two bulbs or *Urginea maritima* as a folk remedy for arthritic pains was sufficient to result in fatal poisoning (4).

There has been a report of coronary vasoconstriction in patients who were given acetyldigoxin 0.8 mg intravenously at angiography (5). Pretreatment with nisoldipine 10 mg, 2 hours before angiography, prevented the digoxin-induced vasoconstriction. These patients all had pre-existing coronary artery disease, but the vasoconstrictor effect occurred in both normal and abnormal coronary segments. However, the effect on high-grade stenoses was more pronounced. There is other evidence that ischemic damage can occur in patients who have been given digoxin intravenously, including an increase in the activity of serum creatinine kinase (6) and impaired left ventricular function after acute myocardial infarction (7).

- A 26-year-old woman who had taken a herbal supplement for stress relief which contained *Scutellaria lateriflora*, *Pedicularis canadensis*, *Cimifuga racemosa*, *Humulus lupulus*, *Valeriana officinalis*, and *Capsicum annuum* developed chest pain of 7 hours duration (8). Her medical history was otherwise unremarkable. Examination of her heart showed no abnormality, but during monitoring her heart rate fell to 39/minute and her blood pressure to 59/36 mmHg. Her serum digoxin concentration was 0.9 ng/ml. The authors therefore concluded that the herbal remedy contained digoxin-like factors that had caused digitalis toxicity.

References

1. Wojtyna W, Enseleit F. A rare cause of complete heart block after transdermal botanical treatment for psoriasis. Pacing Clin Electrophysiol 2004;27(12):1686–8.
2. Haynes BE, Bessen HA, Wightman WD. Oleander tea: herbal draught of death. Ann Emerg Med 1985;14(4):350–3.
3. de Silva HA, Fonseka MM, Pathmeswaran A, Alahakone DG, Ratnatilake GA, Gunatilake SB, Ranasinha CD, Lalloo DG, Aronson JK, de Silva HJ. Multiple-dose activated charcoal for treatment of yellow oleander poisoning: a single-blind, randomised, placebo-controlled trial. Lancet 2003;361(9373):1935–8.
4. Tuncok Y, Kozan O, Cavdar C, Guven H, Fowler J. Urginea maritima (squill) toxicity. J Toxicol Clin Toxicol 1995;33(1):83–6.
5. Nolte CW, Jost S, Mugge A, Daniel WG. Protection from digoxin-induced coronary vasoconstriction in patients with coronary artery disease by calcium antagonists. Am J Cardiol 1999;83(3):440–2.
6. Varonkov Y, Shell WE, Smirnov V, Gukovsky D, Chazov EI. Augmentation of serum CPK activity by digitalis in patients with acute myocardial infarction. Circulation 1977;55(5):719–27.
7. Balcon R, Hoy J, Sowton E. Haemodynamic effects of rapid digitalization following acute myocardial infarction. Br Heart J 1968;30(3):373–6.
8. Scheinost ME. Digoxin toxicity in a 26-year-old woman taking a herbal dietary supplement. J Am Osteopath Assoc 2001;101(8):444–6.

Table 61 Some plants that contain cardiac glycosides

Plant	Common name(s)	Cardiac glycoside(s)	Comments
Adonis vernalis	False hellebore, pheasant's eye	Adonitoxin, strophanthidin	
Antiaris toxicaria	Upas tree	Antiarin	A Javanese tree of the mulberry family, used as an arrow poison
Convallaria majalis	Lily of the valley	Convallamarin	
Erysimum helveticum	Wallflower	Helveticoside	
Helleborus niger	Black hellebore, Christmas rose	Helleborcin	Also called melampodium after the Greek physician Melampus, who used it as a purgative
Nerium oleander	Pink oleander	Neriifolin	
Periploca graeca	Silk vine	Periplocin	
Tanghinia venenifera	Ordeal tree	Tanghinin	At one time used in Madagascar to test the guilt of someone suspected of a crime
Thevetia peruviana	Yellow oleander	Peruvoside, thevetins	Seeds widely used for self-poisoning in Southern India and Sri Lanka
Urginea (Scilla) maritima	Squill	Proscillaridin	Squill was a common remedy for dropsy in ancient times and up to the 19th century

Coumarin

General Information

The plant lactone coumarin (not to be confused with coumarin anticoagulants) is a constituent of some plants, including:

- *Alyxia lucida*
- *Dalea tuberculata*
- *Dipteryx odorata*
- *Dipteryx oppositofolia*
- *Levisticum officinale*
- *Melilotus officinalis*
- *Scabiosa comosa.*

Organs and Systems

Hematologic

Coumarin is devoid of anticoagulant activity, but the molding of sweet clover can give it hemorrhagic potential by transforming coumarin to the anticoagulant dicoumarol. This transformation could explain a case of abnormal clotting function and mild bleeding after the drinking of an herbal tea prepared from tonka beans, sweet clover, and several other ingredients. Unfortunately, this possibility was not investigated, as the reporting physician was not aware of the exact phytochemistry and pharmacology of coumarin-yielding plants.

Liver

Coumarin has hepatotoxic potential in man, when taken in daily doses of 25–100 mg (1). Bile-duct carcinomas have been reported to occur in rats fed coumarin, but the correctness of this diagnosis has been seriously criticized.

Reference

1. Cox D, O'Kennedy R, Thornes RD. The rarity of liver toxicity in patients treated with coumarin (1,2-benzopyrone). Hum Toxicol 1989;8(6):501–6.

Ephedrine and pseudoephedrine

See also Ephedraceae

Ephedrine is a sympathomimetic drug prepared from members of the *Ephedra* family, such as *Ephedra vulgaris*, also called the sea grape. It is available as ordinary pharmaceutical formulations of ephedrine, but also in herbal formulations of *Ephedra*, including a Chinese herbal medicine called Ma huang, which has been used since ancient times as a stimulant and in the treatment of asthma, and in formulations combined with caffeine. It is used to treat asthma, nasal congestion, fever, obesity, and anhidrosis. It is also abused as a recreational drug. Of 140 reports of adverse events related to *Ephedra* supplements submitted to the FDA between June 1997 and March 1999, 31% were definitely or probably causally related to *Ephedra* (1). In 47% of the cases, there were cardiovascular symptoms and in 18% central nervous system effects; 10 patients died.

Both ephedrine and pseudoephedrine remain worryingly popular (2) and are widely available without prescription. Based on increasing evidence of the risks of *Ephedra* self-medication, various national regulatory authorities are currently considering recalling *Ephedra* products from over-the-counter sales. Oral doses of ephedrine 25–30 mg are often prescribed, for example for orthostatic hypotension. Lower oral doses, present in some "cold remedies" in tablet form, are unlikely to be efficacious, although they are risky where the drug is contraindicated, for example in patients with cardiac disease. Ephedrine used in nasal sprays and drops can have systemic effects in the doses normally used. Such products are widely available over the counter in many countries.

Pseudoephedrine is a stereoisomer of ephedrine, in which two of three chiral centers are different.

"Herbal ecstasy" is a term used for many different herbal formulations, none of which contains ecstasy. Some of the names for these herbs (which can be sold in stores) include "The Bomb," "Reds," and "Sublime." In New Zealand analysis of "The Bomb" showed substantial amounts of ephedrine and the Ministry of Health removed it from the market. Some symptoms associated with herbal ecstasy include headache, dizziness, palpitation, tachycardia, and raised blood pressure. Thus, in countries where the term "herbal ecstasy" is commonly used, it is important that those who see patients who have taken Herbal Ecstasy should not confuse it with ecstasy, as toxicity and medical management may be quite different (3).

Organs and Systems

Ephedrine is longer-acting than adrenaline and can exert any type of adrenergic effect, including metabolic changes and dysuria. It has a somewhat smaller effect on the cardiovascular system than adrenaline but a relatively larger effect on the central nervous system; even low doses can, in sensitive patients, cause tremor, insomnia, and anxiety, and such problems are much more marked if ephedrine is given together with caffeine, as it sometimes is for appetite control (SEDA-17, 161). In children, ephedrine sometimes paradoxically induces sedation. Psychosis has resulted from excessive self-medication (4).

The use of ephedrine alone or ephedrine plus caffeine is associated with an increased risk of psychiatric, autonomic, gastrointestinal, and cardiovascular symptoms (5). In Texas some 500 cases of adverse reactions were reported between 1993 and 1995 (SEDA-21, 153). Three of these cases have been highlighted because of their serious nature: one patient died from myocardial infarction, one had a series of generalized absence seizures, and a third had a fatal subarachnoid hemorrhage as the result of taking an unsuitable dose (SEDA-21, 153).

Cardiovascular

In subjects with impaired baroreflex function, oral pseudoephedrine 30 mg or phenylpropanolamine 12.5 mg and 25 mg produced significant increases in blood pressure. When they were taken with water the increase in blood pressure was greater. The maximal increase in systolic blood pressure occurred after 60 minutes and the pressure returned to baseline by 2 hours (6).

Acute myocardial infarction has been attributed to pseudoephedrine (7).

- A 19-year-old male cigarette smoker with normal coronary arteriography had an upper respiratory infection, for which he brought Gripex®, each tablet of which contains paracetamol 325 mg, pseudoephedrine HCl 30 mg, and dextromethorphan HBr 10 mg. He took four tablets twice, about 3 hours apart, a total of 240 mg of pseudoephedrine within 3 hours. He had an acute myocardial infarction 12 hours later. Other drugs, such as cocaine and amphetamines, were excluded. Subsequent coronary angiography and echocardiography showed a non-Q-wave myocardial infarction with normal coronary arteries.

About 5% of patients with a myocardial infarction have normal coronary arteries at angiography. In many cases there is coronary artery spasm and/or thrombosis, perhaps with underlying endothelial dysfunction of the epicardial arteries. It is important in such cases to obtain a complete history of the use of drugs, including over-the-counter drugs.

In a meta-analysis of the effects of *Ephedra* or ephedrine-containing products compared with control, the odds ratio of palpitation was 2.29 (95% CI = 1.27, 4.32) (8).

In a meta-analysis of 24 studies involving 1285 subjects the authors concluded that in healthy individuals pseudoephedrine significant increased heart rate (2.83/minute; CI = 2.0, 3.6) and increased systolic blood pressure (0.99 mmHg; 95% CI = 0.08, 1.9), with no effect on diastolic pressure (9). In patients with controlled hypertension there was an increase of similar magnitude in systolic blood pressure (1.2 mmHg; CI = 0.56, 1.84). Higher doses and immediate-release formulations were associated with greater increases in blood pressure, although it was difficult to be certain what the average dose was in these studies; from the data presented it was usually 60 mg once or twice daily. The authors reported isolated cases of much greater rises in blood pressure (up to 20 mmHg systolic) but again it was difficult to determine the doses used. They concluded that immediate-release formulations had a greater effect than modified-release formulations, and that a dose-response relation could be discerned. Evidently many cases of pseudoephedrine-related cardiovascular effects involved higher doses than have been used in formal studies or are recommended by manufacturers. There appeared to be a smaller effect in women than in men. Shorter duration of use was associated with greater increases in systolic and diastolic blood pressures. There were no clinically significant adverse outcomes. However, a rare adverse event may not be seen with such a small sample size.

Coronary vasospasm has been attributed to pseudoephedrine (10).

- A previously healthy 32-year-old man from Nigeria developed substernal chest pain at rest associated with nausea and sweating 45 minutes after taking two tablets of an over-the-counter cold remedy containing pseudoephedrine 30 mg and paracetamol 500 mg per tablet. He recalled a similar but less severe episode 1 week earlier after taking the same medication. There was no relevant past medical history or family history of coronary artery disease. An electrocardiogram showed ST elevation in the inferolateral leads and the plasma creatine kinase activity and troponin I concentration were both raised. Coronary angiography showed normal arteries.

The authors concluded that this episode had been caused by coronary vasospasm initiated by pseudoephedrine and warned of the dangers of this type of medication, even in otherwise healthy individuals. The temporal association between ingestion of pseudoephedrine and the myocardial infarction suggested a causal relation. The absence of coronary artery disease at catheterization combined with the cardiac magnetic resonance imaging findings were consistent with an acute myocardial infarction caused by vasospasm due to pseudoephedrine.

Hypertensive strokes have been attributed to pseudoephedrine in 22 patients (11c). The effects of pseudoephedrine may be important when considered on a population basis, given their widespread use as decongestants. Although marked rises in blood pressure were uncommon, there were rises above 140/90 mmHg in nearly 3% of the patients. The benefit to harm balance should therefore be evaluated carefully before pseudoephedrine is used in individual patients most at risk of rises in blood pressure and heart rate.

Pseudoephedrine is a component of some non-prescription basal decongestants given by mouth; even quite ordinary doses of such products (for example 120 mg) can cause a hypertensive reaction in sensitive subjects, such as those with a pheochromocytoma and those with at least a family history of hypertension (SEDA-17, 162).

- A 21-year-old man presented with hypertension (blood pressure 220/110 mmHg) and ventricular dysrhythmias after taking four capsules of herbal ecstasy (12). He was treated with lidocaine and sodium nitroprusside and his symptoms resolved within 9 hours.

Severe hypertension has been attributed to pseudoephedrine abuse (13).

- A 36-year-old man with hypertension taking no less than seven antihypertensive drugs had outpatient systolic pressures of over 190 mmHg. Investigations for primary causes of hypertension were negative and there was increasing suspicion of treatment non-compliance or factitious hypertension. Urine screening showed the presence of pseudoephedrine, which the patient could not explain. When he was given his normal antihypertensive drugs under close supervision his systolic blood pressure fell to 70 mmHg and his serum creatinine doubled. His blood pressure became normal when his medication was briefly suspended but he continued to deny any deliberate attempt to alter his blood pressure and discharged himself soon afterwards.

The authors concluded that this represented factitious hypertension due to pseudoephedrine, the first such case reported and a very unusual example of Munchausen's syndrome.

Nervous system

The hazard associated with use of over-the-counter nutritional supplements containing ephedrine by athletes has been illustrated (14).

- A 24-year-old, previously healthy, male university athlete developed a severe right-sided headache and collapsed with left-sided weakness while sprinting (15). Earlier that morning he had taken at least five tablets of Xenadrine, each of which contained ephedrine 34 mg, pseudoephedrine 24 mg, methylephedrine 6 mg, and caffeine 363 mg. He was not taking any other medications and had no history of trauma. His blood pressure was 137/87 mmHg and his heart rate 58/minute. Imaging showed moderate vasospasm in the right middle cerebral artery. He was treated with milrinone and heparin, and after 6 days was discharged to a rehabilitation center taking prophylactic aspirin.

The largest manufacturer of *Ephedra*-containing supplements is Metabolife. The firm has repeatedly insisted that it was not aware of adverse events associated with its products, and claimed that they were "absolutely safe" (16). An investigation by the US Justice Department into the truth of these statements showed that between 1997 and 2002 the company had received over 13 000 reports of suspected adverse drug reactions; they included nearly 2000 reports of significant adverse reactions.

- A 33-year-old man had a stroke while taking a herbal remedy (Thermadrene) containing *Ephedra* alkaloids and guaraná for "boosting his energy" (17). He was immediately treated with alteplase and was referred for rehabilitation 9 days later. At follow-up 5 months later he still had a minor neurological deficit.

Skin

Acute generalized exanthematous pustulosis has been attributed to pseudoephedrine (18).

- A 42-year-old Spanish woman took Vincigrip®, which contains paracetamol, pseudoephedrine, and chlorphenamine, for an upper respiratory tract infection. Ten hours after the first dose she developed fever of over 38°C and a pustular erythematous rash on the face, trunk, and proximal limbs. Skin biopsy showed sterile pustules with papillary edema and focal keratinocyte necrosis, compatible with a diagnosis of acute generalized exanthematous pustulosis. Immunohistochemistry and reverse transcription polymerase chain reaction (PCR), a lymphocyte proliferation assay, and patch testing confirmed the specificity of the reaction due to pseudoephedrine. Immunocytochemistry showed a mononuclear infiltrate consisting of activated memory T cells in addition to polymorphonuclear cells. PCR showed increased expression of IL-8 in the affected skin. She was treated with an oral glucocorticoid, and recovered completely within 10 days.

The authors noted that there had been only report of a similar case, but that this unusual drug reaction is attributed to drugs in 90% of cases, penicillins being most frequently implicated. A previous report of this unusual reaction due to pseudoephedrine was confirmed with a positive patch test (19).

Toxic epidermal necrolysis has been reported in a patient taking pseudoephedrine (20).

- A 57-year-old woman developed a pruritic generalized maculopapular rash with occasional target lesions 8 days after taking two doses of pseudoephedrine, 240 mg in total. She also had oropharyngeal ulceration. Over the next week she developed bullae covering about 15% of the body, which subsided over the next 2 weeks with supportive therapy. Biopsy was consistent with toxic epidermal necrolysis. Three years later she took a single dose of a compound medication containing pseudoephedrine and developed a rash 12 hours later. This was successfully treated with glucocorticoids, topical and systemic.

Drug administration route

Inadvertent epidural injection of ephedrine has been described in a 17-year-old woman in labor (21).

- Ephedrine 50 mg in 10 ml was inadvertently administered via an epidural catheter (instead of the intended ropivacaine and fentanyl) in a 17-year-old nulliparous woman in labor. The error was immediately recognized, and further administration was stopped. For the next 2 hours, her neurological status, blood pressure, heart rate, and temperature and fetal heart rate were monitored. She was comfortable and had no complaints during or after the bolus. The level of analgesia regressed and there was full recovery of sensation. Vital signs in both the mother and fetus remained stable throughout, and there were no changes in uterine contraction. However, because labor did not progress, cesarian delivery was performed, without complications.

Perhaps surprisingly, there were no untoward effects on the mother or baby. A commentary after the report noted that in other similar cases adverse reactions did occur, including hypertension, dysrhythmias, and central nervous system stimulant effects.

Drug overdose

Inadvertent overdose of ephedrine has been reported from Japan (22).

- A 22-year-old man undergoing tonsillectomy was inadvertently given ephedrine 200 mg intravenously instead of neostigmine and atropine, which was intended to reverse residual neuromuscular blockade. His blood pressure rose to 265/165 mmHg with a heart rate of 140/minute. He was immediately given intravenous propranolol 0.4 mg and sublingual nifedipine 5 mg. The hypertension and tachycardia gradually subsided and after 5 hours blood pressure was 120/80 mm Hg with a pulse rate of 80/minute. There were no long term sequelae.

This may be the highest recorded dose of this drug given inadvertently, and so the authors could not refer to previous literature on the management of such a case. They suggested that labetalol may be a preferable remedy.

Drug-drug interactions

Midazolam

Ephedrine 70 micrograms/kg was given 1 minute before induction, and midazolam 7.5 mg, 30-60 minutes before induction (n = 30) (23). Anesthesia was induced with fentanyl 1 micrograms/kg and sodium thiopental 3.5 mg/kg. The patients were intubated with rocuronium 0.9 mg/kg, and the intubation time, determined by Dixon's up and down method, was significantly shorter than in controls. There were no significant hemodynamic changes.

Neuromuscular blocking drugs

The effect of ephedrine on the onset time of neuromuscular blockers has been evaluated in 25 patients who were given ephedrine 70 micrograms/kg 3 minutes before induction with propofol plus remifentanil (24). The onset time of suxamethonium was significantly shortened. Tthere were no significant hemodynamic changes.

References

1. Haller CA, Benowitz NL. Adverse cardiovascular and central nervous system events associated with dietary supplements containing ephedra alkaloids. N Engl J Med 2000;343(25):1833–8.

2. Samenuk D, Link MS, Homoud MK, Contreras R, Theoharides TC, Wang PJ, Estes NA 3rd. Adverse cardiovascular events temporally associated with ma huang, an herbal source of ephedrine. Mayo Clin Proc 2002;77(1):12–6.

3. Yates KM, O'Connor A, Horsley CA. "Herbal Ecstasy": a case series of adverse reactions. NZ Med J 2000;113(1114): 315–7.

4. Herridge CF, a'Brook MF. Ephedrine psychosis. BMJ 1968;2(598):160.

5. Shekelle PG, Hardy ML, Morton SC, Maglione M, Mojica, Suttorp MJ, Rhodes SL, Jungvig L, Gagne J. Efficacy and safety of ephedra and ephedrine for weight loss and athletic performance. A meta-analysis. J Am Med Assoc 2003;289:1537–45.

6. Jordan J, Shannon JR, Diedrich A, Black B, Robertson D, Biaggioni B. Water potentiates the pressor effect of Ephedra alkaloids. Circulation 2004;109:1823–5.

7. Grzesk G, Polak G, Grabczewska Z, Kubica J. Myocardial infarction with normal coronary arteriogram: the role of ephedrine-like alkaloids. Med Sci Monit 2004;10:CS15–21.

8. Shekelle PG, Hardy ML, Morton HC, Maglione M, Mojica WA, Suttorp MJ, Rhodes SL, Jungvig L, Gagne J. Efficacy and safety of Ephedra and ephedrine for weight loss and athletic performance: a meta-analysis. JAMA 2003;289:1537–45.

9. Salerno SM, Jackson JL, Berbano EP. Effect of oral pseudoephedrine on blood pressure and heart rate. A meta-analysis. Arch Intern Med 2005;165:1686–94.

10. Manini AF, Kabrehl C, Thomsen TW. Acute myocardial infarction after over-the-counter use of pseudoephedrine. Ann Emerg Med 2005;45:213–6.

11. Cantu C, Arauz A, Murillo-Bonilla LM, Lopez M, Barinagarrementeria F. Stroke associated with sympathomimetics contained in over-the-counter cough and cold drugs. Stroke 2003;34:1667–72.

12. Zahn KA, Li RL, Purssell RA. Cardiovascular toxicity after ingestion of "herbal ecstacy". J Emerg Med 1999;17(2):289–91.

13. Jacobs KM, Hirsch KA. Psychiatric complications of Mahuang. Psychosomatics 2000;41(1):58–62.

14. Foxford RJ, Sahlas DJ, Wingfied KA. Vasospasm-induced stroke in a varsity athlete secondary to ephedrine ingestion. J Sports Med 2003;13:183–5.

15. Haller CA, Benowitz NL. Adverse cardiovascular and central nervous system effects associated with dietary supplements containing Ephedra alkaloids. New Engl J Med 2000;343:1833–38.

16. Durbin RJ, Waxman HA, Davis SA. Adverse event reports from Metabolife. Oct 2002 House Committee on Government Reform.

17. Kaberi-Otarod J, Conetta R, Kundo KK, Farkash A. Ischemic stroke in a user of thermadrene: a case study in alternative medicine. Clin Pharmacol Ther 2002;72(3): 343–6.

18. Padia MA, Alvarez-Ferreira J, Tapia B, Blanco R, Mañas C, Blanca M, Bellón T. Acute generalized exanthematous pustulosis associated with pseudoephedrine. Br J Dermatol 2004;150:139–42.

19. Assier-Bonnet H, Viguier M, Dubertret L, Revuz J, Roujeau JC. Severe adverse drug reactions due to pseudoephedrine from over-the-counter medications. Contact Dermatitis 2002;47:165–92.

20. Nagge JJ, Knowles SR, Juurlink DN, Shear NH. Pseudoephedrine-induced toxic epidermal necrolysis. Arch Dermatol 2005;141:907–8.

21. Sidi A. Inadvertent epidural injection of ephedrine in labor. J Clin Anesth 2004;16:74–6.

22. Sakuragi T, Yasumoto M, Higa K, Nitahara K. Inadvertent intravenous administration of a high dose of ephedrine. J Cardiothorac Vasc Anesth 2004;18:121.

23. Ittichaikulthol W, Sriswadi S, Nual-On S, Hongpuang S, Sornil A. The effect of ephedrine on the onset time of rocuronium in Thai patients. J Med Assoc Thai 2004;87:264–9.

24. Ganidagli S, Cengiz M, Baysal Z. Effect of ephedrine on the onset time of succinylcholine. Acta Anaesth Scand 2004;48:1306–9.

Haematitium

Haematitium, a component of some herbal preparations, contains zinc, tin, iron, magnesium, and diferric acid.

Liver

Haematitium has been implicated in liver damage.

- A 4-year-old boy was given a herbal mixture containing 20 g of burnt clay, ginger, licorice, mandarin skin, Chinese date, *Inula britannica*, bitter orange, *Codonopsis* root, and haematitium 10 ml tds for 3 days for vomiting (1). The vomiting stopped for a few days but then recurred. He had jaundice and cervical lymphadenopathy. Investigations showed the pattern of acute hepatitis. Liver biopsy showed a severe acute hepatitis with portal-to-portal bridging necrosis and a significant number of eosinophils, raising the possibility of drug-induced hepatitis. He had signs of increasing liver dysfunction, with a worsening coagulopathy and an encephalopathy after 10 days. He underwent orthoptic liver transplant 19 days after the first onset of jaundice.

The authors reported a probable association between the hepatic failure and the herbal preparation, based on the World Health Organization definition of causality assessment.

Reference

1. Webb AN, Hardikar W, Cranswick NE, Somers GR. Probable herbal medication induced fulminant hepatic failure. J Paediatr Child Health 2005;41(9–10):530–1.

HMG coenzyme-A reductase inhibitors

Rhabdomyolysis in a stable renal transplant recipient was attributed to the presence of red yeast rice (*Monascus purpureus*) in a herbal mixture (1). The condition resolved when he stopped taking the product. Rice fermented with red yeast contains several types of mevinic acids, including monacolin-K, which is identical to lovastatin. The authors postulated that the interaction of ciclosporin with these compounds through cytochrome P450 had resulted in the adverse effect. Transplant recipients must be cautioned against using herbal products to lower their lipid concentrations, in order to prevent such complications.

Reference

1. Prasad GV, Wong T, Meliton G, Bhaloo S. Rhabdomyolysis due to red yeast rice (*Monascus purpureus*) in a renal transplant recipient. Transplantation 2002;74(8):1200–1.

Ipecacuanha, emetine, and dehydroemetine

See also Rubiaceae

General Information

Ipecacuanha is an extract of the root of *Psychotria ipecacuanha*, also known as *Cephaelis ipecacuanha*, a member of the Rubiaceae. It contains the emetic alkaloids cephaeline and emetine. It has often been used as a home remedy for various purposes, and not only as an emetic. It is a traditional ingredient of some expectorants, since expectoration often accompanies vomiting. Misuse of ipecacuanha by patients with anorexia nervosa and bulimia has resulted in severe myopathy, lethargy, erythema, dysphagia, cardiotoxicity, and even death. Use in infancy generally seems safe.

Emetine, once the drug of choice for the treatment of amebiasis, despite marked cardiotoxicity, has largely been replaced by metronidazole and related compounds for this indication. Large doses of emetine can damage the heart, liver, kidneys, intestinal tract, and skeletal muscle. Allergic reactions and tumor-inducing effects have not been described. Dehydroemetine is a little less toxic but also less effective than emetine; its adverse effects are similar (SED-11, 594).

Gastrointestinal decontamination in acute toxic ingestion has been reviewed (1).

- Although ipecac generally seems to have a good safety profile, it can be associated with protracted vomiting. Other reported adverse effects include drowsiness, agitation, abdominal cramps, diarrhea, aspiration pneumonia, cerebral hemorrhage, pneumoperitoneum, and pneumomediastinum. Its use is not currently recommended (2).

Gastric lavage can be useful in some patients who have taken life-threatening doses of highly toxic substances. However, toxin absorption can be enhanced by gastric lavage. Reported adverse effects mainly include laryngospasm, hypoxemia, aspiration pneumonia, bradycardia, electrocardiographic ST segment elevation, and rarely mechanical injury to the gastrointestinal tract.

- Activated charcoal has gained popularity as a first choice for gut decontamination, based on its efficacy and relative lack of adverse effects. Poor patient acceptance is a disadvantage. Frequent vomiting can rarely become a problem.
- Saline laxatives (magnesium citrate, magnesium sulfate, sodium sulfate, and disodium phosphate) or saccharide laxatives (sorbitol, mannitol, lactulose) are also used in poisoned patients. Common adverse effects are abdominal cramps, excessive diarrhea, and abdominal distension. Dehydration and electrolyte imbalance in children, and hypermagnesemia and magnesium toxicity (with magnesium-based cathartics) have also been reported.

- Whole bowel irrigation to wash the entire gastrointestinal tract rapidly and mechanically is similar to the methods used by gastroenterologists to prepare patients for colonic investigation or bowel surgery. It is safe, even in children, pregnant women, and patients with cardiac or respiratory failure. Polyethylene glycol isotonic electrolyte solution is commonly used. Complications are usually minor and include nausea, vomiting, abdominal distension and cramps, and anal irritation.

Organs and Systems

Cardiovascular

Cardiotoxicity is the most serious and dangerous adverse effect of emetine. The clinical signs are tachycardia, dysrhythmias, and hypotension. Deaths have been described. Electrocardiographic abnormalities occur in 60–70% of cases; increased T wave amplitude, prolongation of the PR interval, ST segment depression, and T wave changes are all common. It seems possible that emetine influences the cell permeability of sodium and calcium ions, and this could be the basis of its effect on cardiac automaticity and contractility and on the electrocardiogram (SED-11, 594). The symptoms of emetine toxicity suggest that an effect on intracellular magnesium concentrations could be another possible explanation, but there are no data to support this hypothesis (SED-11, 594).

Respiratory

Asthma can occasionally be induced by ipecacuanha; when the compound was more widely used in medicine this was a familiar problem for those compounding medicines (3).

Gastrointestinal

Nausea and vomiting are frequent, perhaps in as many as one-third of all cases. In about half the cases, diarrhea is induced or existing diarrhea aggravated. Melena can occur, but this seems unlikely to be drug-induced (SED-11, 594).

Skin

Dermatitis has been attributed to emetine (4), as has cellulitis at the site of injection (5).

Musculoskeletal

Complaints of weakness, tenderness, and stiffness of skeletal muscles, especially in the neck and shoulder, are common. Following emetine aversion therapy for alcohol abuse, muscle weakness and pathological changes in muscle biopsy specimens have been described (SED-11, 594).

A reversible myopathy secondary to abuse of ipecacuanha by individuals with eating disorders has been noted (SED-12, 945) (SEDA-17, 421); the active alkaloid may have been responsible.

Long-Term Effects

Drug abuse

Munchausen's syndrome by proxy involving syrup of ipecacuanha has been reported in an 18-month-old child who was brought by his mother with persistent vomiting for 4 weeks with generalized myopathy and pneumonia (6). Its over-the-counter availability, low cost, and effective emetic properties give this drug a high appeal for such abuse.

Drug Administration

Drug overdose

There is some suggestion of cardiac impairment with overdosage, presumably also myopathic. Forced emesis can lead to esophageal damage or even complete rupture; pneumomediastinum and pneumoperitoneum have therefore sometimes complicated induced emesis (SEDA-10, 326). A bizarre case of neonatal vomiting, irritability, and hypothermia was attributed to the mother having added ipecacuanha surreptitiously to her baby's feed; at any age, however, emetine in excess can be very irritant to the gastrointestinal tract, resulting, for example, in bloody diarrhea.

Drug–Drug Interactions

Activated charcoal

Activated charcoal, sometimes used to treat self-poisoning, binds the active ingredients of ipecacuanha and inactivates them, at least partly (7). The administration of ipecacuanha also delays the administration of charcoal (8). They should not be co-administered.

References

1. Lheureux P, Askenasi R, Paciorkowski F. Gastrointestinal decontamination in acute toxic ingestions. Acta Gastroenterol Belg 1998;61(4):461–7.
2. Anonymous. Position paper: ipecac syrup. J Toxicol Clin Toxicol 2004;42(2):133–43.
3. Persson CG. Ipecacuanha asthma: more lessons. Thorax 1991;46(6):467–8.
4. Schwank R, Jirasek L. Kožni přecitlivelost na emetin. [Skin sensitivity to emetine.] Cesk Dermatol 1952;27(1–2):50–6.
5. Anonymous. Drugs for parasitic infections. Med Lett Drugs Ther 1986;28(706):17.
6. Cooper C, Kilham H, Ryan M. Ipecac—a substance of abuse. Med J Aust 1998;168(2):94–5.
7. Krenzelok EP, Freedman GE, Pasternak S. Preserving the emetic effect of syrup of ipecac with concurrent activated charcoal administration: a preliminary study. J Toxicol Clin Toxicol 1986;24(2):159–66.
8. Kornberg AE, Dolgin J. Pediatric ingestions: charcoal alone versus ipecac and charcoal. Ann Emerg Med 1991;20(6):648–51.

Limonene

General information

Since the 1980s limonene has been used as an alternative to traditional solvents and degreasing agents. It is a monoterpene found in a variety of fruits, vegetables, herbs, trees, and bushes. Limonene constitutes the main volatile component of some citrus oils and represents a small fraction of the terpenes in pine needle oils and turpentine. Biogenic and anthropogenic sources release large amounts of limonene into the atmosphere, and biogenic emissions may meet or exceed anthropogenic sources (1).

Limonene is used as a substitute for chlorinated hydrocarbons, chlorofluorocarbons, and other solvents. In concentrations of 30–100%, it is used as a metal degreaser before painting, as a paint solvent, and as a cleaner in the printing and electronic industries. It is also used in numerous other products, such as home cleaning products, disinfectants, industrial hand cleansers, odorants, rubber, wax and paint strippers, adhesives, lubricating oils, essential oils, cosmetics, fire extinguishers, and wetting and dispersing agents. For more that 50 years, it has been used as a fragrance and flavoring additive to perfumes, soaps, food, and beverages. Medicinal uses include chemopreventive and chemotherapeutic properties (2)(3).

General adverse effects

Although data are limited, limonene, linoleic acid, and oleic acid are probably of low toxicity. Human dietary and in vitro studies of linoleic and oleic acids have shown alterations in erythrocyte membranes, platelet activity, sperm motility, and thyroxine binding; however, these findings have not been extended to potential hazards in exposed workers. Until more data are available, conclusions regarding occupational health effects remain difficult to determine.

Human exposure to these products is common and almost certainly occurs during daily food intake. Medical management of exposure is symptomatic and supportive. Mild skin irritation can occur, and oxidative products of limonene can produce skin sensitization. Patients with skin or ocular exposure should be decontaminated with water or saline respectively. Treatment of dermal effects involves removal and symptomatic skin care, such as antihistamines for allergic effects. An ophthalmologist should see patients with ocular complaints that do not resolve with irrigation or who have abnormal ocular findings on examination (4).

Organs and Systems

Respiratory

Limonene, and possibly linoleic and oleic acids, can have irritative and bronchconstrictive airway effects. Patients with significant inhalational exposure should be removed from the environment and undergo appropriate decontamination. Inhaled beta2 adrenoceptor agonists can be used for bronchoconstriction.

Urinary tract

Limonene ingested in sufficient quantity can cause proteinuria. However, nephropathy and renal tumors are not expected in humans.

Nicotine and nicotine replacement therapy

Drug overdose

Although 15–39% of the population smoke nicotine-containing products, acute nicotine toxicity is rare and can be difficult to recognize; it can produce rapid and dramatic toxicity (5).

- A 15-year-old boy, who was in generally good health, had a cardiopulmonary arrest with a generalized seizure at home, while resting on a couch, and stopped breathing (6). On arrival in hospital, he was unresponsive, with a Glasgow Coma Scale score of 4, a rectal temperature of 33.8oC, a heart rate of 157/minute, a blood pressure of 170/108 mmHg, and a respiratory rate of 6/minute. The breath sounds were clear but the cardiac rhythm was irregular, with normal heart sounds and normal capillary refill. He had midline fixed pupils, the right slightly larger than the left, with weak corneal reflexes. There was no gag reflex. He was globally hypotonic, with decerebrate posturing to painful stimuli and extensor plantar responses. He had intermittent stiffening of the limbs and arching of the back, with spontaneous resolution, and diaphragmatic spasm. MRI scanning of the brain showed severe hypoxic-ischemic encephalopathy, with multiple areas of cortical and basal ganglia infarction. Despite aggressive intensive care support, he improved minimally and remained dependent on gastric tube feedings and full-time care providers. He had a history of depression and was known to smoke cigarettes and marijuana. In his attic there was a bottle that was nearly empty and had contained an insecticide with the trade name Black Leaf 40, with nicotine sulfate 40% as the active ingredient.

A presumptive diagnosis of nicotine ingestion was made. Laboratory analysis of residual liquid in the bottle showed that it contained nicotine 357 mg/ml.

References

1. IPCS. Concise International Chemical Assessment Document No 5. Limonene. Geneva: WHO, 1998.
2. Crowell PL, Elson CE, Bailey HH, Elegbede A, Haag JD, Could MN. Human metabolism of the experimental cancer therapeutic agent d-limonene. Cancer Chemother Pharmacol 1994;35:31–7.
3. Crowell PL, Could MN. Chemoprevention and therapy of cancer by d-limonene. Crit Rev Oncog 1994;5:1–22.
4. De Witt C, Bebarta V. Botanical solvents. Clin Occup Environ Med 2004;4:445–54.
5. Beenman JA, Hunter WC. Fatal nicotine poisoning. A report of 24 cases. Arch Pathol 1937;24:481–5.
6. Rogers AJ, Denk LD, Wax PM. Catastrophic brain injury after nicotine insecticide ingestion. J Emerg Med 2004;26:169–72.

Nicotine and nicotine replacement therapy

General Information

The introduction of nicotine replacement systems and agents is a significant advance in the treatment of nicotine addiction. Nicotine replacement reduces the severity of nicotine withdrawal symptoms during smoking abstinence and improves smoking cessation rates. During cigarette smoking, nicotine is rapidly absorbed, producing peak blood concentrations of about 15 ng/ml within the first few minutes, while a 21-mg transdermal nicotine patch produces peak blood concentrations of about 18 ng/ml 5–6 hours after application. Thus, nicotine concentrations are markedly increased when the patch and smoking are combined especially within the early minutes of smoking.

Comparative studies

In a meta-analysis of 42 nicotine chewing gum studies and nine patch studies, comprising 17 000 subjects, the odds ratios of long-term success were 1.61 for gum and 2.07 for patch compared with placebo (1). However, effective use of nicotine gum requires careful instructions. Transdermal uptake of nicotine from patches is about 1 mg/hour (2). Large multicenter studies using 16-hour and 24-hour nicotine patches in primary care have shown double the success rate of placebo (3).

The main objectives of the Collaborative European Anti-Smoking Evaluation (CEASE) were to examine whether long-term success rates (that is complete abstinence sustained for 1 year) could be increased by using higher than standard doses of transdermal nicotine, and/or by prolonging the patch treatment period (4). There were four cases of myocardial infarction during the study period, which comprised 950 treatment years and 1700 study years. One 59-year-old man had a myocardial infarct during treatment with the 25 mg nicotine patch, and one 48-year-old man and one 50-year-old man had myocardial infarcts after discontinuing treatment with 15 mg; one 66-year-old man who took placebo had a myocardial infarct. This compared with the 10.2 cases expected during the study period, based on calculations using data from the Framingham study (5). The overall incidence of adverse events was low, and they were generally transient. In order to examine possible dose-related adverse events, the occurrence of events during the first 8 weeks of treatment in each of the three groups (25 mg, 15 mg, and placebo) was analysed, and nausea and vomiting were the only reported symptoms, with a higher frequency in the 25 mg group (7.3%) than the 15 mg group (5.4%).

In a multicenter, community-based, prospective, longitudinal study of the safety of nicotine replacement therapy (NRT; n = 370), bupropion (n = 413), and the combination (n = 121) for smokers seeking to quit, adverse effects were reported by 3.8%, 33%, 22%, and 5.7% of subjects at 15, 30, 60, and 90 days respectively (6).

Adverse effects were significantly more common in those who used the combination therapy or bupropion alone than in those who used NRT alone. A total of 83 smokers (9.3%) withdrew from treatment and 116 (13%) stopped temporarily because of adverse effects. There were no differences in the percentages of withdrawals among the different treatment options. Adverse effects were rarely severe (n = 10). Nevertheless, 41 subjects (4.5%) discontinued drug therapy indefinitely and 55 (6.1%) discontinued temporarily because of mild adverse effects. Pharmacological therapies for smoking cessation are safe as long as they are appropriately prescribed and supervised by clinicians according to clinical practice guidelines. Adverse effects are mostly mild, but may be unacceptable to patients.

General adverse effects

The pharmacokinetics of nicotine transdermal systems have been documented in several studies. In one study, a number of users had headache and transient itching at the site of application (7). In another study in 37 subjects, three reported itching and headache, and one had allergic reactions to the patch, which disappeared after changing the site of application (8). The subjects in a third study had headache (4%), nausea (4%), and vertigo (4%) slightly more often than the placebo group (9). Skin reactions such as mild itching under the patch, lasting for 15–30 minutes, and erythema and acute eczema, persisting for several days after removal, also occurred. Skin reactions necessitating withdrawal of the patches occurred in 10% in another study (10). In the same report sleep disturbances of varying severity was a commonly reported adverse effect.

The effect of transdermal nicotine on concomitant medications has been reported in an open non-comparative study of 80 smokers (42 men and 38 women) who used 24-hour nicotine patches (11). Adverse effects included itching and local erythema, insomnia, and abnormal taste sensations. Two subjects withdrew because of adverse effects. At 12 weeks, 17 were non-smokers. At 26 weeks, 19 were non-smokers and a further 14 had reduced their use of cigarettes significantly.

Nicotine gum used for smoking cessation in 114 smokers caused sore mouth, hiccups, and abdominal pain and vomiting (12). One subject withdrew from the study because of vomiting. Hot flushes have also been reported (13).

Nicotine (15–25 mg/day) has also been used to treat active ulcerative colitis. The most common adverse effects were nausea, light-headedness, headache, sleep disturbance, dizziness, and skin irritation (14,15).

Organs and Systems

Cardiovascular

Nicotine can aggravate cardiovascular disease through the hemodynamic consequences of sympathetic neural stimulation, or systemic release of catecholamines, or both (16). Although several cardiovascular events have been ascribed to nicotine patches, including cerebral arterial narrowing with severe headaches and transient neurological deficits (17), relatively few adverse effects have been reported (16). Suicide attempts by means of overdosage using nicotine patches, presumably with the intention of precipitating myocardial infarction, have failed (18), and it would seem that the risk of nicotine replacement therapy is likely to be much less than that of cigarette smoking, even in patients with coronary heart disease (16).

Nicotine can provoke angina in patients with coronary artery disease and can also trigger myocardial infarction. It has been hypothesized that activities associated with increased catecholamine concentrations can both trigger plaque rupture and promote an occlusive thrombus formation that can result in myocardial infarction or sudden cardiac death. By evoking the release of catecholamines (stimulating platelet aggregation and increasing platelet/vessel wall interactions), nicotine could therefore be a potent trigger of myocardial infarction. However, a randomized, double-blind, placebo-controlled trial in 584 patients with at least one diagnosis of cardiovascular disease clearly showed that transdermal nicotine does not cause a significant increase in cardiovascular events in high-risk patients with cardiovascular disease. On the other hand, the efficacy of transdermal nicotine as an aid to smoking cessation in such patients has proved to be very limited and may not be sustained over time (19). In another study (20) from the Pharmacovigilance Division of the Federal Ministry for Health and Consumer Products of Austria, 41 cases of myocardial infarction were associated with the use of nicotine patches.

Atrial fibrillation was described in association with normal or large doses (21,22), suggesting that accumulation can occur when nicotine gum is taken daily.

Respiratory

In one study, 27% of the participants reported coughing, and 15% had irritation of the mouth or throat after 3–5 days (23). At 3 weeks 7% reported coughing and 10% mouth and throat irritation. In another study, 18 healthy subjects inhaled nicotine for 20 minutes (80 inhalations) every hour for 10 hours at three different environmental temperatures (20, 30, and 40°C) (24). Adverse events were twice as frequent at 40°C as at 20°C. Most often reported were coughing, a lump in the throat, and a sore or irritated throat. Belching, hiccups, pressure over the chest, headache, and heartburn were reported occasionally. About one-fifth of the subjects experienced coughing after the first 5–10 inhalations.

Nervous system

In 53 subjects, two- and three-dimensional analyses of nicotine-induced eye movements were performed to evaluate whether they were primarily of vestibular or oculomotor origin (25). Nicotine-induced nystagmus was detected in 27 subjects (51%). These findings suggest that nicotine causes an imbalance in the vestibulo-ocular reflex.

Psychiatric

A woman developed hallucinations and cerebral arterial narrowing after using nicotine patches (17). In another case delusions and hallucinations occurred when a woman who had been using nicotine to try to stop smoking took a cigarette (26).

Gastrointestinal

Smoking is negatively associated with ulcerative colitis and positively associated with Crohn's disease; it is the most striking and firmly established epidemiological factor associated with the two conditions. The first association of non-smoking with ulcerative colitis was made over 20 years ago (27), and was followed within 2 years by the observation that patients with Crohn's disease were more often smokers (28). Smoking also has opposite effects on the clinical course of the two conditions, with possible benefit in ulcerative colitis (29) and a detrimental effect in Crohn's disease (30). Nicotine is probably the principal active ingredient in smoking responsible for these associations.

In a pilot study of oral nicotine in the treatment of primary sclerosing cholangitis in eight patients, only five completed 1 year of treatment; two had no adverse effects, but three had to temporarily reduce the dose of nicotine because of nausea, dizziness, insomnia, or lightheadedness, but later resumed the full dose (31). One patient completed only 4 months of treatment because of dizziness and bouts of palpitation, even at the lowest dose, requiring permanent discontinuation of nicotine. One patient with ulcerative colitis, who was completely asymptomatic at the time of entry, had insomnia and watery diarrhea shortly after starting oral nicotine; however, his ulcerative colitis later responded to transdermal nicotine. One patient had moderately active Crohn's colitis and bloody diarrhea treated with corticosteroid enemas and mesalazine; his bloody diarrhea increased shortly after starting oral nicotine, requiring drug withdrawal and an increased frequency of corticosteroid enemas; nicotine was not restarted.

Skin

Local skin reactions to transdermal nicotine patches have been reported (32). Other local adverse effects, including erythema with edema, erosions, and systemic skin reactions have also been described (33).

Skin tolerability in patients with a history of eczema, psoriasis, or other skin disorders has been studied in 1481 participants (34). The adverse effects reported were erythema, rash, pruritus, irritation, edema at the site of application, musculoskeletal pain related to the application site, dreaming, and other sleep disturbances.

Musculoskeletal

The acetylcholine receptors involved in myasthenia gravis are nicotinergic, and in myasthenia gravis they are reduced in number at the neuromuscular junction, which can lead to functional overdosage after the use of nicotine, similar to a cholinergic crisis. This is comparable to the myasthenic syndromes that were described during the Second World War in tobacco chewers without any muscle impairment.

- Myasthenia gravis was worsened by a nicotine transdermal system in a man who used to smoke 40 cigarettes a day without an effect on his myasthenia (35). Two hours after applying a nicotine transdermal system he developed increased bilateral ptosis, total ophthalmoplegia, difficulty in chewing, and generalized weakness. The symptoms improved after the patch was removed.

Immunologic

Leukocytoclastic vasculitis has been ascribed to nicotine patches in two patients (36). Two other patients developed vasculitis in association with the use of nicotine patches (37). The authors concluded that it was likely that the reactions in the two patients were related to the nicotine therapy. However, the possibility of a reaction to the vehicle (coconut oil and polymers) could not be excluded.

Susceptibility Factors

Anecdotal reports of paroxysmal atrial fibrillation, myocardial infarction (38,39), dysrhythmias (40), and stroke in patients taking nicotine have made physicians wary of prescribing nicotine gum or transdermal patches for patients with cardiac disease.

Occlusive dressings covering transdermal patches can increase the risk of adverse effects (41).

- A 40-year-old man with renal carcinoma and extensive lung metastases who had smoked three packs per day for 25 years, had two 21 mg transdermal nicotine patches placed on a clean hairless area of his body. Before he took a shower the next afternoon, the nursing staff covered the patches with a polyvinylidrine chloride occlusive dressing (Saranwrap). During the hot shower he felt nauseated, light-headed, and tremorous, and nearly lost consciousness. Removal of the patches and occlusive dressing alleviated the adverse effects.

The adverse effects experienced by this patient could have been due to the use of an occlusive dressing and exposure to a high temperature, in combination with a high dose of nicotine. Absorption of nicotine through the skin is greater at high temperatures, because of increased blood flow to the skin. A high ambient temperature (77–84°C) significantly increases dermal absorption and plasma concentrations of nicotine in the first hour of heat exposure, starting 5 hours after application of nicotine patches (42).

Drug Administration

Drug administration route

Nicotine has also been administered in a nasal spray. In one study, sneezing, irritation in the throat, and headache were reported after application of nasal spray (43).

However, in a larger study (255 subjects, half of whom had been assigned to nicotine nasal spray and the other half to placebo), there were nine cases of sinusitis (five nicotine and four placebo) and two of nose bleeding and occasional spots of blood from the nostrils (one in each group) (44). Two other studies reported irritated throat, hiccups, heartburn, bad taste sensation, dry mouth, and dizziness, nasal irritation, runny nose, watery eyes, sneezing, and coughing as the most frequently reported adverse effects (45,46).

Drug–Drug Interactions

General

The therapeutic effects of many drugs are less predictable in cigarette smokers, since smoking can potentially alter the pharmacokinetics of a drug by reducing its absorption, increasing its plasma clearance, or inducing cytochrome P450 isozymes. Enzyme induction in smokers is probably due to polycyclic aromatic hydrocarbons, as they are potent inducers of cytochrome P450 (47).

Antimalarial drugs

Although rigorous data on the metabolism of antimalarial drugs are lacking, it appears from in vitro and in vivo studies that chloroquine is metabolized by cytochrome P450 (48). It is therefore conceivable that cigarette smoking alters the hepatic metabolism of antimalarial drugs.

Antimalarial drugs (chloroquine, hydroxychloroquine, or quinacrine) were given to 36 patients with cutaneous lupus, 17 smokers and 19 non-smokers (49). The median number of cigarettes smoked was one pack/day, with a median duration of 12.5 years. There was a reduction in the efficacy of antimalarial therapy in the smokers. Patients with cutaneous lupus should therefore be encouraged to stop smoking and consideration may be given to increasing the doses of antimalarial drugs in smokers with refractory cutaneous lupus before starting a cytotoxic agent.

References

1. Silagy C, Mant D, Fowler G, Lancaster T. The effect of nicotine replacement on smoking cessation. In: Lancaster T, Silagy S, Fullerton D, editors. Tobacco Addiction Module of the Cochrane Database of Systematic Reviews. Oxford: The Cochrane Collaboration, 1997:1–8.
2. Palmer KJ, Buckley MM, Faulds D. Transdermal nicotine. A review of its pharmacodynamic and pharmacokinetic properties, and therapeutic efficacy as an aid to smoking cessation. Drugs 1992;44(3):498–529.
3. Imperial Cancer Research Fund General Practice Research Group. Randomized trial of nicotine patches in general practice: results at one year. BMJ 1994;308:1476–7.
4. Tonnesen P, Paoletti P, Gustavsson G, Russell MA, Saracci R, Gulsvik A, Rijcken B, Sawe U. Higher dosage nicotine patches increase one-year smoking cessation rates: results from the European CEASE trial. Collaborative

European Anti-Smoking Evaluation. European Respiratory Society. Eur Respir J 1999;13(2):238–46.

5. US Department of Health Education and Welfare. Public Health Service. NIH. National Heart and Lung Institute. The Framingham Study. An epidemiological investigation of cardiovascular disease. DHEW publication No (NIH) 76–1083, April 1976.

6. Sicari R, Palinkas A, Pasanisi EG, Venneri L, Picano E. Adverse effects of pharmacological therapy for nicotine addiction in smokers following a smoking cessation program. Nicotine Tobacco Res 2005;7:2136–41.

7. Gupta SK, Okerholm RA, Eller M, Wei G, Rolf CN, Gorsline J. Comparison of the pharmacokinetics of two nicotine transdermal systems: Nicoderm and Habitrol. J Clin Pharmacol 1995;35(5):493–8.

8. Saenghirunvattana S. Trial of transdermal nicotine patch in smoking cessation. J Med Assoc Thai 1995;78(9):466–8.

9. Tonnesen P, Norregaard J, Simonsen K, Sawe U. A double-blind trial of a 16-hour transdermal nicotine patch in smoking cessation. N Engl J Med 1991;325(5):311–5.

10. Mant D, Fowler G. Effectiveness of a nicotine patch in helping people stop smoking: results of a randomised trial in general practice. Imperial Cancer Research Fund General Practice Research Group. BMJ 1993;306(6888): 1304–8.

11. Martin PD, Robinson GM. The safety, tolerability and efficacy of transdermal nicotine (Nicotinell TTS) in initially hospitalised patients. NZ Med J 1995;108(992):6–8.

12. Jimenez-Ruiz CA, Florez S, Solano S, Ramos A, Ramos L, Fomies E. Nicotine gum (2 mg) plus counselling as treatment for smokers in a cessation clinic setting J Smok Relat Disord 1994;5:171–5.

13. Sudan BJ. Comment: niacin, nicotine, and flushing. Ann Pharmacother 1994;28(9):1113.

14. Pullan RD, Rhodes J, Ganesh S, Mani V, Morris JS, Williams GT, Newcombe RG, Russell MA, Feyerabend C, Thomas GA, Sawe U. Transdermal nicotine for active ulcerative colitis. N Engl J Med 1994;330(12):811–5.

15. Thomas GA, Rhodes J, Mani V, Williams GT, Newcombe RG, Russell MA, Feyerabend C. Transdermal nicotine as maintenance therapy for ulcerative colitis. N Engl J Med 1995;332(15):988–92.

16. Benowitz NL. Nicotine patches. BMJ 1995;310(6991):1409–10.

17. Jackson M. Cerebral arterial narrowing with nicotine patch. Lancet 1993;342(8865):236–7.

18. Engel CJ, Parmentier AH. Suicide attempts and the nicotine patch. JAMA 1993;270(3):323–4.

19. Joseph AM, Norman SM, Ferry LH, Prochazka AV, Westman EC, Steele BG, Sherman SE, Cleveland M, Antonnucio DO, Hartman N, McGovern PG. The safety of transdermal nicotine as an aid to smoking cessation in patients with cardiac disease. N Engl J Med 1996;335(24):1792–8.

20. Anonymous. Nicotine patches and chewing gum-cardiac risks with concomitant smoking. WHO Newslett 1996;5,6:6.

21. Stewart PM, Catterall JR. Chronic nicotine ingestion and atrial fibrillation. Br Heart J 1985;54(2):222–3.

22. Ottervanger JP, Stricker BH, Klomps HC. Transdermal nicotine: clarifications, side effects, and funding. JAMA 1993;269(15):1940–1.

23. Hjalmarson A, Nilsson F, Sjostrom L, Wiklund O. The nicotine inhaler in smoking cessation. Arch Intern Med 1997;157(15):1721–8.

24. Lunell E, Molander L, Andersson SB. Temperature dependency of the release and bioavailability of nicotine from a nicotine vapour inhaler; in vitro/in vivo correlation. Eur J Clin Pharmacol 1997;52(6):495–500.

25. Pereira CB, Strupp M, Eggert T, Straube A, Brandt T. Nicotine-induced nystagmus: three-dimensional analysis and dependence on head position. Neurology 2000;55(10):1563–6.

26. Foulds J, Toone B. A case of nicotine psychosis? Addiction 1995;90(3):435–7.

27. Harries AD, Baird A, Rhodes J. Non-smoking: a feature of ulcerative colitis. BMJ (Clin Res Ed) 1982;284(6317):706.

28. Somerville KW, Logan RF, Edmond M, Langman MJ. Smoking and Crohn's disease. BMJ (Clin Res Ed) 1984;289(6450):954–6.

29. Green JT, Rhodes J, Ragunath K, Thomas GA, Williams GT, Mani V, Feyerabend C, Russell MA. Clinical status of ulcerative colitis in patients who smoke. Am J Gastroenterol 1998;93(9):1463–7.

30. Cottone M, Rosselli M, Orlando A, Oliva L, Puleo A, Cappello M, Traina M, Tonelli F, Pagliaro L. Smoking habits and recurrence in Crohn's disease. Gastroenterology 1994;106(3):643–8.

31. Angulo P, Bharucha AE, Jorgensen RA, DeSotel CK, Sandborn WJ, Larusso NF, Lindor KD. Oral nicotine in treatment of primary sclerosing cholangitis: a pilot study. Dig Dis Sci 1999;44(3):602–7.

32. Sivyer G, Gardiner J, Hibbins G, Saunders R. Smoking cessation study involving a transdermal nicotine patch. Outcome achieved in a workplace setting and in general practice. Drug Invest 1994;7:244–53.

33. Levin ED, Westman EC, Stein RM, Carnahan E, Sanchez M, Herman S, Behm FM, Rose JE. Nicotine skin patch treatment increases abstinence, decreases withdrawal symptoms, and attenuates rewarding effects of smoking. J Clin Psychopharmacol 1994;14(1):41–9.

34. Gourlay SG, Forbes A, Marriner T, McNeil JJ. Predictors and timing of adverse experiences during trandsdermal nicotine therapy. Drug Saf 1999;20(6):545–55.

35. Moreau T, Vukusic S, Vandenabeele S, Confavreux C. Nicotine et aggravation de la myasthénie. [Nicotine and worsening myasthenia.] Rev Neurol (Paris) 1997;153(2): 141–3.

36. Van der Klauw MM, Van Hillo B, Van den Berg WH, Bolsius EP, Sutorius FF, Stricker BH. Vasculitis attributed to the nicotine patch (Nicotinell). Br J Dermatol 1996;134(2):361–4.

37. Anonymous. Vasculitis attributed to the nicotine patch (Nicotinell). WHO Newslett 1996;7:5.

38. Orleans CT, Ockene JK. Routine hospital-based quit-smoking treatment for the postmyocardial infarction patient: an idea whose time has come. J Am Coll Cardiol 1993;22(6):1703–5.

39. Warner JG Jr, Little WC. Myocardial infarction in a patient who smoked while wearing a nicotine patch. Ann Intern Med 1994;120(8):695.

40. Arnaot MR. Treating heart disease. Nicotine patches may not be safe. BMJ 1995;310(6980):663–4.

41. Kratzer AM, Wolfgang SA. Nicotine patch needs to breathe. Am J Health Syst Pharm 1998;55(13):1413–5.

42. Vanakoski J, Seppala T, Sievi E, Lunell E. Exposure to high ambient temperature increases absorption and plasma concentrations of transdermal nicotine. Clin Pharmacol Ther 1996;60(3):308–15.

43. Lunell E, Molander L, Andersson M. Relative bioavailability of nicotine from a nasal spray in infectious rhinitis and after use of a topical decongestant. Eur J Clin Pharmacol 1995;48(1):71–5.

44. Schneider NG, Olmstead R, Mody FV, Doan K, Franzon M, Jarvik ME, Steinberg C. Efficacy of a nicotine nasal spray in

smoking cessation: a placebo-controlled, double-blind trial. Addiction 1995;90(12):1671–82.

45. Lunell E, Molander L, Leischow SJ, Fagerstrom KO. Effect of nicotine vapour inhalation on the relief of tobacco withdrawal symptoms. Eur J Clin Pharmacol 1995;48(3-4): 235–40.

46. Hjalmarson A, Franzon M, Westin A, Wiklund O. Effect of nicotine nasal spray on smoking cessation. A randomized, placebo-controlled, double-blind study. Arch Intern Med 1994;154(22):2567–72.

47. Schein JR. Cigarette smoking and clinically significant drug interactions. Ann Pharmacother 1995;29(11):1139–48.

48. Ducharme J, Farinotti R. Clinical pharmacokinetics and metabolism of chloroquine. Focus on recent advancements. Clin Pharmacokinet 1996;31(4):257–74.

49. Rahman P, Gladman DD, Urowitz MB. Smoking interferes with efficacy of antimalarial therapy in cutaneous lupus. J Rheumatol 1998;25(9):1716–9.

Phytoestrogens

Phytoestrogens are naturally occurring, polyphenolic, non-steroidal plant compounds that are structurally similar to 17β-estradiol and have estrogenic and/or antiestrogenic effects. These effects are mediated by binding to estrogen receptors, by alterations in the concentrations of endogenous estrogens, and by binding to or stimulation of the synthesis of sex hormone binding globulin.

There are four main groups of phytoestrogens: isoflavonoids, flavonoids, stilbenes, and lignans (1). Of these the most commonly occurring are the flavonoids (of which the coumestans, prenylated flavonoids, and isoflavones have the greatest estrogenic effects) and the lignans. The isoflavonoids include genistein, daidzein, coumestrol, and equol; the prenylated flavonoids include 8-prenylnaringenin, the stilbenes include resveratrol, and the lignans include enterodiol and enterolactone. Mycoestrogens (mycotoxins) are metabolites of *Fusarium*, a fungus that is often found in pastures and in alfalfa and clover.

The sources of phytoestrogens are summarized in Table 62 (2).

In a study of nine common phytoestrogens, the foods with the highest phytoestrogen contents were nuts and oilseeds, followed by soy products, cereals and breads, legumes, meat products, processed foods that contain soy, vegetables, fruits, and alcoholic and non-alcoholic beverages (3). Flax seed and other oilseeds contained the highest total phytoestrogen content, followed by soy bean and tofu. The highest concentrations of isoflavones were in soy bean and soy bean products (for example. tofu) followed by legumes, whereas lignans were the primary source of phytoestrogens found in nuts and oilseeds (for example flax seed) and also in cereals, legumes, fruits, and vegetables.

Reproductive system

In three women abnormal uterine bleeding with endometrial pathology was related to high consumption of soy products (4). The first had postmenopausal bleeding with uterine polyps, a proliferative endometrium, and a leiomyoma. The second had severe dysmenorrhea, abnormal uterine bleeding, endometriosis, and a leiomyoma. The third had severe dysmenorrhea, abnormal uterine bleeding, endometriosis, leiomyomata, and secondary infertility. All three improved after withdrawal of soy from their diet.

Breasts

Phytoestrogens can cause breast changes in men, including gynecomastia (5).

- A 58-year-old otherwise healthy man developed gynecomastia and breast cancer, which was positive for androgen, estrogen, and progesterone receptors. He had taken herbal mixtures high in phytoestrogen content for 6 years, estimated to amount to the equivalent of estriol 6000 IU/day. There was no family history of breast cancer or BRCA1/BRCA2 mutation. He had breast surgery, but the clinical outcome was not reported.

Table 62 Sources of phytoestrogens

Isoflavones	Lignans				Coumestans	
Legumes	Whole-grain cereals	Fruits, vegetables, seeds	Alcohol		Bean sprouts	Fodder crops
Soybeans	Wheat	Cherries	Beer from hops		Alfalfa	Clover
Lentils	Wheat germ	Apples	Bourbon from corn		Soybean sprouts	
Beans	Barley	Pears				
Chick peas	Hops	Stone fruits				
	Rye	Linseed				
	Rice	Sunflower seeds				
	Brans	Carrots				
	Oats	Fennel				
		Onion				
		Garlic				
		Vegetable oils				

Drug-drug interactions

In 20 healthy volunteers who took soy extract containing 50 mg of isoflavones twice daily for 2 weeks there was no evidence of induction of CYP3A, although in vitro studies showed inhibition of all of the CYPs tested (6).

References

1. Cos P, De Bruyne T, Apers S, Vanden Berghe D, Pieters L, Vlietinck AJ. Phytoestrogens: recent developments. Planta Med 2003;69(7):589–99.
2. Murkies AL, Wilcox G, Davis SR. Clinical review 92: phytoestrogens. J Clin Endocrinol Metab 1998;83(2):297–303.
3. Thompson LU, Boucher BA, Liu Z, Cotterchio M, Kreiger N. Phytoestrogen content of foods consumed in Canada, including isoflavones, lignans, and coumestan. Nutr Cancer 2006;54(2):184–201.
4. Chandrareddy A, Muneyyirci-Delale O, McFarlane SI, Murad OM. Adverse effects of phytoestrogens on reproductive health: a report of three cases. Complement Ther Clin Pract 2008;14(2):132–5.
5. Dimitrakakis C, Gosselink L, Gaki V, Bredakis N, Keramopoulos A. Phytoestrogen supplementation: a case report of male breast cancer. Eur J Cancer Prev 2004;13:481–4.
6. Anderson GD, Rosito G, Mohustsy MA, Elmer GW. Drug interaction potential of soy extract and *Panax ginseng*. J Clin Pharmacol 2003;43:643–8.

Pyrrolizidine alkaloids

General Information

Pyrrolizidine alkaloids occur in a large number of plants, notably the genera *Crotalaria* (Fabaceae), *Cynoglossum* (Boraginaceae), *Eupatorium* (Asteraceae), *Heliotropium* (Boraginaceae), *Petasites* (Asteraceae), *Senecio* (Asteraceae), and *Symphytum* (Boraginaceae) (Table 63). Certain representatives of this class and the plants in which they occur are hepatotoxic as well as mutagenic and hepatocarcinogenic. They can produce veno-occlusive disease of the liver with clinical features like abdominal pain with ascites, hepatomegaly and splenomegaly, anorexia with nausea, vomiting, and diarrhea. Sometimes there is also damage to the pulmonary region.

Table 63 Pyrrolizidine alkaloids in various genera

Genus	Pyrrolizidine alkaloids
Crotalaria albida	Croalbidine
Crotalaria anagyroides	Anacrotine, methylpyrrolizidine
Crotalaria aridicola	Various dehydropyrrolizidines
Crotalaria axillaris	Axillaridine, axillarine
Crotalaria barbata	Crobarbatine
Crotalaria burha	Croburhine, crotalarine
Crotalaria candicans	Crocandine, cropodine
Crotalaria crassipes	Retusamine
Crotalaria crispata	Crispatine, fulvine
Crotalaria dura	Crotaline
Crotalaria fulva	Fulvine
Crotalaria globifer	Crotaline, globiferine
Crotalaria goreensis	Hydroxymethylenepyrrolizidine
Crotalaria grahamiana	Grahamine
Crotalaria grantiana	Grantianine
Crotalaria incana	Anacrotine
Crotalaria intermedia	Integerrimine, usaramine
Crotalaria laburnifolia	Crotalaburnine, hydroxysenkirkine
Crotalaria madurensis	Crotafoline, madurensine
Crotalaria mitchelii	Retusamine
Crotalaria mucronata	Mucronitine, mucronitinine
Crotalaria nana	Crotaburnine, crotananine
Crotalaria novae-hollandiae	Retusamine
Crotalaria retusa	Retusamine
Crotalaria semperflorus	Crosemperine
Crotalaria spectabilis	Retronecanol
Crotalaria stricta	Crotastrictine
Crotalaria trifoliastrum	Various alkylpyrrolizidines
Crotalaria usaramoensis	Usaramine, usaramoensine
Crotalaria virgulata	Grantaline, grantianine
Crotalaria walkeri	Acetylcrotaverrine, crotaverrine
Cynoglossum amabile	Amabiline, echinatine
Cynoglossum glochidiatum	Amabiline
Cynoglossum lanceolatum	Cynaustine, cynaustraline
Cynoglossum latifolium	Latifoline
Cynoglossum officinale	Heliosupine
Cynoglossum pictum	Echinatine, heliosupine
Cynoglossum viridiflorum	Heliosupine, viridiflorine
Eupatorium cannabinum	Echinatine, supinine
Eupatorium maculatum	Echinatine, trachelantimidine
Heliotropium acutiflorum	Heliotrine
Heliotropium arguzoides	Trichodesmine
Heliotropium curassavicum	Angelylheliotridine
Heliotropium dasycarpum	Heliotrine
Heliotropium eichwaldii	Angelylheliotrine
Heliotropium europeum	Acetyl-lasiocarpine, helioitrine
Heliotropium indicum	Acetylindicine, indicine, indicinine
Heliotropium lasiocarpum	Heliotrine, lasiocarpine
Heliotropium olgae	Heliotrine, incanine, lasiocarpine
Heliotropium ovalifolium	Heliofoline
Heliotropium ramosissimum	Heliotrine
Heliotropium strigosum	Strigosine
Heliotropium supinum	Heliosupine, supinine
Heliotropium transoxanum	Heliotrine
Petasites japonicus	Fukinotoxin, petasinine, petasinoside
Senecio adnatus	Platyphylline
Senecio alpinus	Seneciphylline
Senecio amphibolus	Macrophylline
Senecio angulatus	Angulatine
Senecio aquaticus	Aquaticine
Senecio argentino	Retrorsine, senecionine
Senecio aureus	Floridanine, florosenine, otosenine
Senecio auricula	Neosenkirkine
Senecio borysthenicus	Seneciphylline
Senecio brasiliensis	Brasilinecine
Senecio campestris	Campestrine

(Continued)

(Continued)

Table 63 (Continued)

Senecio cannabifolius	Senecicannabine
Senecio carthamoides	Carthamoidine
Senecio cineraria	Jacobine, seneciphylline
Senecio cissampelinum	Senampelines
Senecio crucifolia	Jacobine
Senecio doronicum	Doronine
Senecio erraticus	Erucifoline, floridanine
Senecio filaginoides	Retrorsine, ionine
Senecio franchetti	Franchetine, sarracine
Senecio fuchsii	Fuchsisenecionine
Senecio gillesiano	Retrorsine, senecionine
Senecio glabellum	Integerrimine, senecionine
Senecio glandulosus	Retrorsine, senecionine
Senecio glastifolius	Graminifoline
Senecio hygrophylus	Hygrophylline, platyphylline
Senecio ilicifolius	Pterophine, senecionine
Senecio illinitus	Acetylsenkirkine, senecionine
Senecio incanus	Seneciphylline
Senecio integerrimus	Integerrimine, senecionine
Senecio isatideus	Isatidine, retrorsine
Senecio jacobaea	Jacobine, jacoline, jaconine, jacozine, otosenine, renardine
Senecio kirkii	Acetylsenkirkine, senkirkine
Senecio kubensis	Seneciphylline
Senecio latifolius	Senecifolidine, senecifoline
Senecio leucostachys	Retrorsine, senecionine
Senecio longibolus	Integerrimine, longiboline, retronecanol, retrorsine, riddelline, senecionine, seneciphylline
Senecio macrophyllus	Macrophylline
Senecio mikanoides	Mikanoidine
Senecio nemorensis	Nemorensine, oxynemorensine
Senecio othonnae	Floridnine, onetine, otosenine
Senecio othonniformis	Bisline, isoline
Senecio palmatus	Seneciphylline
Senecio paucicalyculatus	Paucicaline
Senecio petasis	Bisline
Senecio phillipicus	Retrorsine, seneciphylline
Senecio platyphylloides	Neoplatyphylline, platyphylline, sarracine, seneciphylline
Senecio platyphyllus	Platyphylline, seneciphylline
Senecio pojarkovae	Sarracine, seneciphylline
Senecio procerus	Procerine
Senecio propinquus	Seneciphylline
Senecio pseudoarnica	Senecionine
Senecio pterophorus	Pterophine
Senecio ragonesi	Retrorsine, senecionine, uspallatine
Senecio renardi	Renardine, seneciphylline
Senecio retrorsus	Senkirkine

(Continued)

Senecio rhombifolius	Isatidine, retrorsine, neoplatyphylline, platyphylline sarracine, seneciphylline
Senecio Riddellii	Riddelline
Senecio rivularis	Angeloyloxyheliotrine
Senecio rosmarinifolius	Rosmarinine
Senecio ruwenzoriensis	Ruwenine, ruzorine
Senecio sarracenicus	Sarracine
Senecio scleratus	Isatidine, scleratine
Senecio seratophiloides	Senecivernine
Senecio spartioides	Seneciphylline, spartoidine
Senecio squalidus	Senecionine, squalidine
Senecio stenocephalus	Seneciphylline
Senecio subalpinus	Seneciphylline
Senecio subulatus	Retrorsine, senecionine
Senecio swaziensis	Swazine
Senecio triangularis	Triangularine
Senecio uspallatensis	Senecionine, uspallatine
Senecio vernalis	Senecivernine
Senecio vira-vira	Anacrotine, neoplatyphylline
Senecio viscosus	Senecionine
Senecio vulgaris	Senecionine
Symphytum asperum	Asperumine, echinatine
Symphytum caucasicum	Heliosupine, echinatine
Symphytum officinalis	Lasiocarpine, symphytine, viridiflorine
Symphytum orientale	Echinatine, heliosupine
Symphytum tuberosum	Lasiocarpine, symphytine
Symphytum uplandicum	Viridflorine, anadoline, heneicisane, tricosane, acetylintermedine, acetyl-lycopsamine, uplandicine

The German Federal Health Office has restricted the availability of botanical medicines containing unsaturated pyrrolizidine alkaloids (1,2). Herbal medicines that provide over 1 mg/day internally or over 100 mg/day externally are not permitted; herbal medicines that provide 0.1–1 mg/day internally or 10–100 mg/day externally may be applied for only a maximum of 6 weeks per year, and they should not be used during pregnancy or lactation.

Organs and Systems

Liver

In a South African study 20 children were identified as suffering from hepatic veno-occlusive disease thought to be caused by the administration of traditional remedies (3). The predominant clinical presentation was ascites and hepatomegaly. Nine children died. The surviving patients progressed to cirrhosis and portal hypertension. In four cases early urine specimens were available, and in all of these the presence of pyrrolizidine alkaloids was confirmed.

Two sets of twins who were treated with traditional remedies developed liver disease (4). One pair survived, albeit with hepatic damage. Of the other pair, one twin died within 24 hours and the second died 1 month after admission, with a diagnosis of veno-occlusive liver disease. In both cases, the pyrrolizidine alkaloid retrorsine was identified.

Second-Generation Effects

Fetotoxicity

It is prudent to avoid exposing unborn or suckling children to herbal remedies containing pyrrolizidine alkaloids. Animal studies have shown that transplacental passage and transfer to breast milk are possible, and there is a human case on record of fatal neonatal liver injury, in which the mother had used a herbal cough tea containing pyrrolizidine alkaloids throughout her pregnancy (5).

References

1. Anonymous. Vorinformation Pyrrolizidinalkaloidhaltige human Arzneimittel. Pharm Ztg 1990;135:2532–31990;135:2623–4.
2. Anonymous. Aufbereitungsmonographien Kommission. E. Pharm Ztg 1990;135:2081–2.
3. Steenkamp V, Stewart MJ, Zuckerman M. Clinical and analytical aspects of pyrrolizidine poisoning caused by South African traditional medicines. Ther Drug Monit 2000;22(3): 302–6.
4. Conradie J, Stewart MJ, Steenkamp V. GC/MS identification of toxic pyrrolizidine alkaloids in traditional remedies given to two sets of twins. Ann Clin Biochem 2005;42(Pt 2):141–4.
5. Rasenack R, Müller C, Kleinschmidt M, Rasenack J, Wiedenfeld H. Veno-occlusive disease in a fetus caused by pyrrolizidine alkaloids of food origin. Fetal Diagn Ther 2003;18(4):223–5.

Sparteine

General Information

Sparteine is a major quinolizidine alkaloid found in a variety of plants:

- *Ammodendron conollyi*
- *Anagyris fetida*
- *Baptisia australis*
- *Baptisia tinctoria*
- *Chelidonium majus*
- *Cytisus scoparius*
- *Genista lucida*
- *Lupinus argentus*
- *Lupinus laxus*
- *Lupinus luteus*
- *Lupinus niger*
- *Lupinus polyphyllus*
- *Sarothamnus scoparius*
- *Sophora pachycarpa*
- *Sophora secundiflora*
- *Spartium junceum*
- *Spartium scoparium.*

Sparteine has been used as a marker of the metabolic activity of CYP2D6 [1]. Pharmacokinetic studies have shown that its metabolic oxidation exhibits genetic polymorphism and that about 6–9% of the Caucasian population are poor metabolizers. Quinidine, haloperidol, and moclobemide are all potent inhibitors of the oxidative metabolism of sparteine [2].

Among the reported effects of sparteine are reduced cardiac conductivity, stimulation of uterine motility, circulatory collapse, and respiratory arrest. Sparteine was contained in a slimming aid ("Herbal Slimming Aid," UK). It was formerly used to induce labor; at the very least it should be considered contraindicated during pregnancy.

References

1. Ozawa S, Soyama A, Saeki M, Fukushima-Uesaka H, Itoda M, Koyano S, Sai K, Ohno Y, Saito Y, Sawada J. Ethnic differences in genetic polymorphisms of CYP2D6, CYP2C19, CYP3As and MDR1/ABCB1. Drug Metab Pharmacokinet 2004;19(2):83–95.
2. De Smet PA. Health risks of herbal remedies. Drug Saf 1995;13(2):81–93.

Vitamin E

There have been two cases of coagulation disorders with the herbal drug cucurbicin, each tablet of which contains vitamin E 10 mg (1). Cucurbicin is an approved herbal drug in Sweden, traditionally used for micturition difficulties. The active components are extracts from the fruit of *Serenoa repens* and the seed of *Cucurbita pepo*.

- A 73-year-old man with a common cold, who had been taking cucurbicin three tablets daily for more than 1 year, developed a coagulation disorder, with an international normalized ratio (INR) of 2.1, despite a normal albumin concentration and no anticoagulant treatment. His INR improved to 1.4 on treatment with vitamin K, but did not normalize until cucurbicin was withdrawn 1 week later.
- A 61-year-old man, who had taken warfarin and simvastatin for a long time and had an INR of 2.4, started to take cucurbicin five tablets daily for micturition difficulties. After 6 days his INR had increased to 3.4.

After withdrawal of cucurbicin the INR returned to the previous value within 1 week.

No anticoagulant effect has been reported with either of the major components of cucurbicin. However, vitamin E antagonizes the effect of vitamin K and can cause an increased risk of bleeding, especially in patients taking oral anticoagulants. The amount of vitamin E in cucurbicin that these two patients consumed was 30–50 mg/day, corresponding to the recommended dose of 40–50 mg daily for vitamin E deficiency. The two cases suggest that caution should be exercised when cucurbicin is used concurrently with warfarin.

Reference

1   Yue QY, Jansson K. Herbal drug curbicin and anticoagulant effect with and without warfarin: possibly related to the vitamin E component. J Am Geriatr Soc 2001;49(6):838.

Yohimbine

General Information

Yohimbine is a major carboline alkaloid in the bark of *Pausinystalia* species, such as *Pausinystalia yohimbe* (*Corynanthe yohimbe*), *Pausinystalia macrocerus*, *Pausinystalia paniculata*, and *Pausinystalia trillesi*. It is also found in *Pseudocinchona africana* and *Rauwolfia canescens*.

Yohimbine is an alpha$_2$-adrenoceptor antagonist and has been used in the treatment of erectile dysfunction. At high doses its most common effects are increased blood pressure, slight anxiety, and increased frequency of micturition (1). In a systematic review of randomized controlled trials (2), the following adverse events were noted: hypertension, allergic skin reactions, anxiety, dizziness, chills, headache, sweating, agitation, tachycardia, gastrointestinal symptoms, diarrhea, loss of energy, and increased urinary frequency.

Organs and Systems

Cardiovascular

In 25 unmedicated subjects with hypertension yohimbine 22 mg increased mean blood pressure by an average of 5 mm Hg, plasma noradrenaline by 66%, and plasma dihydroxyphenylglycol by 25% at 1 hour after administration (3). The magnitude of the pressor response was unrelated to baseline pressure but correlated positively with baseline noradrenaline concentration and with the yohimbine-induced increment in plasma noradrenaline.

In 25 healthy volunteers and 29 sex- and age-matched untreated hypertensive patients yohimbine 10 mg caused a significant increase in diastolic pressure only in the hypertensive patients (4).

In patients taking tricyclic antidepressants, hypertension can occur at a dose of 4 mg tds. The toxicity of yohimbine can be enhanced by other drugs, such as phenothiazines.

Angina pectoris has been attributed to yohimbine (5).

- A patient with CREST syndrome (calcinosis, Raynaud's phenomenon, esophageal dysfunction, sclerodactyly, and telangiectasia) paradoxically experienced worsening of Raynaud's phenomenon when using yohimbine for erectile dysfunction (6).

Respiratory

Bronchospasm has been attributed to yohimbine (7).

Nervous system

In a double-blind, placebo-controlled study in eight patients receiving methadone, yohimbine 4 mg/kg intravenously caused objective and subjective opioid withdrawal symptoms and increased craving for opioids (8). This was attributed to increased sensitivity of postsynaptic responses to noradrenaline.

Sensory systems

In a placebo-controlled, crossover study in seven healthy subjects oral yohimbine 0.4 mg/kg caused a significant but transient reduction in P50 auditory gating (9).

Psychological, psychiatric

Yohimbine commonly causes anxiety; in eight patients with panic disorder this effect was reduced by fluvoxamine (10).

Manic symptoms have been attributed to yohimbine (11).

In a placebo-controlled study in 18 combat veterans with post-traumatic stress disorder and 11 healthy controls, intravenous yohimbine 0.4 mg/kg significantly increased the amplitude, magnitude, and probability of the acoustic startle reflex (used as a model to investigate the neurochemical basis of anxiety and fear states) in the veterans with post-traumatic stress disorder but not in the controls (12).

Hematologic

Agranulocytosis has been attributed to yohimbine (13).

Immunologic

A lupus-like syndrome in conjunction with generalized erythroderma and progressive renal insufficiency has been attributed to yohimbine (14,15).

Drug Administration

Drug overdose

- In a 62-year-old man who took yohimbine 200 mg the only adverse effects were tachycardia, hypertension, and anxiety of brief duration (16).

This case suggests that yohimbine overdose is relatively benign.

- A 16-year-old girl who took yohimbine in the form of an alleged aphrodisiac known as "yo-yo" had an acute dissociative reaction accompanied by weakness, paresthesia, and incoordination, followed by anxiety, headache, nausea, palpitation, and chest pain; she also had hypertension, tachycardia, tachypnea, sweating, pallor, tremor, and an erythematous rash (17). Serum adrenaline and noradrenaline concentrations were raised. Her symptoms lasted about 36 hours and resolved spontaneously.

Drug–Drug Interactions

Phenytoin

Yohimbine can cause loss of the antiepileptic action of phenytoin (2).

Tricyclic antidepressants

Yohimbine has a sialogenic effect in depressed patients with a dry mouth due to tricyclic antidepressants or neuroleptic drugs (18).

However, there is an increased risk of hypertension when tricyclic antidepressants are combined with yohimbine (19).

Yohimbine can cause hypertension in patients taking tricyclic antidepressants. A drug history should include the use of herbal remedies before conventional treatments are prescribed.

References

1. Tam SW, Worcel M, Wyllie M. Yohimbine: a clinical review. Pharmacol Ther 2001;91(3):215–43.
2. Ernst E, Pittler MH. Yohimbine for erectile dysfunction: a systematic review and meta-analysis of randomized clinical trials. J Urol 1998;159(2):433–6.
3. Grossman E, Rosenthal T, Peleg E, Holmes C, Goldstein DS. Oral yohimbine increases blood pressure and sympathetic nervous outflow in hypertensive patients. J Cardiovasc Pharmacol 1993;22(1):22–6.
4. Musso NR, Vergassola C, Pende A, Lotti G. Yohimbine effects on blood pressure and plasma catecholamines in human hypertension. Am J Hypertens 1995;8(6):565–71.
5. Epelde Gonzalo F. Angor inducido por yohimbina. [Yohimbine-induced angina pectoris.] An Med Interna 1998;15(12):676.
6. Johnson S, Iazzetta J, Dewar C. Severe Raynaud's phenomenon with yohimbine therapy for erectile dysfunction. J Rheumatol 2003;30(11):2503–5.
7. Landis E, Shore E. Yohimbine-induced bronchospasm. Chest 1989;96(6):1424.
8. Stine SM, Southwick SM, Petrakis IL, Kosten TR, Charney DS, Krystal JH. Yohimbine-induced withdrawal and anxiety symptoms in opioid-dependent patients. Biol Psychiatry 2002;51(8):642–51.
9. Adler LE, Hoffer L, Nagamoto HT, Waldo MC, Kisley MA, Giffith JM. Yohimbine impairs P50 auditory sensory gating in normal subjects. Neuropsychopharmacology 1994;10(4):249–57.
10. Goddard AW, Woods SW, Sholomskas DE, Goodman WK, Charney DS, Heninger GR. Effects of the serotonin reuptake inhibitor fluvoxamine on yohimbine-induced anxiety in panic disorder. Psychiatry Res 1993;48(2):119–33.
11. Price LH, Charney DS, Heninger GR. Three cases of manic symptoms following yohimbine administration. Am J Psychiatry 1984;141(10):1267–8.
12. Morgan CA 3rd, Grillon C, Southwick SM, Nagy LM, Davis M, Krystal JH, Charney DS. Yohimbine facilitated acoustic startle in combat veterans with post-traumatic stress disorder. Psychopharmacology (Berl) 1995;117(4):466–71.
13. Siddiqui MA, More-O'Ferrall D, Hammod RS, Baime RV, Staddon AP. Agranulocytosis associated with yohimbine use. Arch Intern Med 1996;156(11):1235–8.
14. Sandler B, Aronson P. Yohimbine-induced cutaneous drug eruption, progressive renal failure, and lupus-like syndrome. Urology 1993;41(4):343–5.
15. De Smet PA, Smeets OS. Potential risks of health food products containing yohimbe extracts. BMJ 1994;309(6959):958.
16. Friesen K, Palatnick W, Tenenbein M. Benign course after massive ingestion of yohimbine. J Emerg Med 1993;11(3):287–8.
17. Linden CH, Vellman WP, Rumack B. Yohimbine: a new street drug. Ann Emerg Med 1985;14(10):1002–4.
18. Bagheri H, Schmitt L, Berlan M, Montastruc JL. A comparative study of the effects of yohimbine and anethol-trithione on salivary secretion in depressed patients treated with psychotropic drugs. Eur J Clin Pharmacol 1997;52(5):339–42.
19. Fugh-Berman A. Herb–drug interactions. Lancet 2000;355(9198):134–8.

Index of drug names